Induced Plant Defenses Against Pathogens and Herbivores

Biochemistry, Ecology, and Agriculture

Edited by

Anurag A. Agrawal
University of Toronto
Toronto, Ontario

Sadik Tuzun
Auburn University
Auburn, Alabama

Elizabeth Bent
University of British Columbia
Vancouver

APS PRESS
The American Phytopathological Society
St. Paul, Minnesota

Cover illustration
A cabbage looper larva moving from a damaged leaf onto a leaf infected with downy mildew (illustration by Lynn S. Adler)

This book has been reproduced directly from computer-generated copy submitted in final form to APS Press by the editors of the volume. No editing or proofreading has been done by the Press.

Reference in this publication to a trademark, proprietary product, or company name by personnel of the U.S. Department of Agriculture or anyone else is intended for explicit description only and does not imply approval or recommendation to the exclusion of others that may be suitable.

Library of Congress Catalog Card Number: 99-64871
International Standard Book Number: 0-89054-242-2

Second printing, 2000

Printed in the United States of America on acid-free paper

The American Phytopathological Society
3340 Pilot Knob Road
St. Paul, Minnesota 55121-2097, USA

TABLE OF CONTENTS

LIST OF CONTRIBUTORS

Anurag A. Agrawal
Department of Botany
University of Toronto
25 Willcocks Street
Toronto, ON Canada M5S 3B2

Bobby Bassi
Novartis Crop Protection
PO Box 18300
Greensboro, NC 27419 USA

Elizabeth Bent
Faculty of Agricultural Sciences
University of British Columbia
Vancouver, BC Canada V6T 1Z4

Richard M. Bostock
Department of Plant Pathology
University of California
Davis, CA 95616 USA

Jan Bruin
Section Population Biology
University of Amsterdam
Kruislaan 320, 1098 SM
Amsterdam, The Netherlands

Robin Buell
Institute for Genomic Research
9712 Medical Center Drive
Rockville, MD 20850 USA

C. Peter Constabel
Department of Biological Sciences
University of Alberta
Edmonton, Alberta
Canada T6G 2E9

Bas Drukker
Section Population Biology
University of Amsterdam
Kruislaan 320, 1098 SM
Amsterdam, The Netherlands

Herbert Eichenseer
Department of Entomology
University of Arkansas
Fayetteville, AR 72701 USA

Gary W. Felton
Department of Entomology
University of Arkansas
Fayetteville, AR 72701 USA

Ray Hammerschmidt
Department of Botany and Plant Pathology
Michigan State University
East Lansing, MI 48824 USA

Arne Janssen
Section Population Biology
University of Amsterdam
Kruislaan 320, 1098 SM
Amsterdam, The Netherlands

Richard Karban
Department of Entomology
University of California
Davis, CA 95616 USA

Joseph W. Kloepper
Department of Plant Pathology
Auburn University
Auburn, AL 36849 USA

Joseph Kuć
(Emeritus, University of Kentucky)
700 Front Street #1202
San Diego, CA 92101 USA

Kay Lawton
Novartis Agribusiness Biotech Research Institute
Research Triangle, NC 27709
USA

Casey C. Lehman
Department of Agronomy
University of Nebraska-Lincoln
Lincoln, NE 68583-0915 USA

W. Joe Lewis
Insect Biology and Population Management
Research Laboratory
USDA-ARS
Tifton, GA 31793 USA

Gary D. Lyon
Scottish Crop Research Institute
Invergowrie, Dundee DD2 5DA
Scotland, United Kingdom

John F. Murphy
Department of Plant Pathology
Auburn University
Auburn, AL 36849 USA

Adrian C. Newton
Scottish Crop Research Institute
Invergowrie, Dundee DD2 5DA
Scotland, United Kingdom

Ralph L. Nicholson
Department of Botany and
Plant Pathology
Purdue University
West Lafayette, IN 47907 USA

Michael Oostendorp
Novartis Crop Protection
CH-4002 Basel
Switzerland

Angelo Pallini
Section Population Biology
University of Amsterdam
Kruislaan 320, 1098 SM
Amsterdam, The Netherlands

Paul W. Paré
Center for Medical, Agricultural and Veterinary Entomology
USDA-ARS
1700 SW 23rd Drive
Gainesville, FL 32608 USA

Jane E. Polston
University of Florida, IFAS
Gulf Coast Research and Education Center
Bradenton, FL 34203 USA

Maurice Sabelis
Section Population Biology
University of Amsterdam
Kruislaan 320, 1098 SM
Amsterdam, The Netherlands

David J. Schuster
University of Florida, IFAS
Gulf Coast Research and Education Center
Bradenton, FL 34203 USA

Petru Scutareanu
Section Population Biology
University of Amsterdam
Kruislaan 320, 1098 SM
Amsterdam, The Netherlands

Edward R. Sikora
Department of Plant Pathology
Auburn University
Auburn, AL 36849 USA

Jennifer A. Smith-Becker
Department of Plant Pathology
University of California
Riverside, CA 92521 USA

Paul Staswick
Department of Agronomy
University of Nebraska-Lincoln
Lincoln, NE 68583-0915 USA

Theo Staub
Novartis Crop Protection
CH-4002 Basel, Switzerland

Michael J. Stout
Department of Entomology
Louisiana State University
Baton Rouge, LA 70803 USA

Allison Tally
Novartis Crop Protection
PO Box 18300
Greensboro, NC 27419 USA

Jennifer S. Thaler
Department of Botany
University of Toronto
25 Willcocks Street
Toronto, ON Canada M5S 3B2

James H. Tumlinson
Center for Medical, Agricultural and Veterinary Entomology
USDA-ARS
1700 SW 23rd Drive
Gainesville, FL 32608 USA

Sadik Tuzun
Department of Plant Pathology
Auburn University
Auburn AL 36849 USA

Nora Underwood
Department of Entomology and Center for Population Biology
University of California
Davis, CA 95616 USA

Madelaine Venzon
Section Population Biology
University of Amsterdam
Kruislaan 320, 1098 SM
Amsterdam, The Netherlands

Changbin Yao
Department of Entomology
Auburn University
Auburn, AL 36849 USA

Arthur R. Zangerl
Department of Entomology
University of Illinois
Urbana, IL 61801 USA

Geoffrey W. Zehnder
Department of Entomology
Auburn University
Auburn, AL 36849 USA

Nina K. Zidack
Department of Plant Sciences
Montana State University
Bozeman, MT 59717 USA

Editors' Note on Terminology

There are a variety of terms used to describe the phenomenon of induced plant defenses against herbivores and pathogens, including immunization, induced resistance (IR), induced systemic resistance (ISR), and systemic acquired resistance (SAR). Traditionally, the term IR had been used by entomologists (e.g., Karban and Baldwin 1997), and ISR and SAR had been used by phytopathologists, each essentially having the same meaning (e.g., Kuć 1995, Sticher et al. 1997): systemic protection of a plant after an inducing agent is applied to a single part of the plant.

Some authors have recently distinguished between ISR and SAR (e.g. van Loon et al. 1998), stating that the former is jasmonate-dependent, mediated by non-pathogenic soil bacteria and does not involve the synthesis of pathogenesis-related proteins, while the latter is salicylate-dependent and initiated by a variety of inducing agents. This definition contradicts other literature in which ISR and SAR are synonymous terms (e.g., Kloepper et al. 1992, Hammerschmidt and Kuć 1995). While there may be a fundamental difference between resistance induced by some non-pathogenic soil bacteria and resistance induced by other agents (some of which are also non-pathogenic soil bacteria: van Loon et al. 1998), it seems confusing to distinguish between these phenomena using acronyms that have had an identical meaning. We have encouraged the authors of this book to define terms where possible, and to use mechanistic descriptors (i.e., jasmonate-dependent).

Literature cited

Hammerschmidt, R. and J. A. Kuć, editors. 1995. Induced resistance to disease in plants. Kluwer Academic Publishers, Boston.

Karban, R., and I. T. Baldwin. 1997. Induced responses to herbivory. University of Chicago Press, Chicago.

Kloepper, J.W., S. Tuzun and J. A. Kuć. 1992. Proposed definitions related to induced disease resistance. Biocontrol Science and Technology 2:347-349.

Kuć, J. 1995. Phytoalexins, stress metabolism, and disease resistance in plants. Annual Review of Phytopathology 33:275-297.

Sticher, L., B. Mauch-Mani, and J. P. Métraux. 1997. Systemic acquired resistance. Annual Review of Phytopathology 35:235-270.

van Loon, L.C., P.A.H.M. Bakker, and C.M.J. Pieterse. 1998. Systemic resistance induced by rhizosphere bacteria. Annual Reviews in Phytopathology 36:453-483.

Induced Resistance Against Pathogens and Herbivores: An Overview

Richard Karban and Joseph Kuć

Abstract

For the past several decades entomologists and plant pathologists have investigated induced responses of plants to herbivores and to diseases independently, largely oblivious to the advances and traditions of the other discipline. Recently, intense interest in the transduction signals that plants employ to mediate induced responses has made it clear that these two kinds of plant reactions share much in common. One aim of this overview and of this volume is to allow these two disciplines to converse and to learn from their similarities and differences.

Induced responses to both herbivores and pathogens involve multiple mechanisms that sometimes may be coordinated. Induced resistance to herbivores and pathogens both show a lack of specificity in terms of inducing agents and organisms that are affected by the responses. The transduction pathways that plants use to activate induced responses to many herbivores differ from those that induce responses to many pathogens, although these two pathways clearly interact. Coincident with an improved understanding of the pathways mediating induction have been attempts to artificially induce resistance using chemical elicitors. These techniques will be used more widely in commercial agriculture during the upcoming years. A consideration of induced responses to herbivores and pathogens has led us to develop a list of research priorities and unanswered questions for the future.

Introduction

Induced resistance against pathogens has a history that spans this century (Beauverie 1901, Ray 1901, Chester 1933). By contrast, induced resistance against herbivores has only been discovered more recently (Green and Ryan 1972, Haukioja and Hakala 1975). In both cases, these early descriptions of induced responses were largely ignored by the general scientific community as a tool to manipulate plant resistance to control diseases and herbivores. To be sure, "vaccination" techniques, that involved the release of less damaging pathogens and herbivores in order to gain protection against economically devastating plant parasites, were tried and in some cases found to be successful (reviewed by Karban and Baldwin 1997). However, only recently has a detailed mechanistic understanding of induced responses become available. This understanding has been accompanied by the recent deployment of "elicitors," abiotic treatments that induce resistance, by mainstream agriculture.

Induced resistance against pathogens has now been described for over 30 plant species and induced resistance against herbivores for over 100 plant species. These examples have involved resistance that has been induced by infection by pathogens, herbivores, or their "products" or by a diverse group of inorganic or organic compounds. The activity of the inducing agents is not due to antibiotic activity *per se*, nor to transformation into antibiotic compounds. Rather, these agents have the ability to make the plant more resistant to subsequent challenges by pathogens and herbivores (Kuć 1982, 1987, 1995, Kessman et al. 1994, Karban and Baldwin 1997, Sticher et al. 1997).

Multiple Mechanisms

Two important features of systemic induced resistance against both pathogens and herbivores have been appreciated for some time. First, it has long been clear that several different mechanisms contribute to produce the emergent property that we observe as "induced resistance." For example, induced resistance against many pathogens is initiated by a hypersensitive response and oxidative burst in which plant cells around the site of infection die and, in some instances, effectively trap and kill the pathogen (Dixon et al. 1994, Gilchrist 1998). However, some examples of systemic induced resistance occur without this localized hypersensitive response (e.g., van Loon et al. 1998). Most cases of induced resistance against herbivores lack hypersensitive responses and oxidative bursts, although these two phenomena sometimes characterize effective induced resistance against insects and mites, particularly when the herbivores are small and have low mobility (Fernandes 1990, Dreger-Jauffret et al. 1991, Felton et al. 1994, Bi et al. 1997). Some cases of induced resistance are localized against pathogens (e.g., Hammerschmidt and Nicholson, this volume) and against herbivores (e.g., Zangerl and Berenbaum 1995); others are systemic (e.g., Karban and Carey 1984, Hammerschmidt and Yang-Cashman 1995, Stout and Duffey 1996). Some plant responses are apparent very quickly, within hours, after the induction event (e.g., Zangerl and Berenbaum 1995), some are observed in the season(s) following induction (e.g., Bryant et al. 1991) and some plants exhibit both rapid and delayed responses (e.g., birch, Neuvonen and Haukioja 1991). This diversity of phenomena makes it clear that induced resistance against pathogens and herbivores involves multiple mechanisms which may or may not act in a coordinated manner.

Lack of Specificity

A second feature of many of the induced responses is that they show a surprising lack of specificity. Specificity can be usefully divided into specificity of cues that the plant responds to and specificity of effects on challenge organisms (Karban and Baldwin 1997). Many different cues can cause the induced response. For example, resistance in cucurbits and tobacco can be induced by infection with viruses, bacteria, fungi or treatment with a variety of structurally unrelated inorganic and organic compounds (Doubrava et al. 1988, Gottstein and Kuć 1989, Mucharromah and Kuć 1991, Strobel and Kuć 1995, Kuć 1995, Fought and Kuć 1996). In some systems, herbivores and pathogens elicit the same responses, while in others they elicit distinct responses (Stout and Bostock, this volume). The great structural diversity of agents used to elicit systemic induced resistance makes it clear that activity depends on what elicitors do rather than what they are. Despite this fact, recognition in plant-pathogen interactions is often specific and this event results from specific protein-protein interactions (Hutcheson 1998).

Once induced, many plant responses have activity against a wide variety of organisms. Curiously, this activity often appears to be idiosyncratic. In general, systemic induced resistance is most effective against fungi, less effective against bacteria, and least effective against systemic viruses. In the cucumber system, inoculation with anthracnose fungus provides systemic protection against the same fungus, as well as obligate and parasitic fungi, local lesion and systemic viruses, wilt fungi and bacteria, but not against spider mites, or beet armyworm caterpillars (Kuć 1987, Ajlan and Potter 1991). In cotton, infestation of cotyledons with spider mites protects plants against the same and different species of mites, caterpillars, thrips, whiteflies, true bugs, and a vascular wilt fungus, but not against a bacterial blight (Karban and Baldwin 1997, Agrawal et al. 1999b,c). This lack of specificity appears to be due to the induction of multiple mechanisms which affect a broad spectrum of pests; some of the individual compounds that are produced may be rather specific, although many are not.

The idiosyncratic nature of induced responses can be understood by recognizing that plants have multiple signaling pathways that initiate multiple responses. As with other systems known to be under hormonal control, regulation depends on the internal chemical environment and the precise timing of the various steps in the pathways. These pathways are stimulated by cues associated with some parasites but not others. These are also effective individually against some groups of pathogens and herbivores, but not against others. In the next section we will outline several of these signaling pathways.

Finally we will consider some of the applied consequences of these different pathways and the multiple ways of eliciting them.

Multiple Pathways

The first described and best understood pathway involves induced resistance to a variety of pathogens mediated by salicylic acid. Salicylic acid causes numerous changes, including production of a lignin-like barrier, pathogenesis-related proteins such as chitinase that degrade the structural chitin of fungi, hydrogen peroxide, and strongly reactive oxidative enzymes such as peroxidase. The precise role of each of these putative defense compounds against pathogens is being elucidated using mutants that lack the ability to synthesize or accumulate each of these products (Hammerschmidt and Nicholson, this volume). For example, transformed plants that do not accumulate salicylic acid or pathogenesis-related proteins do not develop resistance against some pathogens (Gaffney et al. 1993). The possible role for these products against herbivores is almost completely unknown and is an important priority for further study (Bi et al. 1997, Inbar et al. 1998, Hammerschmidt and Nicholson, this volume). Although salicylic acid is required for some cases of systemic acquired resistance, it is probably not the signal that moves through the plant (Rasmussen et al. 1991, Vernooij et al. 1994, Hammerschmidt and Becker-Smith, this volume). There is currently considerable research to determine the nature of the translocational signal, and this information is vital to understand the mechanism(s) for induced resistance and the manipulation of these mechanisms for disease control.

Systemic acquired resistance mediated by the salicylic acid pathway can be activated by biotic and abiotic agents. Some nonpathogenic rhizobacteria activate a second, distinct response that also protects plants against pathogenic viruses, bacteria, and fungi (van Loon et al. 1998). Recent evidence suggests that this pathway may also protect plants against some herbivores (Zehnder et al., this volume). This mechanism is distinct in that it is generally not associated with a hypersensitive reaction nor with pathogenesis-related proteins, it often enhances plant growth, and it does not require salicylic acid. Although the details of this signaling pathway are now being determined, it appears to involve jasmonic acid and ethylene (Pieterse et al. 1998).

Another signaling pathway involving jasmonic acid is activated in response to herbivory (sometimes called the octadecanoid pathway). Jasmonic acid has been known for many years, although its role in plant defense has only been appreciated very recently (Staswick 1992). Jasmonic acid is involved in many different plant processes including senescence, storage, growth, and reproduction (Karban and Baldwin 1997, Staswick and Lehman, this volume). Many, if not most, of the putative defensive chemicals that increase following herbivory are induced by jasmonic acid. In contrast, the role of jasmonic acid in induced physical and morphological responses (e.g., hairs, trichomes) is not known. Chemical treatments that inhibited jasmonic acid sythesis failed to induce proteinase inhibitors, a putative plant defense against herbivores, following wounding (Peña-Cortés et al. 1993, Farmer 1994). Mutants that could not mount the jasmonate response did not accumulate defensive compounds following damage and experienced low fitness in laboratory environments with herbivores (Howe et al. 1996, McConn et al. 1997). This indicates that jasmonic acid is required for induced responses to herbivores. Current models posit that systemin, a small polypeptide hormone, rather than jasmonic acid moves through the plant following herbivory, activating the formation of jasmonates, and ultimately the factors that induce resistance (Bergey et al. 1996, Wasternack and Parthier 1997).

Plants also possess other, less well understood, signaling pathways that are involved with induced responses to pathogens and herbivores. Ethylene is produced at the site of wounding by herbivores and pathogens and induces increases in many of the compounds thought to contribute to defense including lignin, pathogenesis-related proteins, chitinase, and phenylalanine ammonia lyase (PAL) (Enyedi et al. 1992). Some plants treated with ethylene become more resistant to fungal, viral, and insect attacks, although some of these responses have been found to occur even when ethylene synthesis is inhibited (Lawton et al. 1993). Abscisic acid is also increased at the wound site and is associated with local and systemic increases in putative defensive chemicals (Peña-Cortés et al. 1989). Abscisic acid responds to stresses such as drought and may operate upstream of the jasmonate pathway (Hildmann et al. 1992, Wasternack and Parthier 1997).

Recognition of these pathways may help explain the idiosyncratic patterns of specificity and cross-resistance observed among different herbivore and pathogen pests (see above). For example, the pathogens that are affected by responses of cucumbers induced by *Colletotrichum lagenarium* are sensitive to salicylate-mediated induced resistance. The mites and insects that were not reduced by plant inoculation with *C. lagenarium* were not affected by this response pathway. Similarly, the mites and insects that were decreased by damage to cotton cotyledons are all affected by jasmonate-mediated induced resistance, although they are affected by different secondary chemicals (Agrawal and Karban, personal observation). Resistance to the verticillium wilt fungus was induced by feeding by mites and has also been found to be affected by induction mediated by jasmonates (Li et al. 1996). The pathogen that was not reduced on cotton plants infested by spider mites, bacterial blight, was not affected by jasmonate-mediated responses.

Interactions Between Pathways

There is little doubt at this point that these pathways interact. These interactions may be synergistic, enhancing the efficacy of the responses, or antagonistic, reducing efficacy when several pathways are simultaneously involved. Many putative defensive reactions can be induced more effectively by a combination of ethylene and jasmonic acid than by either of these elicitors alone (Xu et al. 1994). Jasmonic acid has been found to mediate resistance against pathogens, as well as herbivores, in a variety of systems (Cohen et al. 1993, Penninckx et al. 1996, Thomma et al. 1998, van Loon et al. 1998, Staswick and Lehman, this volume). Salicylic acid may also induce resistance against some herbivores, in addition to pathogens (Hardie et al. 1994, Inbar et al. 1998), although this phenomenon is less well documented.

There are also many examples of these pathways inhibiting one another. Inhibition of the jasmonic acid pathway by salicylic acid is the best documented and best understood of these antagonistic interactions. Treatment with salicylic acid prevents wounded plants from accumulating proteinase inhibitors and polyphenol oxidase, putative defenses against herbivores triggered by jasmonic acid (Doherty et al. 1988, Stout et al. 1998, 1999). This occurred because salicylic acid blocked the pathway that culminates in synthesis of jasmonic acid and transcription of genes coding for defenses against herbivores (Doares et al. 1995). Field applications of BTH, a commercially available inducer of the salicylic acid pathway (see Tally et al., this volume), inhibited the ability of tomato plants to induce resistance against herbivores and caused them to suffer higher levels of herbivory (Thaler et al. 1999a, Thaler, this volume). Salicylic acid also probably interferes with plant responses that are regulated by abscisic acid (Raskin 1992). Conversely, stresses leading to systemically elevated levels of abscisic acid make plants more vulnerable to some pathogens (Bostock, MacDonald, and Duniway, unpublished manuscript). Jasmonic acid can apparently also interfere with resistance mediated by salicylic acid and the plant's ability to protect itself against some pathogens. Plants treated with jasmonic acid produced less pathogenesis-related proteins in response to pathogen infection and were less able to have resistance induced by bacterial infection in the lab and field (Sano and Ohashi 1995, Niki et al. 1998, Fidantsef et al. 1999, Thaler et al. 1999a).

There are also many examples of the "wrong pathway" being stimulated, i.e., pathogens stimulating the pathway mediated by jasmonic acid and herbivores stimulating the pathway mediated by salicylic acid. For example, fungal and bacterial infections caused plants to increase levels of jasmonates (rather than salicylates) (Penninckx et al. 1996, Stout et al. 1998, 1999, Thomma et al 1998). Conversely, responses involving salicylic acid have been implicated in several cases of induced resistance against insect herbivores (Hardie et al. 1994, Inbar et al. 1998). If these interactions between pathways are general, induced responses to different pests (herbivores and pathogens) and to different stresses may not be independently regulated. If so, then response to one may enhance or constrain the plant's ability to respond to others.

Herbivores and pathogens may have evolved mechanisms that prevent plants from recognizing them, or once recognized, that reduce the effectiveness of plant resistance mechanisms. At a macroscopic scale, many folivores sever the vascular connections to a leaf prior to consuming it, minimizing the induced responses to the leaf blade (Carroll and Hoffman 1980, Dussourd and Denno 1991, 1994). If responses mediated by jasmonates and salicylates are generally antagonistic to one another, then perhaps insects that stimulate the salicylate pathway and pathogens that stimulate the jasmonate pathway are indirectly depressing the plant responses that would be most effective, thereby increasing their own fitness. Recent evidence suggests that constituents in insect saliva may inhibit the plant's jasmonate-mediated responses to herbivory (McCloud and Baldwin 1997, Felton and Eichenseer, this volume). Felton and Eichenseer (this volume) have found that a salivary enzyme, glucose oxidase, suppresses induced resistance, perhaps by blocking a key enzyme in the biosynthesis of jasmonic acid.

Agricultural Uses of Induced Resistance

The most widely used and successful technique in western preventive medicine involves immunizing patients by inoculating them with attenuated strains of pathogens or with other chemical elicitors of induced resistance. Similar techniques have been used to protect crop plants against diseases, although vaccinations have been used commercially in only a small percentage of the systems for which they are seemingly amenable (Campbell 1989, Kuć 1995, Tally et al., this volume). Vaccinations have also been used commercially against a few mite pests (Karban and Baldwin 1997). Recently, the pathways and signals that contribute to induce resistance against pathogens and herbivores have become better known (see above). Coincident with this knowledge has come an increased interest in the identification, development, and commercial deployment of chemicals that elicit induced resistance. We will summarize these developments and speculate on their prospects for future use.

Vaccinations involving the actual organisms (pathogens or herbivores) have the advantage of providing all of the signals or cues that the plant has evolved to react to. However, actual microbes or herbivores are difficult to culture, apply, and control. Living organisms may also cause some level of economic loss to the crop. Strains of pathogens, nematodes, and herbivores have been identified which cause trivial or unmeasurable levels of damage under most growing conditions and yet are good inducers. Even when vaccination techniques have provided inexpensive and reliable control, they have not always been adopted widely. For example, reintroductions of native Willamette mites into vineyards with chronically high populations of economically damaging Pacific mites provided consistent control of Pacific mite populations (Karban et al. 1997b). This control was associated with increased yield. However, there is no commercial supplier of Willamette mites and most growers cannot distinguish between the two herbivore species. Vaccinations generally lack a facilitator, someone who can make a profit by encouraging the technique, and this is probably the most insurmountable hurdle to widespread use.

Chemical elicitors of induced resistance offer an attractive alternative to vaccinations. As we learn more about the signal transduction pathways, we should be able to control the defenses that a plant employs without the complications associated with microbial or herbivore inducers. Chemical elicitors should allow growers to express resistance as a constitutive trait or to control the timing and strength of induced responses. For example, resistance could be induced prior to large increases in the pest population and in some cases this resistance is quite persistent. Chemical elicitors may be more effective at inducing resistance than is natural attack. For example, BTH, a synthetic elicitor of salicylate mediated resistance, induced similar responses as did pathogens and was found to be an extremely potent inducer of systemic resistance (Görlach et al. 1996, Tally et al., this volume). Chemical elicitors can be applied exogenously in controlled concentrations to produce graded induced responses, some of which exceed natural responses of the plants to pathogens and herbivores.

Many chemicals have been found to induce resistance against pathogens and herbivores (Doubrava et al. 1988, Gottstein and Kuć 1989, Mucharromah and Kuć 1991, Kessmann et al. 1994, Strobel and Kuć 1995, Ye et al. 1995, Lyon et al. 1995, Fought and Kuć 1996, Karban and Baldwin 1997, Lyon and Newton, this volume). Since plant signal pathways are highly conserved among phylogenetically diverse plants, these elicitors may be adaptable in many diverse crop and pest situations (Constabel and Ryan 1998, Morris et al. 1998). Thus far, very few elicitors have been tested and released for agricultural use (Kendra et al. 1989, Ye et al. 1995, Lyon and Newton, this volume). The most exciting of these endeavors is the commercial production of BTH which has been marketed in Europe to protect wheat against powdery mildew, although it confers broad spectrum resistance against diverse pathogens (Morris et al. 1998, Tally et al., this volume). This large-scale experiment will soon provide information about the difficulties and possibilities of using elicitors for induced resistance in mainstream agriculture. Elicitors against herbivores are not as well developed as those against pathogens. Recent results indicate that exogenously applied jasmonic acid induces resistance against all the common herbivores of tomato plants under conditions of production agriculture (Thaler et al. 1999b, Thaler, this volume). Leaves of treated plants were 60% less likely to receive herbivore damage compared to controls, although this did not translate into increased yields (Thaler 1999).

Much of the commercial interest in elicitors has involved attempts to permanently express resistance mechanisms that are naturally expressed only following attack by pathogens or herbivores. This would essentially make induced resistance into a trait that was expressed constitutively. Such a manipulation has several potential advantages as well as disadvantages. Constitutive resistance pre-empts the attacking pathogen or herbivore and may reduce the likelihood of infection. Constitutive resistance also eliminates the time lag between initial attack and plant response inherent in induced responses.

There are many benefits of induced resistance that constitutive resistance may lack (Karban et al. 1997a, Agrawal and Karban 1999). Pathogens and herbivores may be more likely to evolve resistance to constitutively expressed plant traits compared to induced responses that are expressed inconsistently, only after infection. Since elicitors can activate multiple plant responses, these presumably will be more durable to counter adaptation by pathogens and herbivores and this issue may not become a problem. Defensive plant phenotypes that change may be more difficult for pathogens and herbivores to deal with physiologically (in ecological rather than evolutionary time) and this advantage may be lost if resistance is expressed constitutively (Stockhoff 1993, Karban et al. 1997a). Finally, resistance traits may become autotoxic to plants when expressed constitutively. For example, in the mid 1980s, Kuć (1987) developed the ability to elicit production of phytoalexins in unattacked green bean and soybean plants by exogenous applications of glucans derived from fungi. Unfortunately, frequent applications of this elicitor were required to maintain resistance and these caused severe necrotization and stunting. Similarly, an isonicotinic acid was identified by scientists at Ciba as a promising elicitor of systemic resistance against many pathogens, although it caused unacceptably high levels of phytotoxicity when resistance was expressed constitutively (Kessman et al. 1994).

It has been argued that induced defenses may be favored over constitutive ones if defenses are costly (Rhoades 1979). Induced defenses are deployed in nature only when needed and the plant can save the costs when defenses are not needed and not deployed (Agrawal, this volume). These costs may take a variety of forms. Becoming better defended against one organism may make the plant more attractive or susceptible to another. The antagonisms between the jasmonate and salicylate pathways for resistance described above may represent such a cost. Autotoxicity may be another cost of defense. Defense may be expensive to plants in terms of energy or precursors, although these "allocation costs" have proven difficult to demonstrate, especially in agricultural situations (but see Tally et al., this volume). However, several recent examples suggest that costs of defense may be real. By stimulating wild radish and wild tobacco plants to induce resistance, plant lifetime fitness was increased (Agrawal 1998, Baldwin 1998). This enhanced fitness only occurred in environments that contained herbivores. In environments without herbivores, plants that were forced to induce resistance suffered reduced fitness (Baldwin 1998, Agrawal et al. 1999a). Experiments inducing barley plants in environments with and without virulent mildew strains suggest that induced resistance to pathogens may also be costly in terms of reduced fitness (Smedegaard-Petersen and Stolen 1981). If defenses are costly to plants, then converting resistance that is now inducible to constitutive resistance may be associated with reduced yields. These costs may be offset by agricultural practices such as fertilization and irrigation. Similarly, costs that reduce male fitness (e.g., Agrawal et al. 1999a) may have little relevance to production agriculture if crops are not pollen or pollinator limited.

Another potential cost of expressing host plant resistance is the possibility that the quality or safety of the food crop will be reduced. However, there is no evidence that the metabolites which accumulate in plants induced with elicitors pose any more of a hazard than those present in constitutively resistant plants (Lyon and Newton, this volume).

Economics will probably determine whether techniques for induced resistance will ever become widely used. For elicitors to be accepted they must be economically beneficial for both the grower and for the chemical companies that will produce and market them. Preliminary studies suggest that elicitors can provide very effective control of many pests. They are easy to apply and can provide long term control at very low concentrations. Elicitors will be most enthusiastically received in those cropping systems where growers are currently experiencing problems controlling pests. For example, elicitors have the potential to control plant viruses, a market in which no antibiotic chemicals are currently available. Their lack of specificity may allow elicitors to be marketed for many pest problems in many different crops, which is probably an absolute requirement for commercialization.

Research Priorities for the Future

There can be little question that a thorough understanding of the biochemical basis of induced resistance as well as ecological considerations will greatly advance our ability to use induced resistance to manage pest populations. Until these mechanisms are understood, it is difficult to fully appreciate when and where these techniques will work or how to manage them most effectively. The following is a list of research priorities for the future that we envision will advance this field:

1. Which of the putative defensive compounds and the timing of their appearance are important to induced resistance?

2. What are the metabolic similarities and differences between constitutive and induced resistance?

3. To what extent are the pathways for induced resistance conserved? Are there other pathways that are also involved in plant defense? What are the translocated signals?

4. Are the pathways interactive? Does the plant coordinate its responses against herbivores and pathogens?

5. What effect will the products induced by pathogens (e.g., PR proteins, phytoalexins) have on herbivores?

6. Sensitization, the process in which attacked plants respond more rapidly or more effectively to challenge after initial damage, has been reported numerous times for pathogens. How does sensitization work? Does a similar process occur for induced responses against herbivores (see Karban and Niiho 1995, Baldwin and Schmelz 1996)?

7. Plants may become more tolerant of attacks by pathogens and herbivores instead of more resistant (e.g., Wittmann and Schoenbeck 1996). Tolerant plants experience the same amount of damage but it causes less of a reduction in fitness compared to less tolerant plants. Is tolerance generally inducible and how is it related to resistance? Since we often screen the effects of induced responses on the herbivores or pathogens rather than by evaluating the plant, we may be missing methods that induce tolerance and chasing methods that induce resistance but fail to benefit plants.

8. What are the benefits and costs of induced resistance against pathogens and herbivores? How do these occur? Are there particular conditions that allow plants to accrue benefits from phenotypically plastic defenses as opposed to constitutively expressed defenses?

9. How do induced responses fit into the natural or agricultural environments involving multiple species interactions and multiple trophic levels? Are there antagonisms and synergisms between responses to various ecological participants - pathogens, herbivores, pollinators, mycorrhizal fungi, neighboring plants?

10. Have we recognized the potential contribution of simple, non patented, environmentally safe compounds as inducers of resistance?

11. There have been several elegant recent examples of induced responses that affect predators and parasites of pathogens and herbivores (e.g., De Moraes et al. 1998). Are these interactions generally important in natural and agricultural systems?

Acknowledgments

We thank Anurag Agrawal and Jennifer Thaler for improving the manuscript. This work was supported by grants from the USDA NRI 9606025 and 9802362.

Literature Cited

Agrawal, A. A. 1998. Induced responses to herbivory and increased plant performance. Science 279:1201-1202.

Agrawal, A. A., and R. Karban. 1999. Why induced defenses may be favored over constitutive strategies in plants. Pages 45-61 *in* R. Tollrian and C. D. Harvell, editors. The Ecology and Evolution of Inducible Defenses. Princeton University Press, Princeton, N.J.

Agrawal, A. A., S. Y. Strauss, and M. J. Stout. 1999a. Costs of induced responses and tolerance to herbivory in male and female fitness components of wild radish. Evolution (in press).

Agrawal, A. A., C. Kobayashi, and J. S. Thaler. 1999b. Influence of prey availability and induced host plant resistance on omnivory by western flower thrips. Ecology 80:518-523.

Agrawal, A. A., R. Karban, and R. Colfer. 1999c. How leaf domatia and induced plant resistance affect herbivores, natural enemies and plant performance. Submitted to Oikos.

Ajlan, A. M., and D. A. Potter. 1991. Does immunization of cucumber against anthracnose by *Colletotrichum lagenarium* affect host suitability for arthropods? Entomologia Experimentalis et Applicata 58:83-91.

Baldwin, I. T. 1998. Jasmonate-induced responses are costly but benefit plants under attack in native populations. Proceedings of the National Academy of Sciences USA 95:8113-8118.

Baldwin, I. T., and E. A. Schmelz. 1996. Immunological "memory" in the induced accumulation of nicotine in wild tobacco. Ecology 77:236-246.

Beauverie, J. 1901. Essais d'immunization des vegetaux contre les maladies cryptogamiques. Comptes Rendus Hebdomadaires des Seances de l'Acadamie des Sciences, Paris 133:107-110.

Bergey, D. R., G. A. Howe, and C. A. Ryan. 1996. Polypeptide signaling for plant defensive genes exhibits analogies to defense signaling in animals. Proceedings of the National Academy of Sciences USA 93:12053-12058.

Bi, J. L., J. B. Murphy, and G. W. Felton. 1997. Antinutritive and oxidative components as mechanisms of induced resistance in cotton to *Helicoverpa zea*. Journal of Chemical Ecology 23:97-117.

Bryant, J. P., I. Heitkonig, P. Kuropat, and N. Owen-Smith. 1991. Effects of severe defoliation on the long-term resistance to insect attack and on leaf chemistry in six woody species of the southern African savanna. American Naturalist 137:50-63.

Campbell, R. 1989. The use of microbial inoculants in the biological control of plant diseases. Pages 67-77 *in* R. Campbell and R. M. Macdonald, editors. Microbial Inoculation of Crop Plants. Oxford University Press, Oxford.

Carroll, C. R. and C. A. Hoffman. 1980. Chemical feeding deterrent mobilized in response to insect herbivory and counteradaptation by *Epilachna tredecimnotata*. Science 209:414-416.

Chester, K. S. 1933. The problem of acquired physiological immunity in plants. Quarterly Review of Biology 8:129-154, 275-324.

Cohen, Y., U. Gisi, and T. Niderman. 1993. Local and sytemic protection against *Phytophthora infestans* induced in potato and tomato plants by jasmonic acid and jasmonic methyl ester. Phytopathology 83:1054-1062.

Constabel, C. P., and C. A. Ryan. 1998. A survey of wound- and methyl jasmonate-induced leaf polyphenol oxidase in crop plants. Phytochemistry 47:507-511.

De Moraes, C. M., W. J. Lewis, P. W. Pare, H. T. Alborn, and J. H. Tumlinson. 1998. Herbivore-infested plants selectively attract parasitoids. Nature 393:570-573.

Dixon, R. A., M. J. Harrison, and C. J. Lamb. 1994. Early events in the activation of plant defense responses. Annual Review of Phytopathology 32:479-501.

Doares, S., T. Syrovets, E. Weiler, and C. A. Ryan. 1995. Oligogalacturonides and chitosan activate plant defense genes through the octadecanoid pathway. Proceedings of the National Academy of Sciences USA 92:4095-4098.

Doherty, H. M., R. R. Selvendran, and D. J. Bowles. 1988. The wound response of tomato plants can be inhibited by aspirin and related hydroxy-benzoic acids. Physiological and Molecular Plant Pathology 33:377-384.

Doubrava, N., R. Dean, and J. Kuć. 1988. Induction of systemic resistance to anthracnose caused by *Colletotrichum lagenarium* in cucumber by oxalates and extracts from spinach and rhubarb leaves. Physiological and Molecular Plant Pathology 33:69-79.

Dreger-Jauffret, F., R. Bronner, and E. Westphal. 1991. Caracterisation de la resistance aux Eriophyides (Acariens) chez la pomme de terre (Var. Nicola), p. 267-273 *in* ANPP - Deuxieme Conference Internationale Sur les Ravageurs en Agriculture, Versailles.

Dussourd, D. E., and R. F. Denno. 1991. Deactivation of plant defense: correspondence between insect behavior and secretory canal architecture. Ecology 72:1383-1396.

Dussourd, D. E., and R. F. Denno. 1994. Host range of generalist caterpillars: trenching permits feeding on plants with secretory canals. Ecology 75:69-78.

Enyedi, A., N. Yalpani, P. Silverman, and I. Raskin. 1992. Signal molecules in systemic plant resistance to pathogens and pests. Cell 70:879-886.

Farmer, E. E. 1994. Fatty acid signalling in plants and their associated microorganisms. Plant Molecular Biology 26:1423-1437.

Felton, G. W., J. L. Bi, C. B. Summers, A. J. Mueller, and S. S. Duffey. 1994. Potential role of lipoxygenases in defense against insect herbivory. Journal of Chemical Ecology 20:651-666.

Fernandes, G. W. 1990. Hypersensitivity: a neglected plant resistance mechanism against insect herbivores. Environmental Entomology 19:1173-1182.

Fidantsef, A. L., M. J. Stout, J. S. Thaler, S. S. Duffey, and R. M. Bostock. 1999. Signal interactions in pathogen and insect attack: expression of lipoxygenase, proteinase inhibitor II, and pathogenesis-related protein P4 in the tomato, *Lycopersicon esculentum*. Physiological and Molecular Plant Pathology.

Fought, L. and J. Kuć. 1996. Lack of specificity in plant extracts and chemicals as inducers of systemic resistance in cucumber plants to anthracnose. Journal of Phytopathology 144:1-6.

Gaffney, T., L. Friedrich, B. Vernooij, D. Negrotto, G. Nye, S. Uknes, E. Ward, H. Kessmann, and J. Ryals. 1993. Requirement of salicylic acid for the induction of systemic acquired resistance. Science 261:754-756.

Gilchrist, D. G. 1998. Programmed cell death in plant disease: the purpose and promise of cellular suicide. Annual Review of Phytopathology 36:393-414.

Görlach, J., S. Volrath, G. Knauf-Beiter, U. Beckhove, K.-H. Kogel, M. Oostendorp, T. Staub, E. Ward, H. Kessmann, and J. Ryals. 1996. Benzothiadiazole, a novel class of inducers of systemic acquired resistance, activates gene expression and disease resistance in wheat. Plant Cell 8:629-643.

Gottstein, H. D. and J. Kuć. 1989. The induction of systemic resistance to anthracnose in cucumber by phosphates. Phytopathology 79:176-179.

Green, T. R., and C. A. Ryan. 1972. Wound-induced proteinase inhibitor in plant leaves: A possible defense mechanism against insects. Science 175:776-777.

Hammerschmidt, R., and P. Yang-Cashman. 1995. Induced resistance in cucurbits. Pages 63-85 *in* R. Hammerschmidt and J. Kuć, editors. Induced Resistance to Disease in Plants. Kluwer, Amsterdam.

Hardie, J., R. Isaacs, J. A. Pickett, L. J. Wadhams, and C. M. Woodcock. 1994. Methyl salicylate and (-)-(1R,5S)-myrtenal are plant-derived repellents for black bean aphid, *Aphis fabae* Scop. (Homoptera: Aphididae). Journal of Chemical Ecology 20:2847-2855.

Haukioja, E., and T. Hakala. 1975. Herbivore cycles and periodic outbreaks. Formulation of a general hypothesis. Report of the Kevo Subarctic Research Station 12:1-9.

Hildmann, T., M. Ebneth, H. Peña-Cortés, J. J. Sanchez-Serrano, L. Willmitzer, and S. Prat. 1992. General roles of abscisic and jasmonic acids in gene activation as a result of mechanical wounding. Plant Cell 4:1157-1170.

Howe, G. A., J. Lightner, J. Browse, and C. A. Ryan. 1996. An octadecanoid pathway mutant (JL5) of tomato is compromised in signaling for defense against insect attack. Plant Cell 8:2067-2077.

Hutcheson, S. W. 1998. Current concepts of active defense in plants. Annual Review of Phytopathology 36:59-90.

Inbar, M., H. Doostdar, R. M. Sonoda, G. L. Leibee, and R. T. Mayer. 1998. Elicitors of plant defensive systems reduce insect densities and disease incidence. Journal of Chemical Ecology 24:135-149.

Karban, R., A. A. Agrawal, and M. Mangel. 1997a. The benefits of induced defenses against herbivores. Ecology 78:1351-1355.

Karban, R., and I. T. Baldwin. 1997. Induced responses to herbivory. University of Chicago Press, Chicago.

Karban, R., and J. R. Carey. 1984. Induced resistance of cotton seedlings to mites. Science 225:53-54.

Karban, R., G. English-Loeb, and D. Hougen-Eitzman. 1997b. Mite vaccinations for sustainable management of spider mites in vineyards. Ecological Applications 7:183-191.

Karban, R., and C. Niiho. 1995. Induced resistance and susceptibility to herbivory: plant memory and altered plant development. Ecology 76:1220-1226.

Kendra, D. F., D. Christian, and L. A. Hadwiger. 1989. Chitosan oligomers from *Fusarium solani*/pea interactions, chitosan/b glucanase digestion of sporelings and from fungal wall chitin actively inhibit fungal growth and enhance disease resistance. Physiological and Molecular Plant Physiology 35:215-230.

Kessmann, H., T. Staub, C. Hofmann, T. Maetzke, J. Herzog, E. Ward, S. Uknes, and J. Ryals. 1994. Induction of systemic acquired disease resistance in plants by chemicals. Annual Review of Phytopathology 32:439-459.

Kuć, J. 1982. Induced immunity to plant disease. BioScience 32:854-860.

Kuć, J. 1987. Plant immunization and its applicability for disease control. Pages 255-274 *in* I. Chet, editor. Innovative Approaches to Plant Disease Control. John Wiley, New York.

Kuć, J. 1995. Induced systemic resistance - an overview. Pages 169-175 *in* R. Hammerschmidt and J. Kuć, editors. Induced Resistance to Diseases in Plants. Kluwer, Amsterdam.

Lawton, K., B. Vernooij, L. Friedrich, T. Gaffney, D. Alexander, D. Negrotto, J. P. Métraux, H. Kessmann, M. G. Rella, S. Uknes, E. Ward, and J. Ryals. 1993. Signal transduction in systemic acquired resistance. Pages 126-133 *in* J. C. Schultz and I. Raskin, editors. Plant Signals in Interactions with Other Organisms. American Society of Plant Physiologists, Rockville, Maryland.

Li, J., I. Zingen-Sell, and H. Buchenauer. 1996. Induction of resistance of cotton plants to Verticillium wilt of tomato plants to Fusarium wilt by 3-aminobutyric acid and methyl jasmonate. Zeitschrift fur Pflanzenkrankheiten und Pflanzenschutz 103:288-299.

Lyon, G. D., T. Reglinski, and A. C. Newton. 1995. Novel disease control compounds: the potential to 'immunize' plants against infection. Plant Pathology 44:407-427.

McCloud, E. S., and I. T. Baldwin. 1997. Herbivory and caterpillar regurgitants amplify the wound-induced increases in jasmonic acid but not nicotine in *Nicotiana sylvestris*. Planta 203:430-435.

McConn, M., R. A. Creelman, E. Bell, J. E. Mullet, and J. Browse. 1997. Jasmonate is essential for insect defense in *Arabidopsis*. Proceedings of the National Academy of Sciences USA 94:5473-5477.

Morris, S. W., B. Vernooij, S. Titatarn, M. Starrett, S. Thomas, C. C. Wiltse, R. A. Frederiksen, A. Bhandhufalck, S. Hulbert, and S. Uknes. 1998. Induced resistance responses in maize. Molecular Plant-Microbe Interactions 11:643-658.

Mucharromah, E. and J. Kuć. 1991. Oxalates and phosphates induce systemic resistance against diseases caused by fungi, bacteria, and viruses in cucumber. Crop Protection 10:265-270.

Neuvonen, S., and E. Haukioja. 1991. The effects of inducible resistance in host foliage on birch-feeding herbivores. Pages 277-291 *in* D. W. Tallamy and M. J. Raupp, editors. Phytochemical Induction by Herbivores. John Wiley, New York.

Niki, T., I. Mitsuhara, S. Seo, N. Ohtsubo, and Y. Ohashi. 1998. Antagonistic effect of salicylic acid and jasmonic acid on the expression of pathogenesis-related (PR) protein genes in wounded mature tobacco leaves. Plant Cell Physiology 39:500-507.

Peña-Cortés, H., T. Albrecht, S. Prat, E. W. Weiler, and L. Willmitzer. 1993. Aspirin prevents wound -induced gene expression in tomato by blocking jasmonic acid biosynthesis. Planta 19:123-128.

Peña-Cortés, H., J. J. Sánchez-Serrano, R. Mertens, L. Willmitzer, and S. Prat. 1989. Abscisic acid is involved in the wound-induced expression of the proteinase inhibitor II gene in potato and tomato. Proceedings of the National Academy of Sciences USA 86:9851-9855.

Penninckx, I. A. M. A., K. Eggermont, F. R. G. Terras, B. P. H. J. Thomma, G. W. De Samblanx, A. Buchala, J.-P. Métraux, J. M. Manners, and W. F. Broekaert. 1996. Pathogen-induced systemic activation of a plant defensin gene in *Arabidopsis* follows a salicylic acid-independent pathway. Plant Cell 8:2309-2323.

Pieterse, C. M. J., S. C. M. van Wees, J. A. van Pelt, M. Knoester, R. Laan, H. Gerrits, P. J. Weisbeek, and L. C. van Loon. 1998. A novel signaling pathway controlling induced systemic resistance in *Arabidopsis*. Plant Cell 10:1571-1580.

Raskin, I. 1992. Role of salicylic acid in plants. Annual Review of Plant Physiology and Plant Molecular Biology 43:439-463.

Rasmussen, J. B., R. Hammerschmidt, and M. N. Zook. 1991. Systemic induction of salicylic acid accumulation in cucumber after inoculation with *Pseudomonas syringae* pv. syringae. Plant Physiology 97:1342-1347.

Ray, J. 1901. Les maladies cryptogamiques des vegetaux. Revue Generale de Botanique 13:145-151.
Rhoades, D. F. 1979. Evolution of plant chemical defense against herbivores. Pages 3-54 *in* G. A. Rosenthal and D. H. Janzen, editors. Herbivores: Their Interaction with Secondary Plant Metabolites. Academic Press, New York.
Sano, H., and Y. Ohashi. 1995. Involvement of small GTP-binding proteins in defense signal-transduction pathways of higher plants. Proceedings of the National Academy of Sciences USA 92:4138-4144.
Smedegaard-Petersen, V., and O. Stolen. 1981. Effects of energy-requiring defense reactions on yield and grain quality in a powdery mildew-resistant barley cultivar. Phytopathology 71:396-399.
Staswick, P. E. 1992. Jasmonate, genes, and fragrant signals. Plant Physiology 99:804-807.
Sticher, L., B. Mauch-Mani, and J. P. Métraux. 1997. Systemic acquired resistance. Annual Review of Phytopathology 35:235-270.
Stockhoff, B. A. 1993. Diet heterogeneity: implications for growth of a generalist herbivore, the gypsy moth. Ecology 74:1939-1949.
Stout, M. J., and S. S. Duffey. 1996. Characterization of induced resistance in tomato plants. Entomologia Experimentalis et Applicata 79:273-283.
Stout, M. J., A. L. Fidantsef, S. S. Duffey, and R. M. Bostock. 1999. Signal interactions in pathogen and insect attack: systemic plant-mediated interactions between pathogens and herbivores of the tomato, *Lycopersicon esculentum*. Physiological and Molecular Plant Physiology (in press).
Stout, M. J., K. V. Workman, R. M. Bostock, and S. S. Duffey. 1998. Stimulation and attenuation of induced resistance by elicitors and inhibitors of chemical induction in tomato (*Lycopersicon esculentum*) foliage. Entomologia Experimentalis et Applicata 86:267-279.
Strobel, N. E. and J. Kuć. 1995. Chemical and biological inducers of systemic resistance to pathogens protect cucumber and tobacco plants from damage caused by paraquat and cupric chloride. Phytopathology 85:1306-1310.
Thaler, J. S. 1999. Induced resistance in agricultural crops: effects of jasmonic acid on herbivory and yield in tomato plants. Environmental Entomology 28:30-37.
Thaler, J. S., A. L. Fidantsef, S. S. Duffey, and R. M. Bostock. 1999a. Tradeoffs in plant defense against pathogens and herbivores: a field demonstration using chemical elicitors of induced resistance. Journal of Chemical Ecology (in press).
Thaler, J.S., M. J. Stout, R. Karban, and S.S. Duffey. 1999b. Linking biochemical mechanisms of induced plant resistance and herbivore population dynamics. Submitted to Ecology.
Thomma, B. P. H. J., K. Eggermont, I. A. M. A. Penninckx, B. Mauch-Mani, R. Vogelsang, B. P. A. Cammue, and W. F. Broekaert. 1998. Separate jasmonate-dependent and salicylate-dependent defense-response pathways in *Arabidopsis* are essential for resistance to distinct microbial pathogens. Proceedings of the National Academy of Sciences USA 95:15107-15111.
van Loon, L. C., P. A. H. Bakker, and C. M. J. Pieterse. 1998. Systemic resistance induced by rhizosphere bacteria. Annual Review of Phytopathology 36:453-483.
Vernooij, B., L. Friedrich, A. Morse, R. Reist, R. Kolditz-Jawhar, E. Ward, S. Uknes, H. Kessman, and J. Ryals. 1994. Salicylic acid is not the translocated signal responsible for inducing systemic acquired resistance but is required for signal transduction. Plant Cell 6:959-965.
Wasternack, C., and B. Parthier. 1997. Jasmonate-signalled plant gene expression. Trends in Plant Science 2:302-307.
Wittmann, J., and F. Schonbeck. 1996. Studies of tolerance in wheat infested with powdery mildew or aphids. Zeitschrift fur Pflanzenkrankheiten und Pflanzenschutz 103:300-309.
Xu, Y., P. L. Chang, D. Llu, M. L. Naraslmhan, K. G. Raghothama, P. M. Hasegawa, and R. A. Bressan. 1994. Plant defense genes are synergistically induced by ethylene and methyl jamonate. Plant Cell 6:1077-1085.

Ye, X. S., N. Strobel, and J. Kuć. 1995. Induced systemic resistance (ISR): activation of natural defense mechanisms for plant disease control as part of integrated pest management (IPM). Pages 95-113 *in* R. Reuveni, editor. Novel Approaches to Integrated Pest Management. CRC Press, Boca Raton, Florida.

Zangerl, A. R., and M. R. Berenbaum. 1995. Spatial, temporal, and environmental limits on xanthotoxin induction in wild parsnip foliage. Chemoecology 5/6:37-42.

Part I

Biochemistry and Mechanisms

Herbivore Saliva and its Effects on Plant Defense Against Herbivores and Pathogens

Gary W. Felton and Herbert Eichenseer

Abstract

There has been considerable interest in understanding the role of oral secretions of herbivores as specific damage cues for induced plant responses. We provide needed clarification on the meanings of terms such as oral secretions, regurgitant and saliva due to the inconsistencies in their usage. Evidence from several seminal studies on the elicitation of plant responses is reviewed. Our recent work on the role of the salivary protein, glucose oxidase, in eliciting induced resistance to pathogens is discussed. Finally, we hypothesize that an important function of herbivore saliva may be to suppress jasmonate-mediated induced defenses against herbivores.

Introduction

Until recently it was assumed that plant responses to real herbivory were indistinguishable from mechanically-induced damage. Although mechanical damage may elicit plant responses similar to insect feeding (e.g., Hartley and Lawton 1991); in many instances herbivores may activate unique responses (e.g., Dyer and Bokhari 1976, Hartley and Lawton 1987, Lin et al. 1990; Hanhimäki and Senn 1992, Mattiacci et al. 1995, Alborn et al. 1997, Korth and Dixon 1997, De Moraes et al. 1998, Stout et al. 1998, Bernasconi et al. 1998). Unique damage cues may originate from herbivores with differing feeding patterns/habits (e.g., chewing vs. sucking insects) that differentially induce resistance by causing distinct types of cellular damage (Hartley and Lawton 1987, Felton et al. 1994, Stout et al. 1994, Turlings et al. 1998a). Microbes (e.g., fungi and bacteria) which may contain elicitors may also contaminate insect mouthparts (Hartley and Lawton 1991). Finally, qualitative and quantitative differences in elicitors in the oral secretions of herbivores may be responsible for responses unique to herbivory. In this chapter, we highlight recent evidence from several laboratories that demonstrates the importance of saliva in understanding host responses to plant attack by herbivores and phytopathogens.

Insect salivary glands. Several reviews (e.g., House 1980, Sehnal and Akai 1990, Ribeiro 1995, Ali 1997) describe the anatomy and physiology of insect salivary glands; however, we present additional information for purposes of clarification. Considering the diversity of mouthparts and feeding habits of insect herbivores, it is not surprising that there are diverse salivary gland morphologies. Salivary glands can be divided into three broad morphological types: 1) acinar, 2) tubular, and 3) reservoir (Ribeiro 1995). Salivary glands are paired, associated with at least one mouthpart and are of ectodermal origin (Sehnal and Akai 1990, Chapman 1998). Acinar (or aveolar) glands are found in chewing hemimetabolous insects (e.g., cockroaches, grasshoppers). Acinar glands are composed of several cell types organized like grapes around a common duct that coalesce with other acini to form branching salivary ducts eventually merging with the main salivary duct (House 1980, Ribeiro 1995, Ali 1997). The different cell types within individual acini are specialized for either fluid or protein secretions and directly innervated by neurons originating from the subesophageal ganglion or stomatogastric nervous system. Fluid and protein secretion is regulated by release of the biogenic amines dopamine or serotonin, respectively (Ali and Orchard 1996, Just and Walz 1996). Tubular glands are found in holometabolous insects (e.g., caterpillars, beetles and flies) where the glands are comprised of a single cell layer covered with a basement membrane (Srivastava 1959, House 1980, Sehnal and Akai 1990, Chapman 1998). These glands are not directly innervated, but secretion is controlled by release of serotonin from the nervous system (Bay 1978 a, b, Trimmer 1985). The complex, many-lobed reservoir glands found in Heteropterans specialize in storing and secreting a variety of substances (Taylor and Miles 1994, Cohen and Wheeler 1998).

In caterpillar species, there are two pairs of salivary glands, the mandibular and labial glands. The mandibular glands secrete their fluids along the inner surface of the cutting edge of the mandibles and contain unknown lipophilic substances that readily leak out when these organs are cut and cohere in the dissection media. Labial glands appear to contain more hydrophilic substances in their lumen than the mandibular glands. During larval development the role of the labial glands may change from one that is primarily digestive or salivary to one that produces silk (Sehnal and Akai 1990, Chapman 1998).

The salivary gland secretions are normally "extra-oral" for insects such as caterpillars. The labial gland secretions are released through the spinneret whereas the mandibular gland secretions are released through pores in the mandibles. Both secretions are external to the buccal cavity. These secretions may be found in the alimentary canal due to ingestion of saliva during feeding. Hence, salivary components can be found in the regurgitant.

Definitions. The term "oral secretions" has been used in a manner synonymous with "regurgitant". Both materials are collected by disturbing the insect (e.g., caterpillars) through handling, squeezing with forceps or even removal of the head. We suggest that these secretions are not normal secretions released during feeding, but result from trauma. Thus, the collected material may arise from the salivary glands, although much of it may be from the alimentary canal. We obtained saliva from substrates chewed upon by insects. For example, application of a sucrose solution to glass filter disks encourages many insects to feed upon these substrates. Following signs of feeding, salivary proteins may be washed off these materials and concentrated. Fig. 1 depicts proteins collected from the regurgitant, labial glands, and saliva of the noctuid caterpillar, *Helicoverpa zea*.

The protein composition is quite dissimilar between collected materials. Thus, we suggest the term saliva should only refer to secretions released during feeding. For instance, proteins obtained from the functional salivary glands of blood-feeding arthropods are assumed to be the same proteins released during feeding. However, we are aware of few studies to focus on the identification and quantification of salivary components actually released into or on the host. Furthermore, the full chemical composition of saliva is unknown for any insect. Describing these are essential to properly understand the ecological role of saliva in plant-parasite interactions.

Saliva performs multiple functions, including digestion, lubrication of mouthparts, detoxication, excretion, defense against predators, pH regulation, suppression of host responses, or elicitation of host responses (Ribeiro 1995). In the rest of this chapter, we focus on how saliva affects host defensive responses.

Caterpillar Saliva and Induction of Host Resistance

Studies on elicitors of plant resistance in herbivorous insects were remarkably meager until the mid-1990's. Lin et al. (1990) found that application of soybean looper, *Pseudoplusia includens*, regurgitant to mechanical wounds greatly enhanced phytoalexin production in soybeans compared to mechanical injury alone. Application of regurgitant from the caterpillar, *Apocheima pilosaria*, to wounds on birch (*Betula pendula*) foliage, induced phenylalanine ammonia lyase (PAL) and subsequent phenolic biosynthesis (Hartley and Lawton 1987, 1991). To date, the most convincing evidence involves studies of induced plant volatiles that attract natural enemies of phytophagous insects (Turlings et al. 1990, 1998a, b, Turlings and Tumlinson 1992, Horikoshi et al. 1997). Turlings et al. (1993, 1995) showed that regurgitant from several insect species (i.e., *Spodoptera exigua*, *Spodoptera frugiperda*, *Trichoplusia ni*, *Anticarsia gemmatilis*, *H. zea*, and *Schistocerca americana*) applied to artificial wounds induced corn seedlings to emit specific volatile attractants. Although

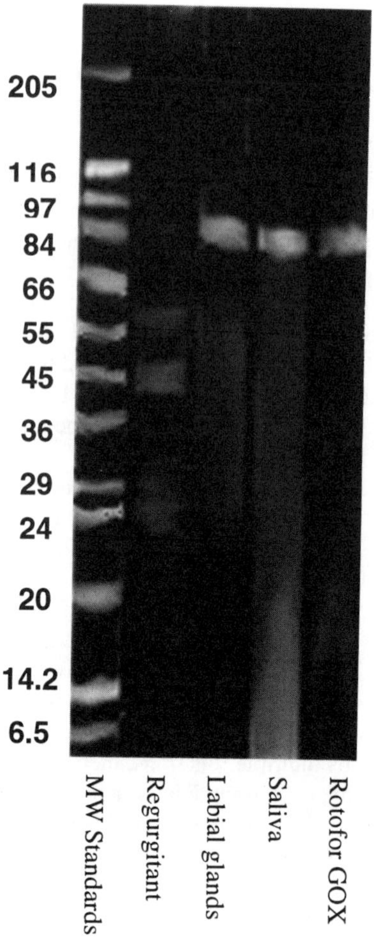

Figure 1: SDS-PAGE of proteins in oral secretions of *H. zea* larvae. Regurgitant was collected by gently holding the larva behind the head over parafilm. Labial gland protein was obtained from homogenates of labial glands. Saliva was obtained as described in the text. Purified glucose oxidase (Rotofor GOX) was obtained after two successive isoelectric-focusing runs on a Biorad Rotofor. The left lane represents molecular weight (MW) standards with molecular weights (in KDa) indicated to the left of each band

quantitative differences were found in response to these various secretions, the blend of volatiles was identical. However, De Moraes et al. (1998) found that cotton, corn and tobacco plants each produced distinct blends of volatiles in response to herbivory by the noctuid caterpillars *Heliothis virescens* and *H. zea*. These studies clearly show that herbivore damage cues are distinct from artificial damage and raise the exciting possibility of herbivore-specific elicitors.

The first clue to the identity of an elicitor from herbivores was reported by Boland et al. (1992), who found that applying almond β-glucosidase to wounded lima bean leaves triggered the release of two homoterpenes that are also produced when plants are infested with spider mites. The similarity of the emitted volatiles suggested that β-glucosidase from the mites may be considered the elicitor for odor induction (Hopke et al. 1994). The homoterpenes could be stored as inactive β-glucosides, but converted to active forms upon hydrolysis by β-glucosidases, and emitted as attractants to predatory mites (Boland et al. 1992). However, more recent evidence indicates that these compounds may be synthesized *de novo* in response to insect feeding (Hopke et al. 1994). Later, another research group first identified an insect elicitor as β-glucosidase (Mattiacci et al. 1995). This enzyme was obtained from the regurgitant and head extracts of the caterpillar, *Pieris brassicae*, and found it to elicit cabbage plant volatiles that attract parasitoid wasps.

More recently the work of Tumlinson and colleagues led to the identification of a non-protein elicitor, N-(17-hydroxylinolenoyl)-L-glutamine (termed volicitin), from the regurgitant of the beet armyworm, *S. exigua* (Alborn et al. 1997). Volicitin applied to damaged corn leaves induced volatiles attractive to parasitoids of *S. exigua,* whereas mechanical damage alone did not induce the same suite of volatiles (Alborn et al. 1997). Ironically, the linolenic acid moiety of volicitin is plant-derived and is esterified to larval-derived glutamine (Paré et al. 1998). Both volicitin and β-glucosidase are putative elicitors of *de novo* synthesis of volatiles. Insect feeding also causes the release of volatiles not associated with natural enemy attraction (i.e., C_6-aldehydes), which are known to induce several defensive genes including lipoxygenases and phenylpropanoid related genes (Bate and Rothstein 1998, see Paré et al., this volume). For example, specific potato volatiles are induced by Colorado potato beetle, *Leptinotarsa decemlineata*, feeding; when regurgitant was applied to mechanical damage, plants attracted more potato beetles to potatoes infested with conspecifics, indicating that herbivores also use induced plant volatiles (Schütz et al. 1997, reviewed by Sabelis et al., this volume).

Korth and Dixon (1997) identified a heat-stable factor in the regurgitant of the caterpillar, *Manduca sexta,* which elicited the rapid induction of proteinase inhibitor transcripts in potato. Accumulation of these transcripts occurred more rapidly in leaves damaged by herbivory or with regurgitant applied to wounds than with mechanical damage alone. Regurgitant was forced through a 0.2-μm filter to prevent microbial contamination. McCloud and Baldwin (1997) found that regurgitant from *M. sexta* dramatically amplified the wound signal jasmonic acid in damaged leaves of *Nicotiana sylvestris*. However, neither root jasmonic acid or whole-plant nicotine levels were affected by regurgitant applied to the wound site.

These studies indicate that distinct signal transduction pathways exist in plants that may specifically distinguish between arthropod herbivory and abiotic damage. However, current studies on caterpillar elicitors are limited to regurgitant (e.g., Lin et al. 1990, Turlings et al. 1993, 1995, 1998b, Mattiacci et al. 1995, Korth and Dixon 1997, Alborn et al. 1997) and/or extracts of heads (Mattiacci et al. 1995). Nonetheless, regurgitant contains components from the salivary glands, foregut, and midgut, and thus may not fully represent the typical saliva deposited on the plant during feeding by insects.

Aphid Saliva and Host Utilization

The best understood interactions between insect saliva and host plants are known from aphids, where saliva plays multiple roles in facilitating host utilization. Aphids secrete at least two types of saliva: one that gels soon after secretion, forming a sheath around the stylet, and the other, a "watery saliva". In the case of the sheath saliva, the sulfhydryl groups of lipoproteins, catalyzed by oxidases and O_2, form cross-links causing the material to gel and form a sheath. Watery saliva contains hydrolytic enzymes for digestion (e.g., pectinase, amylase) and oxidative enzymes. In the grain aphid, *Sitobion avenae*, both sheath and watery saliva contain polyphenol oxidase activity, but peroxidase activity was detected only in sheath material (Urbanska et al. 1998). Control of these particular salivary components is unknown, perhaps the watery saliva and sheath saliva are secreted at the same time, but as the sheath material gels, watery saliva and their enzymes remain (Urbanska et al. 1998).

The stylet sheath may help aphids by: 1) holding stylets in place as it penetrates the substrate's surface and various tissues, 2) sealing wounds and fluid loss when individual cells are punctured, 3) preventing ingestion of unacceptable fluids, 4) preventing signals produced by aphid feeding from diffusing out of the wound, and 5) adsorbing antifeedant phenolics to stylet sheath material by oxidation to toxic quinones which bind to the sheath (Miles 1987a, Peng and Miles 1988, Miles and Peng 1989, Miles 1990).

Plant responses to aphid feeding share common features with the hypersensitive responses to certain phytopathogens (Fernandes 1990, Miles 1990, Campbell and Dreyer 1990, see also Stout and Bostock, this volume). In order to reach phloem where aphids mostly feed, the stylets primarily move through intercellular spaces of mesophyll cells. By secreting pectinesterases and polygalacturonases into the middle lamellae, cell wall polysaccharides are digested and promote stylet movement (Madhusudhan and Miles 1998). Digestion of intercellular pectin may be an important element of compatible aphid-plant interactions (Dreyer and Campbell 1984, 1987, Campbell and Dreyer 1985, 1990). The subsequent release of oligosaccharide products may be possible signals initiating plant hypersensitive responses (Ma et al. 1990, Campbell and Dreyer 1990, Hammerschmidt 1993) or activating the expression of proteinase inhibitor genes (Ryan and Farmer 1991). The specific composition of these oligosaccharides has not been reported and there is no evidence that these products are translocated from the wound site (Campbell and Dreyer 1985). Weevils with chewing mouthparts also secrete polygalacturonases and other cell wall degrading enzymes that are responsible for premature abscission of their fruit hosts (Levine and Hall 1978).

Although sheath formation and secretion of oxidative enzymes may protect aphids from detrimental effects of phenolics and other allelochemicals, oxidation of *o*-dihydroxyphenolics initiates redox cycling in host tissues to generate reactive oxygen species (ROS). This may lead to cell necrosis in some aphid-plant interactions (Jiang and Miles 1993, Miles and Oertli 1993, Jiang 1996) that is a component of induced resistance to phytopathogens. Salivary polyphenol oxidase and peroxidase have only been partially characterized in aphids. They may be responsible for eliciting biochemical responses in plants (Madhusudhan and Miles 1998, Urbanska et al. 1998). For example, both the addition of saliva collected from spotted alfalfa aphid, *Therioaphis trifolii maculata*, and mushroom tyrosinase enhanced phenolic synthesis in alfalfa cell cultures (Jiang 1996), suggesting that the salivary polyphenol oxidase is responsible for the production of phenolics. Alfalfa leaves infested with these aphids also show an increase in ROS, which can be reversed by reducing agents such as ascorbate and glutathione and accelerated with addition of mushroom tyrosinase and peroxidase (Jiang and Miles 1993). Damaged tissues with higher amounts of oxidized phenolics are more acceptable and support more aphids than undamaged alfalfa leaves (Miles and Oertli 1993, Jiang and Miles 1993, Jiang, 1996). For *Aphis fabae*, previous feeding by other conspecifics enhanced host plant acceptability of previously infested plants, but the effect was opposite for *Rhopalosiphum padi*. Improved quality of phloem or enhanced sylet movement on infested plants was believed to enhance host plant acceptability on the plants infested by *A. fabae* (Prado and Tjallingii 1997). The generation of ROS may ultimately interfere with the activation of induced resistance to insect herbivores.

Translocation of pectinases and oxidative enzymes secreted by the spotted alfalfa aphid may cause hypersensitive reactions in alfalfa leaves distal to aphid feeding. Oxidative enzyme activities were higher in the tips of infested plants and detection of C^{14} glutamate derived from aphids was found in alfalfa leaf tips; however, incorporation of radiolabel into secreted protein or secretion of free glutamate was not determined. This is particularly important because relatively high amounts of glutamate are secreted by these aphids (Madhusudhan and Miles 1998).

Saliva and Regulation of Plant Growth

In some cases herbivore saliva can regulate plant regrowth after herbivory. Biomass and tiller production of clipped blue grama grass was reduced, compared to controls, when salivary gland extracts of the grasshopper, *Brachystola magna*, were applied (Detling and Dyer 1981). Bioactive polypeptides extracted from midguts and crops of *Romalea guttata* enhanced growth of sorghum coleoptiles (Dyer 1995). Similar growth enhancing activity in some plants has been attributed to $NaHCO_3$ in the saliva of American bison

Bison bison (Detling et al. 1981), but not on blue grama grass, a grass that bison would naturally defoliate (Detling et al. 1980).

Salivary secretions injected into plant tissue by piercing-sucking insects are also responsible for abnormal plant growth and gall formation. Indoleacetic acid (IAA) found in some salivary secretions or enzymes that inactivate IAA action appear to regulate plant growth (Hori 1974, 1976). Amino acid secretion may potentiate the action of plant growth-regulating substances (Hori 1992). Miles (1987b) and Hori (1992) provide recent reviews of this literature. The influence of salivary secretions on plant growth has not been studied in recent years. To our knowledge, how these substances may modify host plant signal transduction pathways involved in induced responses to herbivory has not been investigated.

Saliva and Induced Resistance to Phytopathogens

Arthropod herbivory may activate a salicylate-dependent plant defense pathway involving the systemic induction of resistance against a broad spectrum of phytopathogens (Hatcher 1995), a phenomenon often called systemic acquired resistance (SAR). SAR involves a necrotic response on the infected leaf, with subsequent translocation of an uncharacterized signal to systemic leaves that activates a suite of genes that encode antimicrobial defenses, including pathogenesis-related proteins, lignification and phytoalexins (Sticher et al. 1997, Hammerschmidt and Nicholson, Hammerschmidt and Smith-Becker, this volume). For example, SAR against the fungal pathogen, *Verticillium dahliae,* was induced by spider mites infesting cotton (Karban et al. 1987). Similarly, the corn earworm, *H. zea*, induced SAR against the bacterial pathogen, *Pseudomonas syringae*, on tomato plants (Stout et al. 1998). However, in other cases, herbivory does not trigger SAR. Herbivory by *M. sexta* failed to elicit SAR against tobacco mosaic virus in tobacco (Ajlan and Potter 1992) and feeding by spider mites, *Tetranychus utricae*, or fall armyworms, *S. frugiperda*, did not induce resistance to anthracnose, *Colletotrichum lagenarium*, in cucumber (Ajlan and Potter 1991). These contradictory results suggest that herbivore-specific elicitors may induce SAR in some systems.

We are unaware of any previous studies to demonstrate that insect saliva may contribute to resistance to phytopathogens, although indirect evidence supported the possibility. Mechanical damage alone does not induce SAR (Sticher et al. 1997), although in soybean, defoliation by the soybean looper induced SAR against stem canker disease, *Diaporthe phaseolorum* var. *caulivora* and red crown rot, *Calonectria crotalariae* (Russin et al. 1989, Padgett et al. 1994). Regurgitant from this insect stimulates the production of phytoalexins, compounds typically associated with SAR (Lin et al. 1990).

Our studies began with observations that feeding by the caterpillar, *H. zea*, induces SAR in soybean to *P. syringae* pv. *glycinea* (Fig. 2). Mechanical damage was insufficient to induce resistance, but the application of saliva to mechanically damaged plants induced resistance comparable to insect feeding

(Fig. 2). SAR was not observed if saliva was heated by autoclaving (Fig. 2). We have since determined that salivary proteins produced H_2O_2 and cell death in soybean cell cultures (unpublished data). Cell necrosis is often associated with SAR and necrosis is mediated, in part, by an oxidative burst of H_2O_2 (Sticher et al. 1997). Addition of catalase to the cells (to decompose H_2O_2) prevented cell death. We found that a 78 kD glycoprotein from salivary glands triggered rapid accumulation of H_2O_2 and subsequent death of soybean cells (unpublished data). Further characterization revealed the protein to be a glucose oxidase that oxidizes glucose in the presence of O_2, to produce gluconic acid and H_2O_2. Based upon an N-terminal sequence, the *H. zea* glucose oxidase (GOX) bears little homology to fungal GOXs (unpublished data).

To determine if *H. zea* GOX was responsible for inducing systemic resistance to *P. syringae*, we applied the purified protein to mechanical wounds on soybean leaves and assayed for resistance 1, 3 and 7 days later. Treatments assayed for resistance one day after application of GOX did not affect resistance (data not shown), but 3 and 7 days later the GOX application induced SAR at levels consistent with that induced by herbivory (Fig. 3).

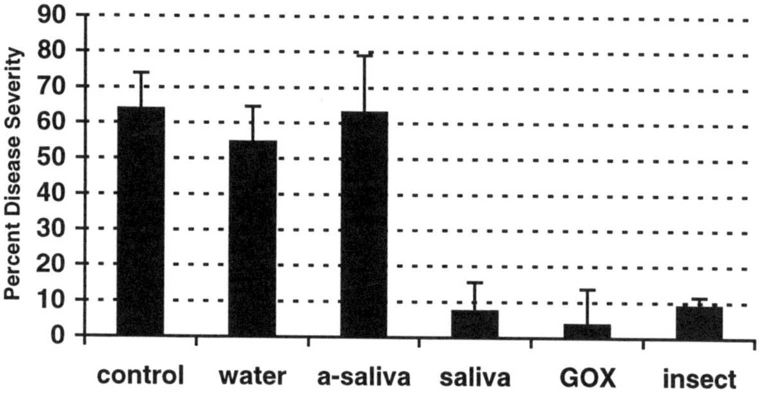

Figure 2: Effect of herbivory and saliva on severity of bacterial blight in soybean. Soybean plants (cv. Williams) at the five-node stage were inoculated with *Pseudomonas syringae* pv. *glycinea* three days after treatment. Disease severity was assessed using the Horsfall-Barratt scale with ratings converted to percent severity using Elanco conversion tables. Treatments included **control**, no treatment; **water,** two 0.5 cm diameter holes were removed with a cork borer on each trifoliate of the first true leaf, water was brushed on wound surface with a camel hair brush; **a-saliva,** same treatment as water except 10 μg of autoclaved saliva prepared from labial glands of 6th instar *H. zea*; **saliva,** same treatment as a-saliva except saliva was not autoclaved prior to treatment; **GOX,** same treatment as saliva except 5 μg of purified glucose oxidase from *H. zea* was applied instead of saliva; **insect**, one fourth instar *H. zea* was confined to the first trifoliate leaf for three days prior to assay. The saliva, GOX and insect treatments were significantly lower than control ($P < 0.05$). Similar responses were seen 7 days after treatment (data not shown).

We also analyzed for biochemical markers of systemic resistance including the phytoalexin daidzein. GOX stimulates systemic increases in total daidzein (free and glycoside forms) above that found for mechanical damage alone and again, comparable to damage inflicted by *H. zea* (Fig. 3). Our results indicate that a specific factor, probably glucose oxidase, in herbivore saliva induces SAR against a bacterial pathogen. More recent findings indicate glucose oxidase also induces resistance to the fungus *Cercospora sojinae* (unpublished).

Herbivore Saliva and Immunosuppression of Induced Resistance

Plants and animals deploy a rich diversity of constitutive and inducible defenses against invading organisms (Wikel 1996, Karban and Baldwin 1997, Lamb and Dixon 1997). Recent evidence points to analogies in the defense signaling pathways of plants and animals (Bergey et al. 1996, Farmer 1997). The ability of organisms to effectively utilize a specific host depends, in part, upon their ability to evade host defenses. The prevailing view in insect-plant interactions is that insects adapt to their host plants through evolution of detoxification pathways against plant defenses (e.g., Berenbaum and Zangerl 1998). We present a different hypothesis: that insect herbivores adapt to host plants by suppression of early events associated with plant recognition and defense signal transduction.

Our evidence supporting this hypothesis is based upon several observations. Several salivary enzymes (e.g., glucose oxidase, peroxidase) of herbivores produce reactive oxygen species that may disrupt the octadecanoid

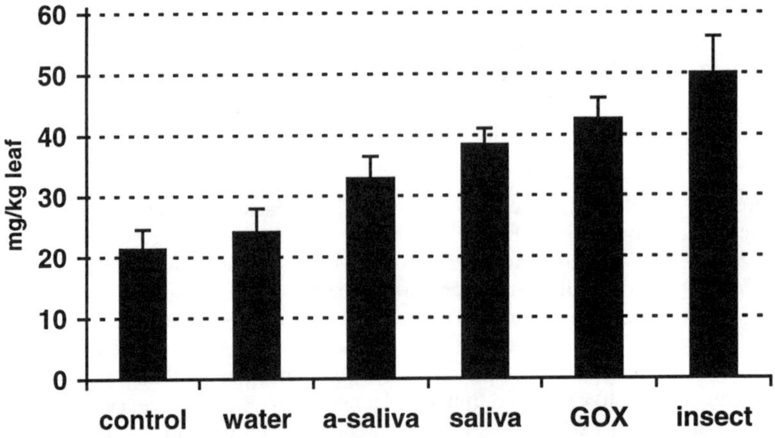

Figure 3: Effect of herbivory and saliva on phytoalexin (daidzein) levels in soybean leaves. Levels of daidzein glycosides were determined in the fifth soybean leaf three days following treatment. See Fig. 2. for explanation of treatments. The saliva, GOX and insect treatments were significantly higher than the control ($P < 0.05$).

pathway associated with induction of insect resistance. Systemin is a peptide signal in tomato plants that is a key component of the octadecanoid pathway. Wounding produces a rapid oxidative burst that oxidizes the peptide systemic signal, systemin, to an inactive peptide (Narvaez-Vasquez et al. 1995). An even greater fraction of the peptide may become oxidized when glucose oxidase is secreted at the feeding sites. We have found that the H_2O_2–scavenging enzyme, catalase, applied to wound sites (e.g., tomato, tobacco plants) enhances insect resistance (unpublished data), whereas application of glucose oxidase to wounded tissue in tobacco plants reduces the magnitude of induced resistance to the tobacco budworm *H. virescens* (unpublished data). In other words, components of insect saliva suppress induced resistance to insects.

Second, reactive oxygen species such as H_2O_2 or nitric oxide can induce salicylic acid production (Niki et al. 1998, Durner et al. 1998), an essential compound for expression of SAR, and an inhibitor of the octadecanoid pathway (Doares et al. 1995, Niki et al. 1998). In fact, either application of salicylic acid or analogues inhibited induced resistance to insects (Thaler et al. 1999). Similarly, we have recently reported that genetic overexpression of salicylic acid levels in tobacco reduces induced resistance to insects (Felton et al. 1999). We used tobacco plants with genetic co-suppression or overexpression of phenylalanine ammonia lyase (PAL), a key enzyme in salicylate biosynthesis, to find that jasmonate and salicylate levels are inversely related. Thus in PAL-suppressed plants, salicylate production is compromised, but there is greater wound-induced production of jasmonate with a concomitant increase in induced resistance to *H. virescens* (Felton et al. 1999). Conversely, plants that were compromised in salicylate biosynthesis exhibited increased susceptibility to tobacco mosaic virus, but plants with increased salicylate accumulation were more resistant to the virus. These results demonstrate cross-talk between pathways for systemic induction of anti-microbial and anti-insect responses. Rather than viewing the octadecanoid and salicylate-dependent SAR pathways as distinct and separate pathways, a more realistic view is that they function as interconnected signal networks for defense (Stout and Bostock, this volume).

We suggest that the saliva of insect herbivores may produce cross-talk between two signal transduction pathways that partially triggers SAR, at the expense of the octadecanoid signal cascade targeted at defense against herbivores. In fact, glucose oxidase treatment induces salicylate accumulation in wounded leaves (unpublished data), which may partly explain the decrease in induced resistance we observed.

In addition to the possible effects of glucose oxidase on systemin oxidation and salicylic acid biosynthesis, our preliminary data suggests that glucose oxidase inhibits lipoxygenase activity, a key enzyme in the biosynthesis of the octadecanoid signal, jasmonic acid. Consistent with this finding is that glucose oxidase treatment of tobacco foliage inhibits jasmonic acid production (unpublished data). In summary, our findings to date indicate that the secretion of glucose oxidase by herbivores may interfere with several integral components of the octadecanoid pathway, several of which are associated with SAR.

Our hypothesis is analogous to insect-host interactions involving blood-feeding arthropods (e.g., mosquitoes, ticks, biting flies), where a multitude of salivary components circumvent the hemostatic processes of their vertebrate hosts (Champagne 1994, Ribeiro 1995, Wikel 1996). Salivary proteins act as anticoagulants, antiplatelet factors, and vasodilators to facilitate the blood meal (Ribeiro 1995, Wikel 1996). Hard ticks differ from other hematophagous arthropods because they remain attached to their host for an extended period, thus allowing for host immune defenses to be mounted. Nevertheless, these ticks suppress the early stages of the immune response by host T-cells (Wikel 1996). In stark contrast, the saliva of phytophagous, chewing insects (e.g., Lepidopteran larvae) has been virtually ignored as a factor mediating host defensive responses (Ribeiro 1995). We propose, like the saliva of hematophagous arthropods, saliva from herbivores can perform a *physiological* role in evasion of host defenses, via the suppression of constitutive defenses and disruption of host signals that mediate inducible defenses. We base our hypothesis on the following observations: 1) there is striking similarity in the signal cascades for systemic wound or trauma responses between animals and plants (e.g., Ryan and Pearce 1998), 2) the saliva of blood-feeding arthropods possesses many factors associated with suppressing host defensive responses, and 3) there are many similarities among the salivary enzymes of blood-feeding arthropods, herbivorous homopterans and lepidoptera. Oxidative enzymes are common to each feeding guild (Table 1).

The ROS-producing enzymes from blood-feeding arthropods may be involved with suppression or inhibition of host defense responses in vertebrates to blood-feeding (Ribeiro 1995). The conserved elements of the immune responses of plants and animals and the similarities in salivary enzymes among plant and animal-feeding arthropods suggests that an important function of insect saliva is to suppress host defenses. The goal of our work is to provide a better understanding of how saliva mediates host defensive responses.

Synthesis

Insect oral secretions or saliva are produced by salivary glands that have multiple functions. For example, the labial glands of actively feeding caterpillars may secrete saliva, but begin to produce silk as the insect completes its immature development. There are multiple enzymes secreted and produced by the salivary glands and we are just now beginning to characterize these proteins. Other material, amino acids, lipophilic compounds, etc. have also been characterized; how these materials interact with each other to produce a distinct signal remains to be tested. A single salivary substance may act on multiple pathways. Our studies with salivary GOX suggests the production of ROS triggers SAR that inhibits the octadecanoid pathway, which is more specifically targeted to resistance against insect herbivores.

Table 1. Similarities in Salivary Enzymes among Blood-Feeding and Plant-Feeding Arthropods.

Enzyme (ROS product)	Hematophagous Arthropods	Herbivorous Homoptera	Herbivorous Lepidoptera
Catechol oxidase/ peroxidase (O_2^-, H_2O_2)	Yes[1]	Yes[2]	?
NADPH oxidase (O_2^-, H_2O_2)	Yes	Possibly	?
thiol oxidase (H_2O_2)	Yes	?	Yes[3]
glucose oxidase (H_2O_2)	?	?	Yes
Nitric oxide synthase (NO)	Yes	?	?

[1]References from top to bottom in this column, Ribeiro et al. 1994, Ribeiro 1996a, b, and Yuda et al. 1996.
[2]References from top to bottom in this column, Madhusudhan and Miles 1998 and Jiang and Miles 1993.
[3]References from top to bottom in this column, Felton et al., unpublished data, Eichenseer et al., unpublished data.
? = No information available.

Future Directions

The composition of herbivore saliva is unknown. Saliva contains myriad proteins with unknown functions. Development of transgenic approaches will have a major impact on studying salivary gland function. The identification of salivary gland-specific promoters or enhancers of direct transcription to salivary cells, and the development of vectors to transfer genes into the glands would be a great aid to understanding gland function. Overexpression or suppression of endogenous genes would aid in understanding the function of specific salivary proteins. For example, Wang et al. (1999) have used polyclonal antibodies to establish that a cyst nematode salivary protein is secreted into its host.

There is lack of information on the dose-response of saliva on plant responses. In most instances, the amounts of salivary elicitors secreted into plant tissues are unknown. This is an important question that needs to be addressed in order to establish the role of saliva in mediating ecological interactions.

Little is known about the genotypic variability of plant responses to salivary signals. Likewise, virtually nothing is known regarding the intraspecific and interspecific variation in herbivore salivary signals. These important issues will aid in understanding the evolutionary interplay between plants, herbivores and their natural enemies. Finally, there is essentially nothing known regarding the perception systems for salivary signals of herbivores. Plant pathologists have been very active in investigating how plants specifically perceive pathogens; entomologists need to make similar efforts for studying how plants perceive insect signals.

Acknowledgements

We thank the USDA-NRI Entomology and NSF-Evolutionary and Physiological Ecology Programs for support. Gratitude is expressed to J. L. Bi, D. Hua, J. B. Murphy, D. Tebeest, J. Rupe, R. Dixon, K. Korth, C. Lamb and R. Plymale for research cooperation.

Literature Cited

Ajlan, A.M. and D.A. Potter. 1991. Does immunization of cucumber against anthracnose by *Colletotrichum lagenarium* affect host suitability for arthropods? Entomologia Experimentalis et Applicata 58:83-91.

Ajlan, A.M. and D.A. Potter. 1992. Lack of effect of tobacco mosaic virus-induced systemic acquired resistance on arthropod herbivores in tobacco. Phytopathology 82:647-651.

Alborn. H.T., T.C.J. Turlings, T.H. Jones, G. Stenhagen, J.H. Loughrin and J.H. Tumlinson. 1997. An elicitor of plant volatiles from beet armyworm oral secretion. Science 276:945-948.

Ali, D.W. 1997. The aminergic and peptidergic innervation of insect salivary glands. Journal of Experimental Biology 200:1941-1949.

Ali, D.W. and I. Orchard. 1996. The uptake and release of serotonin and dopamine associated with locust (*Locusta migratoria*) salivary glands. Journal of Experimental Biology 199:699-709.

Bate, N.J. and S.J. Rothstein. 1998. C_6-volatiles derived from the lipoxygenase pathway induce a subset of defense-related genes. Plant Journal 16:561-569.

Bay, C.M.H. 1978a. The control of enzyme secretion from fly salivary glands (*Calliphora erythrocephala*). Journal of Physiology 274:421-435

Bay, C.M.H. 1978b. The secretion and action of the digestive enzymes of the salivary glands of the blowfly, *Calliphora erythrocephala*. Journal of Insect Physiology 24:141-149.

Berenbaum, M.R. and A.R. Zangerl. 1998. Chemical phenotype matching between a plant and its herbivore. Proceedings of the National Academy of Sciences USA 95:13743-13748.

Bergey, D., G. Howe, and C.A. Ryan. 1996. Polypeptide signaling for plant defensive genes exhibits analogies to defense signaling in animals. Proceedings of the National Academy of Science USA 93:12053-12058.

Bernasconi, M.L., T.C.J. Turlings, L. Ambrosetti, P. Bassetti and S. Dorn. 1998. Herbivore-induced emissions of maize volatiles repel the corn leaf aphid, *Rhopalosiphum maidis*. Entomologia Experimentalis et Applicata 87:133-142.

Boland, W., Z. Feng, J. Donath and A. Gabler. 1992. Are cyclic C11 and C16 homoterpenes plant volatiles indicating herbivory? Naturwissenschaften 79:368-371.

Campbell, B.C., and D.L. Dreyer. 1985. Host-plant resistance of sorghum: Differential hydrolysis of sorghum pectic substances by polysaccharases of greenbug biotypes (*Schizaphis graminum*, Homoptera: Aphididae). Archives of Insect Biochemistry and Physiology 2:203-215.

Campbell, B.C. and D.L. Dreyer. 1990. The role of plant matrix polysaccharides in aphid-plant interactions. Pages 149-170 *in* R. K. Campbell and R. D. Eikenbary, editors. Aphid-Plant Genotype Interactions. Elsevier, Amsterdam.

Champagne, D.E. 1994. The role of salivary vasodilators in bloodfeeding and parasite transmission. Parasitology Today 10:430-433.

Chapman, R.F. 1998. The Insects: Structure and Function, 4th ed. Cambridge University Press, Cambridge.
Cohen, A.C. and A.G. Wheeler, Jr. 1998. Role of saliva in the highly destructive fourlined plant bug (Hemiptera: Miridae: Mirinae). Annals of the Entomological Society of America 91:94-100.
De Moraes, C.M., W.J. Lewis, P.W. Paré, H.T. Alborn and J.H. Tumlinson. 1998. Herbivore-infested plants selectively attract parasitoids. Nature 393:570-573.
Detling, J.K. and M.I. Dyer. 1981. Evidence for potential plant growth regulators in grasshoppers. Ecology 62:485-488.
Detling, J.K., M.I. Dyer, C. Gregg-Procter and D.T. Winn. 1980. Plant herbivore interactions: examination of potential effects of bison saliva on regrowth of *Bouteloua gracilis* (H.B.K.) Lag. Oecologia 45:26-31.
Detling, J.K., C.W. Ross, M.H. Walmsley, D.W. Hilbert, C.A. Bonilla and M.I. Dyer. 1981. Examination of North American bison saliva for potential plant growth regulators. Journal of Chemical Ecology 7:239-246.
Doares, S.H., J. Narvaez-Vasquez, A. Conconi. and C.A. Ryan. 1995. Salicylic acid inhibits synthesis of proteinase inhibitors in tomato leaves induced by systemin and jasmonic acid. Plant Physiology 108:1741-1746.
Dreyer, D.L. and B.C. Campbell. 1984. Association of the degree of methylation of intercellular pectin with plant resistance to aphids and with induction of aphid biotype (*Schizaphis graminum*). Experientia 40:224-226.
Dreyer, D.L. and B.C. Campbell. 1987. Chemical basis of host-plant resistance to aphids. Plant Cell and Environment 10:353-361.
Durner, J., D. Wendehenne and D.F. Klesig. 1998. Defense gene induction in tobacco by nitric oxide, cyclic GMP, and cyclic ADP-ribose. Proceedings of the National Academy of Sciences USA 95:10328-10333.
Dyer, M.I. 1995. Grasshopper crop and midgut extract effects on plants: an example of reward feedback. Proceedings of the National Academy of Sciences USA 92:5475-5478.
Dyer, M.I. and U.G. Bokhari. 1976. Plant-animal interactions: studies of the effects of grasshopper (*Melanoplus sanguinipes*) grazing on blue grama grass (*Bouteloua gracilis*). Ecology 57:762-772.
Farmer, E.E. 1997. New fatty acid-based signals: a lesson from the plant world. Science 276:912-913
Felton, G.W., C.B. Summers, A.J. Mueller. 1994. Oxidative responses in soybean to herbivory bean leaf beetle and three-cornered alfalfa hopper. Journal of Chemical Ecology 20:639-650.
Felton, G.W., K.L. Korth, J.L. Bi, S.V. Wesley, D.V. Huhman, M.C. Mathews, J.B. Murphy, C. Lamb and R.A. Dixon. 1999. Inverse relationship between systemic resistance of plants to microorganisms and to insect herbivory. Current Biology 9:317-320.
Fernandes, G.W. 1990. Hypersensitivity: a neglected plant resistance mechanism against insect herbivores. Environmental Entomology 19:1173-1182.
Hammerschmidt, R. 1993. The nature and generation of systemic signals induced by pathogens, arthropod herbivores and wounds. Advances in Plant Pathology 10:307-337.
Hanhimäki, S. and J. Senn. 1992. Sources of variation in rapidly inducible responses to leaf damage in the mountain birch-insect herbivore system. Oecologia 91:318-331.
Hartley, S.E. and J.H. Lawton. 1987. Effects of different types of damage on the chemistry of birch foliage, and the responses of birch feeding insects. Oecologia 74:432-437.
Hartley, S.E. and J.H. Lawton. 1991. Biochemical aspects and significance of the rapidly induced accumulation of phenolics in birch foliage. Pages 105-132 *in* D. W. Tallamy and M. J. Raupp (editors). Phytochemical Induction by Herbivores. John Wiley, New York.
Hatcher, P.E. 1995. Three-way interactions between plant pathogenic fungi, herbivorous insects and their host plants. Biological Reviews 70:639-694.
House, C.R. 1980. Physiology of invertebrate salivary glands. Biological Reviews 55:417-473.
Hopke, J., J. Donath, S. Blechert and W. Boland. 1994. Herbivore-induced volatiles: the emission of acyclic homoterpenes from leaves of *Phaseolus lunatus* and *Zea mays* can be triggered by a β-glucosidase and jasmonic acid. FEBS Letters 352:146-150.
Hori, K. 1974. Plant growth-promoting factor in the salivary gland of the bug, *Lygus disponsi*. Journal of Insect Physiology 20:1623-1627.

Hori, K. 1976. Plant growth-regulating factor in the salivary gland of several heteropterous insects. Comparative Biochemistry and Physiology 53B:435-438.
Hori, K. 1992. Insect secretions and their effect on plant growth, with special reference to Hemipterans. Pages 157-170 *in* J.D. Shorthouse and O. Rohfritsch, editors. Biology of Insect-induced Galls. Oxford University, Oxford.
Horikoshi, M., J.Takabayashi, S. Yano, R. Yamaoka, N. Ohsaki and Y. Sato. 1997. *Cotesia glomerata* female wasps use fatty acids from plant herbivore complex in host searching. Journal of Chemical Ecology 23:1505-1515.
Jiang, Y. 1996. Oxidative interactions between the spotted alfalfa aphid (*Therioaphis trifolii maculata*) (Homoptera: Aphididae) and the host plant *Medicago sativa*. Bulletin of Entomological Research 86:533-540.
Jiang, Y. and P.W. Miles. 1993. Responses of a compatible lucerne variety to attack by spotted alfalfa aphid: changes in the redox balance in affected tissues. Entomologia Experimentalis et Applicata 67:263-274.
Just, F. and B. Walz. 1996. The effects of serotonin and dopamine on salivary secretion by isolated cockroach salivary glands. Journal of Experimental Biology 199:407- 413.
Karban, R., R. Adamchak and W.C. Schnathorst. 1987. Induced resistance and interspecific competition between spider mites and a vascular wilt fungus. Science 235:678-680.
Karban R. and I.T. Baldwin. 1997. Induced Responses to Herbivory. University of Chicago Press, Chicago, IL.
Korth, K.L. and R.A. Dixon. 1997. Evidence for chewing insect-specific molecular events distinct from a general wound response in leaves. Plant Physiology 115:1299-1305.
Lamb, C. and R. Dixon. 1997. The oxidative burst in plant disease resistance. Annual Review of Plant Physiology and Plant Molecular Biology 48:251-175.
Levine, E. and F.R. Hall. 1978. Pectinases and cellulases from plum curculio larvae: possible causes of apple and plum fruit abscission. Entomologia Experimentalis et Applicata 23:259-268.
Lin, H., M. Kogan, and D. Fischer. 1990. Induced resistance in soybean to the Mexican bean beetle (Coleoptera: Coccinellidae): comparisons of inducing factors. Environmental Entomology 19:1852-1857.
Ma, R., J.C. Reese, W.C.I. Black and P. Bramel-Cox. 1990. Detection of pectinesterase and polygalacturonase from salivary secretions of living greenbugs, *Schizaphis graminum* (Homoptera: Aphididae). Journal of Insect Physiology 36:507-512.
Madhusudhan, V. V. and P.W. Miles. 1998. Mobility of salivary components as a possible reason for differences in the responses of alfalfa to the spotted alfalfa aphid and pea aphid. Entomologia Experimentalis et Applicata 86:25-39.
Mattiacci, L., M. Dicke and M.A. Posthumus. 1995. β-Glucosidase: an elicitor of herbivore-induced plant odor that attracts host-searching parasitic wasps. Proceedings of the National Academy of Sciences USA 92:2036-2040.
McCloud, E.S. and I.T. Baldwin. 1997. Herbivory and caterpillar regurgitants amplify the wound-induced increases in jasmonic acid but not nicotine in *Nicotiana sylvestris*. Planta 203:430-435.
Miles, P.W. 1987a. Feeding process of Aphidoidea in relation to effects on their food plants. Pages 321-339 *in* A. K. Minks and P. Harrewijn, editors. Aphids, their Biology, Natural Enemies and Control, vol. A. Elsevier, Amsterdam.
Miles, P.W. 1987b. The responses of plants to the feeding of Aphidoidea: Principles. Pages 1-21 *in* A. K. Minks and P. Harrewijn, editors. Aphids, their Biology, Natural Enemies and Control, vol. B. Elsevier, Amsterdam.
Miles, P.W. 1990. Aphid salivary secretions and their involvement in plant toxicoses. Pages 131-147 *in* R. K. Campbell and R. D. Eikenbary, editors. Aphid-Plant Genotype Interactions. Elsevier, Amsterdam.
Miles, P.W. and Z. Peng. 1989. Studies on the salivary physiology of plant bugs: detoxification of phytochemicals by the salivary peroxidase of aphids. Journal of Insect Physiology 35:865-872
Miles, P.W. and J.J. Oertli. 1993. The significance of antioxidants in the aphid-plant interaction: the redox hypothesis. Entomologia Experimentalis et Applicata 67:275-283.

Narvaez-Vasquez, J., G. Pearce, M. Orozco-Cardenas, V. Fracheschi and C.A. Ryan. 1995. Autoradiographic and biochemical evidence for the systemic translocation of systemin in tomato plants. Planta 195:593-600.

Niki, T., I. Mitsuhara, S. Seo, N. Ohtsubo and Y. Ohashi. 1998. Antagonistic effect of salicylic acid and jasmonic acid on the expression of pathogenesis-related (PR) protein genes in wounded mature tobacco leaves. Plant Cell Physiology 39:500-507.

Padgett, G.B., J.S. Russin, J.P. Snow, D.J. Boethel and G.T. Berggren. 1994. Interactions among the soybean looper (Lepidoptera: Noctuidae), threecornered alfalfa hopper (Homoptera: Membracidae), stem canker, and red crown rot in soybean. Journal of Entomological Science 29:110-119.

Paré, P.W. and J.H. Tumlinson. 1997. De novo synthesis of volatiles induced by insect herbivory in cotton plants. Plant Physiology 114:1161-1167.

Paré, P.W., H.T. Alborn and J.H. Tumlinson. 1998. Concerted biosynthesis of an insect elicitor of plant volatiles. Proceedings of the National Academy of Sciences USA 95:13971-13975.

Peng, Z. and P.W. Miles. 1988. Studies on the salivary physiology of plant bugs: function of the catechol oxidase of the rose aphid. Journal of Insect Physiology 34:1027-1033

Prado, E. and W.F. Tjallingii. 1997. Effects of previous plant infestation on sieve element acceptance by two aphids. Entomologia Experimentalis et Applicata 82:189-200.

Ribeiro, J.M.C. 1995. Insect saliva: function, biochemistry, and physiology. Pages 74-98 *in* R. F. Chapman and G. de Boer, editors. Regulatory Mechanisms in Insect Feeding. Chapman and Hall, New York.

Ribeiro, J.M.C. 1996a. NAD(P)H-dependent production of oxygen reactive species by the salivary glands of the mosquito *Anopheles albimanus*. Insect Biochemistry and Molecular Biology 26:715-730.

Ribeiro, J.M.C. 1996b. Salivary thiol oxidase activity of *Rhodnius prolixus*. Insect Biochemistry and Molecular Biology 26:899-905.

Ribeiro, J.M.C., R.H. Nussenzveig and G. Tortorella. 1994. Salivary vasodilators of *Aedes triseriatus* and *Anopheles gambiae* (Dipetera: Culicidae). Journal of Medical Entomology 31:747-753.

Russin, J.S., M.B. Layton and D.J. Boethel. 1989. Severity of soybean stem canker disease affected by insect-induced defoliation. Plant Disease 73:144-147.

Ryan, C.A. and E.E. Farmer. 1991. Oligosaccharide signals in plants: a current assessment. Annual Review of Plant Physiology and Plant Molecular Biology 42:651-674.

Ryan, C.A. and G. Pearce. 1998. Systemin: a polypeptide signal for plant defensive genes. Annual Review of Cell and Developmental Biology 14:1-17.

Schütz, S., B. Weissbecker, A. Klein and H.E. Hummmel. 1997. Host plant selection of the Colorado potato beetle as influenced by damage induced volatiles of the potato plant. Naturwissenschaften 84:212-217.

Sehnal, F. and H. Akai. 1990. Insect silk glands: their types, development and function, and effects of environmental factors and morphogenetic hormones on them. International Journal of Insect Morphology and Embryology 19:79-132.

Sticher, L., B. Mauch-Mani and J.P. Métraux. 1997. Systemic acquired resistance. Annual Review of Phytopathology 35:235-270.

Stout, M.J., J. Workman and S.S. Duffey. 1994. Differential induction of tomato foliar proteins by arthropod herbivores. Journal of Chemical Ecology 20:2575-2594.

Stout, M.J., K.V. Workman, R.M. Bostock and S.S. Duffey. 1998. Specificity of induced resistance in the tomato, *Lycopersicon esculentum*. Oecologia 113:74-81.

Srivastava, U.S. 1959. The maxillary glands of some Coleoptera. Proceedings of Royal Entomological Society London (A). 34:57-62

Taylor, G.S. and P.W. Miles. 1994. Composition and variability of the saliva of coreids in relation to phytoxicoses and other aspects of the salivary physiology of phytophagous Heteroptera. Entomologia Experimentalis et Applicata 73:265-277.

Thaler, J. S., A. L. Fidantsef, S. S. Duffey, and R. M. Bostock. 1999. Tradeoffs in plant defense against pathogens and herbivores: a field demonstration using chemical elicitors of induced resistance. Journal of Chemical Ecology (in press).

Trimmer, B.A. 1985. Serotonin and the control of salivation in the blowfly *Calliphora*. Journal of Experimental Biology 114:307- 328.
Turlings, T.C.J. and J.H. Tumlinson. 1992. Systemic release of chemical signals by herbivore-injured corn. Proceedings of the National Academy of Sciences USA 89:8399-8402.
Turlings, T.C.J., J.H. Tumlinson and W.J. Lewis. 1990. Exploitation of herbivore-induced plant odors by host-seeking wasps. Science 250:1251-1253.
Turlings, T.C.J., P.J. McCall, H.T. Alborn and J.H. Tumlinson. 1993. An elicitor in caterpillar oral secretions that induces corn seedlings to emit chemical signals attractive to parasitic wasps. Journal of Chemical Ecology 19:411-425.
Turlings, T.C.J., J.H. Loughrin, U. Röse, P.J. McCall, W.J. Lewis and J.H. Tumlinson. 1995. How caterpillar-damaged plants protect themselves by attracting parasitic wasps. Proceedings of the National Academy of Sciences USA 92:4169-4174.
Turlings, T.C.J., M.L. Bernasconi, R. Bertossa, F. Bigler, G. Caloz and S. Dorn. 1998a. The induction of volatile emissions in maize by three herbivore species with different feeding habits: possible consequences for their natural enemies. Biological Control 11:122-129.
Turlings, T.C.J., U.B. Lengwiler, M.L. Bernasconi and D. Wechsler. 1998b. Timing of induced volatile emissions in maize seedlings. Planta 207:146-152.
Urbanska, A., W.F. Tjallingii, A.F.G. Dixon, and B. Leszczynski. 1998. Phenol oxidising enzymes in the grain aphid's saliva. Entomologia Experimentalis et Applicata 86:197-203.
Wang, X., D. Meyers, Y. Yan, T. Baum, G. Smant, R.S. Hussey and E. Davis. 1999. *In planta* localization of a β-1,4-endoglucanase secreted by *Heterodera glycines*. Molecular Plant-Microbe Interactions 12:64-67.
Wikel, S.K. 1996. Host immunity to ticks. Annual Review of Entomology 41:1-22.
Yuda, M., M. Hirai, K. Miura, H. Matsumura, K. Ando and Y. Chinzei. 1996. cDNA cloning, expression and characterization and nitric-oxide synthase from the salivary glands of the blood-sucking insect *Rhodnius prolixus*. European Journal of Biochemistry 242:807-812.

The Role of Salicylic Acid in Disease Resistance

Ray Hammerschmidt and Jennifer A. Smith-Becker

Abstract

Plant resistance against infection is mediated through the coordinated expression of defenses. Over the last decade, it has become clear that one of the factors involved in regulating the expression of these defenses is salicylic acid, a simple hydroxybenzoic acid. Salicylic acid has been shown to play an important role in the expression of both local resistance controlled by major genes and systemic induced resistance that develops after an initial pathogen attack. The biosynthesis, role in defense, and recent advances in the genetic analysis of the salicylic acid dependent expression of induced resistance is discussed.

Introduction

Phenolic compounds have long been associated with the defense of plants to both pathogens and herbivores. In the earliest studies, simple phenols such as catechol and chlorogenic acid were associated with defense against pathogens as pre-formed antibiotics (Hammerschmidt and Schultz 1996). As a result of the phytoalexin hypothesis of Muller and Borger (see Kuć 1995), infection-induced phenolic compounds have also received attention as factors in defense. These would include phenolic phytoalexins such as the isoflavonoids and pterocarpans found in legumes and deoxyanthocyanidins found in some member of the poaceae, as well as constitutive phenols like chlorogenic acid that accumulate after infection or wounding (Nicholson and Hammerschmidt 1992, Kuć 1995). Phenolic compounds also may contribute to defense as lignins and wall bound phenolic esters that crosslink and strengthen cell walls (Nicholson and Hammerschmidt 1992). Phenolic compounds are also implicated in defense against herbivores and include preformed and induced compounds such as the tannins and furanocoumarins (Hammerschmidt and Schultz 1996). Oxidation products of the phenols may also contribute to defense (Appel 1993).

Phenolic compounds may also contribute to the outcome of plant-parasite interactions by acting indirectly as signals that condition the final reaction. For example, flavonoids are important in the establishment of mutualistic interactions that develop between legume roots and *Rhizobium* and *Bradyrhizobium* species (Lynn and Chang 1990, Harborne 1993) as well as the interaction of plant roots and endomycorrhizal fungi (Nair et al. 1991, Fries et al. 1997). Phenolic compound derived benzoquinones are important in the successful establishment of parasitism by the parasitic seed plant *Striga* in its hosts (Lynn and Chang 1990).

One very simple phenolic compound, 2-hydroxybenzoic or salicylic acid (SA), has been shown to be an important signal molecule for several plant responses. For example, salicylic acid is important in the triggering of thermogenesis in voodoo lily, which plays an important role in attracting pollinators (reviewed in Harborne 1993). However, most of the attention on the signaling properties of SA has involved its role in disease resistance (Durner et al. 1997). The first hint that SA may be involved in the regulation of plant resistance was based on White's (1979) observation that exogenous SA would induce resistance in tobacco to TMV. Further work in cucumber and tobacco showed that an increase in SA occurred just prior to the onset of systemic resistance (known by various terms, including "systemic acquired resistance" or SAR) (Malamy et al. 1990, Métraux et al. 1990, Rasmussen et al. 1991). For the rest of this chapter SAR is used to indicate SA-mediated systemic resistance responses.

Biosynthesis of Salicylic Acid

Phenolic compound biosynthesis from phenylalanine is one of the biochemical changes that almost always occurs in infected plants (Nicholson and Hammerschmidt 1992). Because of this and the wide distribution of SA in the plant kingdom (Raskin 1992), it is likely that infection in most plants will elicit SA biosynthesis. Raskin and co-workers have shown that phenylalanine, cinnamic acid (the product of phenylalanine ammonia lyase action on phenylalanine), and benzoic acid are precursors of SA (Leon et al. 1993, 1995b, Yalpani et al. 1993). They also suggest that benzoic acid, the immediate precursor of SA, is most likely produced via the β-oxidation of the CoA-ester of cinnamic acid (Ribnicky et al. 1998). Benzoic acid is then converted to SA by the enzyme benzoic acid 2-hydroxylase. Cinnamic and benzoic acids are also precursors of SA in potato (Coquoz et al. 1998). The proposed biosynthetic pathway to SA from phenylalanine is shown in Fig. 1.

Bacteria can also produce SA (Haslam 1993). Because some of these bacteria fall into a group of plant growth-promoting rhizobacteria that induce systemic resistance (van Loon et al. 1998), the biosynthesis of SA in these

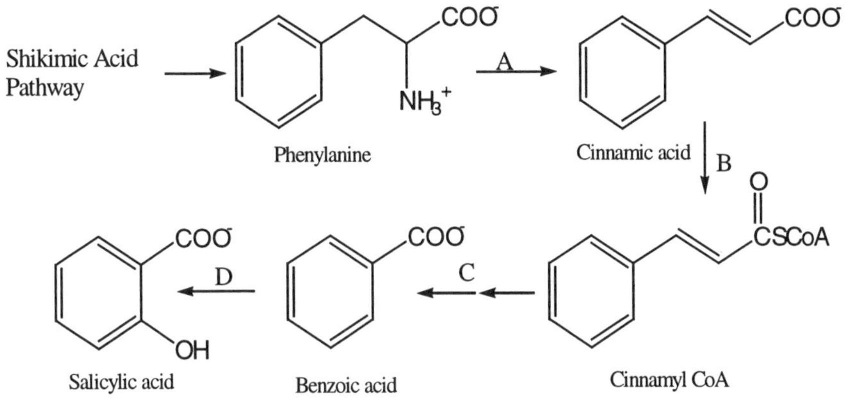

Figure 1: Biosynthesis of salicylic acid in plants from phenylalanine. A= phenylalanine ammonia lyase, B= cinnamyl CoA ligase, C= beta oxidation of cinnamyl CoA to benzoic acid, D= benzoic acid-2-hydroxylase.

organisms is briefly addressed. In bacteria, the shikimic acid pathway also provides the precursor to SA. Unlike the plant biosynthetic pathway, SA is produced from isochorismic acid that is derived from chorismic acid, an intermediate of the shikimic acid pathway (Fig. 2) (Haslam 1993).

Metabolism

The rapid accumulation of SA in plant tissues after infection is followed by a rapid decline in this phenol. As the amount of the free acid declines, there is a concomitant increase in a 2-β-D-glucoside of SA. This glucosylation is mediated by a UDP-glucose-glucosyl transferase that is activated in tissues in response to free SA (Enyedi and Raskin 1993). Although the function of the glucoside is not known, it has been suggested to act as a storage form of free SA that can be released over time (Enyedi and Raskin 1993, Hennig et al. 1993).

Salicylic Acid as a Signal in Resistance

The ability of exogenously applied SA to induce resistance and the expression of associated genes (Raskin 1992, Klessig and Malamy 1994, Pierpont 1994) suggested that this compound might be a natural signal for SAR expression in plants. The systemic increase in SA in tobacco and cucumber during the onset of SAR was the first evidence that SA might play a role as a resistance signal (Malamy et al. 1990, Métraux et al. 1990, Rasmussen et al. 1991).

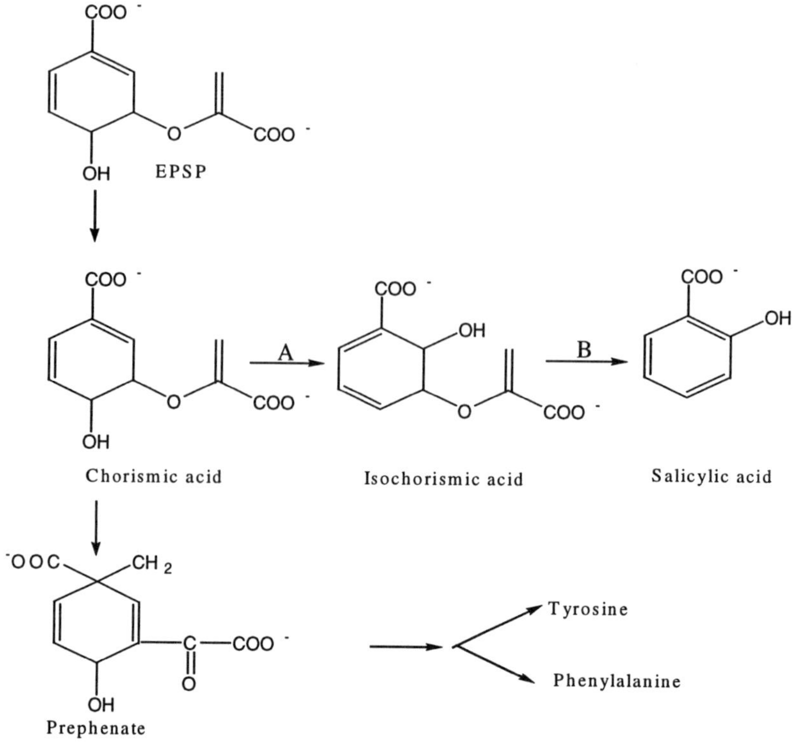

Figure 2: Prokaryotic biosynthesis of salicylic acid from the shikimic acid pathway intermediate chorismic acid. A= isochorismate synthase, B= salicylate synthase

Salicylic acid and SAR. One possible function for SA is as the systemic translocated signal hypothesized to be involved in SAR (e.g., Dean and Kuć, 1986a,b). If SA is the translocated signal, it must be produced in the infected leaves and then translocated throughout the plant. Rasmussen et al. (1991) tested this hypothesis using the cucumber-*Pseudomonas syringae* pv. *syringae* system. Smith et al. (1991) previously demonstrated that infection of cucumber with this bacterium would result in expression of SAR within 24 hours. These authors also showed that the signal for SAR was generated within 4 to 6 hours after inoculation with *P. syringae* pv. *syringae*. Using the time course of resistance induction established by Smith et al. (1991), Rasmussen et al. (1991) detached *Pseudomonas* inoculated leaves of cucumber at intervals after inoculation in order to determine the time of SA transport out of the infected leaf. SA could be detected in the phloem exudates of the *Pseudomonas* inoculated leaf by 8 hours after inoculation, and increases in SA occurred in the phloem of the leaf above the inoculated leaf by 12 hours. These data suggested that SA might be moving from one leaf to another. However, in plants that had the inoculated leaf detached before there was a detectable increase in SA, the entire plant contained greatly increased levels of SA when analyzed 24 hours after the initial inoculation. Based on these results, Rasmussen et al. (1991) concluded that SA was a secondary signal that was induced by a primary, translocated signal generated at the infection site.

The conclusion of Rasmussen et al. (1991) that SA was not the primary translocated resistance signal was supported by Vernooij et al. (1994). They used tobacco plants transformed with the *nahG* gene that codes for the enzyme salicylate hydroxylase as a means of reducing the amount of SA in the plants. This enzyme converts SA to catechol which has no disease resistance inducing activity, and the *nahG* plants did not locally or systemically accumulate SA in response to TMV (Gaffney et al. 1993). In addition, *nahG* transformed plants do not express a SAR response (Gaffney et al. 1993). The role of SA as the translocated signal was tested by Vernooij et al. (1994) by grafting wild type tobacco plants onto *nahG* rootstocks and vice versa. They found that TMV inoculated leaves on the *nahG* rootstock were able to generate a signal that resulted in the expression of resistance and SA accumulation in the wild type scion. Thus, they also concluded that salicylate was not the primary translocated signal. Similarly, tobacco plants that were suppressed in PAL activity (and also did not accumulate SA) also supported, albeit indirectly, the same conclusions (Pallas et al. 1995).

The above observations suggest that the synthesis of SA may be systemically induced. Using labeled phenylalanine, Meuwly et al. (1995) demonstrated a greater rate of synthesis of SA in systemically induced cucumber as compared to controls. In addition, Smith-Becker et al. (1998) have recently shown that cucumber plants induced with *P. syringae* pv. *syringae* have elevated levels of PAL in tissues at a distance from the inducer leaf. The timing of PAL activity increases correlates well with the appearance in phloem exudates of SA and the related compound 4-hydroxy benzoic acid. The greatest increase in PAL activity was in the petiole of the inoculated leaf and the stem above the inoculated leaf, leading to the conclusion that the majority of the SA is synthesized in vascular tissue. The simplest explanation for this observation is that the signals involved in local SA synthesis move out of the site of infection into the vascular tissue where they induce additional SA synthesis.

The presence of salicylic acid in phloem exudates (Métraux et al. 1990, Rasmussen et al. 1991, Smith-Becker et al. 1998) supports the hypothesis that SA is translocated throughout the plant for activation of PR-protein expression and subsequent SAR. Working with tobacco and cucumber, respectively, Shulaev et al. (1995) and Molders et al. (1996) demonstrated that some SA is transported to the upper leaves of these plants from the infected leaf used for resistance induction. Thus, the SA accumulating in upper leaves for SAR induction is most likely a combination of SA synthesized by the inoculated leaf and SA synthesized systemically, possibly in the vascular tissue.

Salicylic acid and local resistance. The resistance-inducing properties of exogenous application of SA and the synthesis of this compound in infected plants suggests that SA may play a role in host resistance. Excellent evidence for a role for endogenous SA in the expression local resistance was reported by Gaffney et al. (1993) using tobacco transformed with *nah*G. TMV lesions formed on the *nahG* transformants were larger than on the wild type controls. Tobacco and *Arabidopsis* transformed with *nahG* were later shown to be more susceptible to other pathogens (Delaney et al. 1994). Potato has a higher constitutive level of SA than tobacco. However, reduction of SA levels via transformation with *nahG* did not increase the susceptiblity of late blight susceptible potato plants to infection by *Phytophthora infestans* (Yu et al. 1997).

Maher et al. (1994) previously reported that tobacco plants with suppressed PAL activity were more susceptible to fungal infection. Although the initial explanation was that the plants were more susceptible because of a lack of induced phenolic defenses, a more recent report suggests that the plants were more susceptible because they could not produce sufficient quantities of SA to be used in regulating the defense response (Pallas et al. 1995). Further evidence for a role of SA was provided by Mauch-Mani and Slusarenko (1996) who reported that part of the function of increased PAL activity in infected *Arabidopsis* was for SA production.

A set of *Arabidopsis* mutants that are deficient in camalexin (a phytoalexin) production have been described (Glazebrook and Ausubel 1994, Glazebrook et al. 1997). One these mutants, *pad4*, has been is a regulatory gene that is affected by SA. Although *PAD4* was not necessary for camalexin accumulation in response to the maize pathogen *Cochliobolus carbonum*, *pad4* plants accumulated reduced levels of camalexin in response to the virulent pathogen *P. syringae* pv. *maculicola* ES4 326 (*Psm*) (Glazebrook et al. 1997). Infection of *pad4* plants with *Psm* resulted in lower levels of SA as compared to the wild type (Zhou et al. 1998), and *pad4* plants also expressed lower amounts of PR1 in response to *Psm* infection. These results and the ability of exogenous SA application to partially restore PR1 and camalexin production in *Psm* infected *pad4* plants illustrates another function of salicylate in local defense.

Treatment of plants with various inhibitors of phenolic compound accumulation have been used to provide evidence for the role of phytoalexins and other phenolic compounds in defense. The research described above may, however, provide an alternative explanation for the enhanced susceptibility of plants such as soybean and bean after treatment with glyphosate activity via inhibition of SA synthesis (e.g., Keen et al. 1982, Liu et al. 1997).

Methyl salicylate as a volatile signal for resistance. Methyl salicylate is a common, volatile derivative of SA. This compound has been shown to be involved in plant-herbivore interactions as an attractant to anthocorid predators of psyllids on pear (e.g., Scutareanau et al. 1997, Sabelis et al., this volume). Recently, the production of methyl salicylate in TMV-infected plants has been described (Shulaev et al. 1997). These authors found that TMV-infected tobacco released volatile methyl salicylate into the air surrounding the plants and this suggested that perhaps this compound could serve as an inter-plant signal for defense. Exposure of plants to methyl salicylate was shown to activate PR-1 expression and increased resistance of tobacco to TMV infection. Methyl salicylate applied as an airborne volatile appeared to act by first being converted to SA after being absorbed by the plant tissues.

SA and Insect Resistance. Although SA is an effective inducer of broad based pathogen resistance, application of SA to plant tissues has limited evidence as an inducer of enhanced resistance to insect herbivores (e.g., Bi et al. 1997, Inbar et al. 1998). SA treatment of wounded tomato has also been shown to enhance herbivory (Stout et al. 1998), and this may be due to the inhibition of jasmonic acid stimulated defenses by SA (Doares et al. 1995, see also Felton et al. 1999, Thaler, this volume).

Salicylic Acid: Resistance Expression and Perception

Exogenous treatment of plant tissues with SA results in the induction of disease resistance. Associated with the induction of resistance is the expression of several putative defense genes that are also systemically induced by pathogens. For example, treatment of tobacco with SA induces the same set of pathogenesis related (PR) protein genes as are induced by TMV (Ward et al. 1991). In cucumber, the class III acidic chitinase and a group of acidic peroxidases are induced systemically by pathogens (Métraux et al. 1988, Smith et al. 1991) and locally by SA treatment (Métraux et al. 1989, Rasmussen et al. 1991, 1995).

SA treatment of plant tissues has also been shown to condition these tissues to respond more quickly to pathogen attack. For example, inhibition of penetration is a characterized resistance response of cucumber leaves to *Colletotrichum orbiculare* (reviewed in Hammerschmidt and Smith-Becker 1997), and exogenous SA treatment of cucumber cotyledon tissues conditions this tissue to also resist infection by inhibiting penetration (Rasmussen et al. 1991). SA treatment of cucumber hypoctyls also enhanced their ability to produce chitinase (Kastner et al. 1998) and hydrogen peroxide (Kauss and Jeblick 1996) in response to elicitor treatment and to deposit lignin-like material in response to pathogen attack (Siegrist et al. 1994). SA pre-treatment of parsley suspension cells enhanced their ability to deposit wall bound phenolic compounds after elicitor treatment (Kauss et al. 1993, Kauss and Jeblick 1995), and SA treated soybean cells responded to an avirulent strain of *P. syringae* by a more rapid expression of the hypersensitive response and H_2O_2 production (Shirasu et al. 1997).

In order for SA to induce resistance and the associated biochemical changes, it must be perceived by the plant. One logical approach to elucidate the mode of action of SA is to identify proteins or other cellular factors that interact with salicylic acid and, as a result, induce the expression of resistance. The identification of catalase as a SA-binding protein led to the hypothesis that inhibition of catalase by SA led to the accumulation of H_2O_2 which then induced the resistance response (Chen and Klessig 1991, Chen et al. 1993, Conrath et al. 1995). Several alternative lines of investigation, however, suggested that SA acted downstream of H_2O_2 rather than upstream in the induction of resistance in tobacco and *Arabidopsis* (Bi et al. 1995, Neuenschwander et al. 1995) and that H_2O_2 could also induce the synthesis of SA (Leon et al. 1995a). Similar conflicting reports on the interaction of SA and an ascorbate peroxidase in the induction of resistance have also been reported (Durner and Klessig 1995, Kvaratskhelia et al. 1997).

To test the role of catalase inhibition in the expression of resistance, transgenic tobacco plants that are suppressed in catalase production have been produced (Chamnongpol et al. 1996, 1998, Takahashi et al. 1997). In all cases, plants with reduced catalase activity exhibited induced levels of resistance along with tissue necrosis and the accumulation of PR1. Although these experiments show that catalase inhibition will result in resistance, it does not offer an explanation of the position of H_2O_2 in relation to SA in resistance signaling. Du and Klessig (1997b), however, have provided further evidence that SA is acting downstream of H_2O_2. They crossed *nahG* tobacco with tobacco that was suppressed in catalase production. They found that progeny that were *nahG* and suppressed in catalase were unable to express enhanced resistance even though the necrotic phenotype associated with H_2O_2 production was present. Thus, they were also able to conclude that SA was acting downstream of H_2O_2.

A 25kD soluble protein has recently been identified as another SA binding protein (Du and Klessig 1997a). This protein has a much higher affinity for SA than either catalase or ascorbate peroxidase. SA analogs and the synthetic resistance activator BTH (see Lyon and Newton, Tally et al., this volume) were able to compete for binding with SA in relation to their ability to induce resistance. Further evaluation of this protein is needed to determine what role it plays in SAR.

Nitric oxide (NO) has recently been shown to be an additional putative signal for the expression of defense. Unlike SA, application of NO releasing agents induces the accumulation of phytoalexins in potato tuber tissue (Noritake et al. 1996). Similary, inhibition of NO synthesis increases susceptibility of *Arabidopsis* to *P. syringae* and is involved in the expression of the hypersenstive response in soybean cells (Delledonne et al. 1998). Durner et al. (1998) also showed that NO and SA both induce PR1 but only NO will induce PAL and the accumulation of SA. Using *nahG* tobacco, Durner et al. (1998) further showed that NO acts via the induction of SA.

Genetic Analysis of Induced Resistance Signaling in *Arabidopsis*

As in other biochemical systems, *Arabidopsis* has proven to be useful as a genetic model system for SAR (Cameron et al. 1994, Mauch-Mani and Slusarenko 1993, Uknes et al. 1992). The observation that induced resistance occurs in *Arabidopsis* has led to an effort to isolate mutants in the induced resistance signaling pathway (reviewed by Delaney 1997, Buell, this volume). The majority of mutants identified so far fall into one of several categories. The first, the lesion simulating disease (*lsd*) mutants, all display spontaneous lesions during development (Deitrich et al. 1994, Greenberg et al. 1994, Weymann et al. 1995, Hunt et al. 1996). Regulation of the *lsd* phenotype is complex, since the presence of the mutation in a nahG backround, in which SA is removed, suppresses lesion formation in certain mutants (Weymann et al. 1995). Five of six *lsd* mutants isolated by Dietrich et al. (1994) exhibited elevated levels of PR-1a mRNA and enhanced resistance to *Peronospora parasitica* when grown under lesion permissive conditions. The ability of some *lsd* mutants to express SAR suggests that the mutations identify signaling components normally activated during pathogen induced necrosis. Identification and functional analysis of the disrupted genes in *lsd* mutants should help describe the components of pathogen induced necrosis responsible for early signaling in induced resistance. Mutants that exhibit constitutive induced resistance without visible lesions have been termed *cim*, for constitutive immunity (Lawton et al. 1993), and *cpr* for constitutive expression of PR genes (Bowling et al. 1994). The *cpr1* mutant contains elevated levels of SA and is resistant to both *P. parasitica* and *P. syringae* pv. *maculicola.* Since the mutation is recessive, it was proposed that the *CPR1* gene product is a negative regulator of induced resistance.

Perhaps the most exciting of the induced resistance mutants was independently isolated by at least three separate groups. This mutant, known as *npr1* (Cao et al. 1994), *nim1* (Delaney et al. 1995), and *sai1* (Shah et al. 1997) was selected for its inability to express induced resistance in response to SA or the salicylic acid analogs INA and BTH. Interestingly, the mutant accumulates salicylic acid upon pathogen infection but is incapable of transducing the salicylic acid signal into PR gene expression (Delaney et al. 1995). Sequence analysis of the *NPR1* gene revealed the presence of ankyrin repeats, suggesting it may function as a transcription-regulating protein (Cao et al. 1997). Visual comparison of the predicted protein product of the gene with 70 known ankyrin-containing proteins revealed similarities between *NPR1* and the IκBα class of transcription regulators (Ryals et al. 1997). This finding is particularly intriguing, since IκB-NF-κB signaling is well documented in pathogen-host interactions in animal systems (reviewed by Baeuerle and Baltimore 1996).

It appears, however, that the mechanism of SA action in plants is very different from its activity in animal systems. In animal systems, IκBα and NF-κB form an inactive cytosolic complex. A signal associated with pathogen infection triggers phosphorylation and degradation of IκBα (the putative NPR1 homolog) and release of NF-κB to the nucleus, where it activates genes associated with inflammation. SA in this system prevents the degradation of IκBα and thus prevents the release of NF-κB and subsequent gene activation. In contrast, SA in plants acts as an activator of the signal transduction pathway. This has led Ryals et al. (1997) to conclude that the NF-κB homolog in plants, assuming one exists, acts as a repressor of induced resistance gene expression.

It is tempting to speculate that the common point of action of salicylic acid in animal and plant signal transduction pathways is a vestige of a pathway that developed before the divergence of the two kingdoms. These recent findings also suggest that compounds selected for activity in one system, for example BTH, may have activity in other systems.

Salicylate-Mediated Defense and Disease Control

Two synthetic compounds, 2,6-dichloronicotinic acid (CGA-41396) and the benzothiadiazole derivative BTH (CGA-245704) have been shown to be effective inducers of resistance (Kessmann et al. 1994, Friedrich et al 1996, Görlach et al. 1996, Lawton et al. 1996, Tally et al., this volume). Although these compounds are not structurally related to SA, they appear to function as SA analogs and induce the same spectrum of defenses as does SA (Friedrich et al. 1996, Görlach et al. 1996, Lawton et al. 1996). Thus, a solid understanding of the way that salicylate functions may lead future improvements in using induced resistance in disease control. Fungicide-mediated control of disease may also benefit from the study of SA. Molina et al. (1998) reported that fungicide control of *Peronospora* in *nahG* and *nim1 Arabidopsis* plants was not as effective as in wild type plants, thus suggesting that there is an interaction of the fungicide-mediated control with the plants' natural defenses.

Synthesis and Future Directions

Over the last decade, the important role of SA in plant-pathogen defense responses has been recognized, and studies on this phenol have provided new information on secondary metabolism and the expression of local and systemic defenses. Most of this work has been done with *Arabidopsis*, cucumber and tobacco, but many other systems which demonstrate induced resistance (e.g., Hammerschmidt and Smith 1997) remain unstudied with regard to SA. One of the challenges that remains is to determine how SA functions to induce resistance as well as determining if the mechanisms that are being elucidated in the model systems such as *Arabidopsis* are universal in the plant kingdom.

Due to the nature of the screens employed for induced resistance mutant analysis, most of the mutants identified to date either express induced resistance constitutively or are deficient in their perception of SA and its analogs. Notably absent from the mutant collections are isolates incapable of producing SA in response to pathogen inoculation. Although some evidence suggests that H_2O_2 induces SA synthesis (Bi et al. 1995, Leon et al.1995a, Neuenschwander et al. 1995, Sharma et al. 1996), it has not been verified that pathogen induced necrosis produces sufficient peroxide to induce the high levels of SA that accumulate locally and systemically. Furthermore, with the exception of PAL, none of the SA biosynthetic enzymes have been cloned. Leon et al. (1993) have described the activity of a benzoic acid 2-hydroxylase in tobacco, but the activity of this enzyme has not been reported in other species. The survival of *nahG* transformants, in which SA is rapidly converted to catechol, suggests that SA is not essential for plant survival and that it should be possible to isolate mutants incapable of producing SA. In order to select for mutations in early signaling events it may be necessary to screen for mutants using pathogens, or specific elicitors derived from pathogens, as systemic resistance inducers. The study of early signaling in induced resistance will undoubtedly overlap with the vast body of information available on R gene function (Buell, this volume).

SA mediated resistance is clearly not the only mechanism for resistance induction. Studies on resistance induced by rhizobacteria have shown that some of these bacteria induce resistance via a non-salicylate pathway which appears to utilize jasmonic acid and ethylene as modulators or resistance (Hoffland et al. 1995, Pieterse et al. 1998, van Loon et al. 1998). In addition, recent studies with *Arabidopsis* have shown that salicylate signaling is important in resistance to *Peronospora parasitica* while jasmonate is needed for resistance to *Alternaria brassicicola* (Thomma et al. 1998). Similarly, jasmonic acid is required for resistance of *Arabidopsis* to *Pythium* (Vijayan et al. 1998). Thus, it will become increasingly important to evaluate the role of these other signals in addition to or in concert with the action of SA in defense. For example, there is evidence that SA interferes with JA signalling (e.g., Doares et al. 1995, Niki et al. 1998, Thaler, this volume). Thus, the practical use of plant activating materials for disease control must be evaluated in relation to effects of other defense signals in the plant.

Acknowledgements

The support of the Michigan Agricultural Experiment Station, the Novartis Corporation and the USDA/NRICGP is gratefully acknowledged.

Literature Cited

Appel, H.M. 1993. Phenolics in ecological interactions: The importance of oxidation. Journal of Chemical Ecology 19:1521-1552.

Baeuerle, P. and D. Baltimore. 1996. NF-6B: Ten years later. Cell 87:13-20.

Bi, Y.-M., P. Kenton, L. Mur, R. Darby, and J. Draper. 1995. Hydrogen peroxide does not function downstream of salicylic acid in the induction of PR protein expression. The Plant Journal 8:235-245.

Bi, J.L., J.B. Murphy, and G.W. Felton. 1997. Does salicylic acid act as a signal in cotton for induced resistance to *Helicoverpa zea*? Journal of Chemical Ecology. 23 :1805-818.

Bowling, S. A., A. Guo, H., Cao, A.S. Gordon, D.F. Klessig, and X. Dong. 1994. A mutation in *Arabidopsis* that leads to constitutive expression of systemic acquired resistance. The Plant Cell 6:1845-1857.

Cameron, R.K., R. Dixon, and C.J. Lamb. 1994. Biologically induced systemic acquired resistance in *Arabidopsis thaliana*. The Plant Journal 5:715-725.

Cao, H., S.A. Bowling, A.S. Gordon, and X. Dong. 1994. Characterization of an *Arabidopsis* mutant that is nonresponsive to inducers of systemic acquired resistance. The Plant Cell 6:1583-1592.

Cao, H., J. Glazebrook, J.D. Clarke, S. Volko, and X. Dong. 1997.The *Arabidopsis* NPR1 gene that controls systemic acquired resistance encodes a novel protein containing ankyrin repeats. Cell 88:57-63.

Chamnongpol, S., H. Willekens, C. Langebartels, M. van Montagu, D. Inze, and W. van Camp, W. 1996. Transgenic tobacco with a reduced catalase activity develops necrotic lesions and induces pathogenesis-related expression under high light. The Plant Journal 10:491-503.

Chamnongpol, S., H. Willekens, W. Moeder, C. Langebartels, H. Sandermann Jr., M. van Montagu, D. Inze, and W. van Camp. 1998. Defense activation and enhanced pathogen tolerance induced by H_2O_2 in transgenic tobacco. Proceedings of the National Academy of Sciences USA 95:5818-5823.

Chen, Z. and D.F. Klessig. 1991. Identification of a soluble salicylic-acid binding protein that may function in signal transduction in the plant disease resistance response. Proceedings of the National Academy of Sciences USA 88:8179-8183.

Chen, Z., H. Silva, and D.F. Klessig. 1993. Active oxygen species in the induction of plant systemic acquired resistance by salicylic acid. Science 262:1883-1885.

Conrath, U., Z.X. Chen, J.R. Ricigliano, and D.F. Klessig. 1995. Two inducers of plant defense responses, 2,6-dichloroisonicotinic acid and salicylic acid, inhibit catalase activity in tobacco. Proceedings of the National Academy of Sciences USA 92:7143-7147.

Coquoz, J.L.; A. Buchala, and J.P. Métraux. 1998. The biosynthesis of salicylic acid in potato plants. Plant Physiology 117:1095-1101

Dean, R.A. and J. Kuć. 1986a. Induced systemic protection in cucumber: The source of the "signal". Physiological and Molecular Plant Pathology. 28:227-233.

Dean, R.A. and J. Kuć. 1986b. Induced systemic protection in cucumber: Time of production and movement of the signal. Phytopathology 76:966-970.

Delaney, T. P. 1997. Genetic dissection of acquired resistance to disease. Plant Physiology 113: 5-12.

Delaney, T. P., L. Friedrich, and J. Ryals. 1995. *Arabidopsis* signal transduction mutant defective in chemically and biologically induced disease resistance. Proceedings of the National Academy of Sciences USA 92:6602-6606.

Delaney, T.P., S. Uknes, B. Vernooij, L. Friedrich, K. Weymann, D. Negrotto, T., Gaffney, M. Gut-Rella, H. Kessmann, E. Ward, and J. Ryals. 1994. A central role of salicylic acid in plant disease resistance. Science 266:1247-1250.

Delledonne, M, Y.J. Xia, R.A. Dixon, and C. Lamb. 1998. Nitric oxide functions as a plant disease resistance. Nature 394:585-588.

Dietrich, R. A., T.P. Delaney, S.J. Uknes, E.R. Ward, J.A. Ryals, and J.L. Dangl. 1994. *Arabidopsis* mutants simulating disease response. Cell 77:565-577.

Doares, S.H., J. Narvaez-Vasquez, A. Conconi, and C.A. Ryan. 1995. Salicylic acid inhibits synthesis of proteinase inhibitors in tomato leaves induced by systemin and jasmonic acid. Plant Physiology 108:1741-1746.

Du, H. and D.F. Klessig. 1997a. Identification of a soluble, high-affinity salicylic acid-binding protein in tobacco. Plant Physiology 113:1319-1327.

Du, H. and D.F. Klessig. 1997b. Role for salicylic acid in the activation of defense responses in catalase-deficient transgenic tobacco. Molecular Plant-Microbe Interactions 10:922-925.

Durner, J. and D.F. Klessig. 1995. Inhibition of ascorbate peroxidase by salicylic acid and 2,6-dichloroisonicotinic acid, two inducers of plant defense responses. Proceedings of the National Academy of Sciences USA 92:11312-11316.

Durner, J., J. Shah, and D. F. Klessig. 1997. Salicylic acid and disease resistance in plants. Trends In Plant Science 2:266-274.

Durner J., D. Wendhenne, and D.F. Klessig. 1998. Defense gene induction by nitric oxide, cyclic GMP, and cyclic ADP-ribose. Proceedings of the National Academy of Sciences USA. 95:10328-10333.

Enyedi, A.J. and I. Raskin. 1993. Induction of UDP-glucose:salicylic acid glucosyltransferase activity in tobacco moasaic virus-inoculated tobacco (*Nicotiana tabacum*) leaves. Plant Physiology 101:1375-1380.

Felton, G .W., K. L. Korth, J. L. Bi, S. V. Wesley, D. V. Huhman, M. C. Mathews, J. B. Murphy, C. Lamb, R. A. Dixon. 1999. Inverse relationship between systemic resistance of plants to microorganisms and to insect herbivory. Current Biology 9:317-320.

Friedrich, L., K. Lawton, W. Ruess, P. Masner, N. Specker, M.G. Rella, B. Meier, S. Dincher, T. Staub, Uknes, J.P. Métraux, H. Kessmann, and J. Ryals. 1996. A benzothiadiazole derivative induces systemic acquired resistance in tobacco. The Plant Journal 10:61-70.

Fries, L.L.M. R.S. Pacovsky, G.R. Safir, and J.O. Siqueira. 1997. Plant growth and arbuscular mycorrhizal fungal colonization affected by exogenously applied phenolic compounds. Journal of Chemical Ecology. 23:1755-1767.

Gaffney, T., L. Friedrich, B. Vernooij, D. Negrotto, G. Nye, S. Uknes, E. Ward, H. Kessmann, and J. Ryals. 1993. Requirement of salicylic acid for the induction of systemic acquired resistance. Science 261:754-766.

Glazebrook, J. and F.M. Ausubel. 1994. Isolation of phytoalexin deficient mutants of *Arabidopsis thaliana* and characterization of their interactions with bacterial pathogens. Proceedings of the National Academy of Sciences USA 91:8955-8959.

Glazebrook, J., M. Zook, F. Mert, I. Kagan, E.E. Rogers, I.R. Crute, E.B. Holub, R. Hammerschmidt, and F.M. Ausubel. 1997. Phytoalexin-deficient mutants of *Arabidopsis* reveal that *PAD4* encodes a regulatory factor and that four PAD genes contribute to downy mildew resistance. Genetics 146:381-392.

Görlach, J., S. Volrath, G. Knauf-Beiter, G., Hengy, U. Beckhove, K.H. Kogel, M. Oostendorp, T. Staub, E. Ward, H. Kessmann, and J. Ryals. 1996. Benzothiadiazole, a novel class of inducers of systemic acquired resistance, activates gene expression and disease resistance in wheat. The Plant Cell 8:629-643.

Greenberg, J., A. Guo, D. Klessig, and F. Ausubel. 1994. Programmed cell death in plants: a pathogen triggered response activated coordinately with multiple defense functions. Cell 77:551-563.

Hammerschmidt, R. and J. Smith-Becker. 1997 Acquired resistance to disease. Horticultural Reviews. 18:247-289.

Hammerschmidt, R. and J.C. Schultz. 1996. Multiple defenses and signals in plant defense against pathogens and herbivores. Recent Advances in Phytochemistry 30:121-154.

Harborne, J.B. 1993. Introduction to Ecological Biochemistry, 4th Ed. Academic Press, London.

Haslam, E. 1993. Shikimic Acid. John Wiley and Sons, Chichester, UK.

Hennig J., J. Malamy, G. Grynkiewicz, J. Indulski, and D.F.Klessig. 1993. Interconversion of the salicylic acid signal and its glucoside in tobacco. The Plant Journal 4:593-600.

Hoffland, E., C. M. J. Pieterse, L. Bik, and J. A. van Pelt. 1995. Induced systemic resistance in radish is not associated with accumulation of pathogenesis-related proteins. Physiological and Molecular Plant Pathology 46:309-320.

Hunt, M. D., U.H. Neuenschwander, T.P. Delaney, K.B. Weymann, L.B. Friedrich, K.A. Lawton, H.-Y. Steiner, and J.A. Ryals. 1996. Recent advances in systemic acquired resistance research. Gene 17:89-95.

Inbar, M., H. Doostdar, R.M. Sonoda, G.L. Leibee, and R.T. Mayer. 1998. Elicitors of plant defensive systems reduce insect densities and disease incidence. Journal of Chemical Ecology 24 :135-149.

Kastner, B., R. Tenhaken, and H. Kauss. 1998. Chitinase in cucumber hypocotyls is induced by germinating fungal spores and by fungal elicitor in synergism with inducers of acquired. The Plant Journal13:447-454.

Kauss, H., R. Franke, K. Krause, U. Conrath, W. Jeblick, B. Grimming, and U. Matern. 1993. Conditioning of parsley (*Petroselinum crispum* L.) suspension cells increases elicitor-induced incorporation of cell wall phenolics. Plant Physiology 102:459-466.

Kauss, H. and W. Jeblick. 1995. Pretreatment of parsley suspension cultures with salicylic acid enhances spontaneous and elicited production of H_2O_2. Plant Physiology 108:1171-1178.

Kauss, H. and W. Jeblick. 1996. Influence of salicylic acid on the induction of competence for H_2O_2 elicitation: comparison of ergosterol with other elicitors. Plant Physiology 111:755-763.

Keen, N.T., M.J. Holliday, and M. Yoshikawa. 1982. Effects of glyphosate on glyceollin production and the expression of resistance to *Phytophthora megasperma* f. sp. *glycinea* in soybean. Phytopathology 72:1467-1470.

Kessmann, H., T. Staub, C. Hofmann, T. Maetzke, J. Herzog, E. Ward, S. Uknes, S. and J. Ryals. 1994. Induction of systemic acquired disease resistance in plants by chemicals. Annual Review of Phytopathology 32:439-459.

Klessig, D.A. and J. Malamy. 1994. The salicylic acid signal in plants. Plant Molecular Biology 26:1439-1458.

Kuć, J. 1995. Phytoalexins, stress metabolism and disease resistance in plants. Annual Review of Phytopathology 33:275-297.

Kvaratskhelia, M., S.J. George, and R.N.F. Thorneley. 1997. Salicylic acid is a reducing substrate and not an effective inhibitor of ascorbate peroxidase. Journal of Biological Chemistry 272:20998-21001.

Lawton, K., S. Uknes, L. Friedrich, T. Gaffney, D. Alexander, R. Goodman, J.-P. Métraux, H. Kessmann, P. Ahl-Goy, M. Gut-Rella, E. Ward, and J. Ryals. 1993 The molecular biology of systemic acquired resistance. Pages 410-420 *in* B. Fritig and M Legrand, editors. Mechanisms of Defense Responses in Plants. Kluwer Academic, Dordrecht, The Netherlands.

Lawton, K.A., L. Friedrich, M. Hunt, K. Weymann, T. Delaney, H. Kessmann, T. Staub, and J. Ryals. 1996. Benzothiadiazole induces disease resistance in *Arabidopsis* by activation of the systemic acquired resistance signal transduction pathway. The Plant Journal 10:71-82.

Leon, J., N. Yalpani, I. Raskin, and M.A. Lawton. 1993. Induction of benzoic acid 2-hydroxylase in virus-inoculated tobacco. Plant Physiology 103:323-328.

Leon, J., M.A. Lawton, and I. Raskin. 1995a. Hydrogen peroxide stimulates salicylic acid biosynthesis in tobacco. Plant Physiology 108:1673-1678.

Leon, J., V. Shulaev, N. Yalpani, M.A. Lawton, and I. Raskin. 1995b. Benzoic acid 2-hydroxylase, a soluble oxygenase from tobacco, catalyzes salicylic acid biosynthesis. Proceedings of the National Academy of Science USA 92:10413-10417.

Liu, L., Z.K. Punja, and J.E. Rahe. 1997. Altered root exudation and suppression of induced lignification as mechanisms of predisposition by glyphosate of bean roots (*Phaseolus vulgaris* L.) to colonization by *Pythium* spp. Physiological and Molecular Plant Pathology. 51:110-127.

Lynn, D.G. and M. Chang. 1990. Phenolic signals in cohabitation: Implications for plant development. Annual Review of Plant Physiology and Plant Molecular Biology 41:497-526.

Maher, E.A., N.J. Bate, W. Ni, Y. Elkind, R.A. Dixon, and C.J. Lamb.1994. Increased disease susceptibility of transgenic tobacco plants with suppressed levels of preformed phenylpropanoid products. Proceedings of the National Academy of Sciences USA 91:7802-7806.

Malamy, J., J.P. Carr, D.F. Klessig, and I. Raskin. 1990, Salicylic acid - a likely endogenous signal in the resistance response of tobacco to tobacco mosaic virus. Science 250:1002-1004.

Mauch-Mani, B. and A. Slusarenko. 1993. *Arabidopsis* as a model host for studying plant-pathogen interactions. Trends in Microbiology 1:265-267.

Mauch-Mani, B., and A.J. Slusarenko 1996. Production of salicylic acid precursors is a major function of phenylalanine ammonia-lyase in the resistance of *Arabidopsis* to *Peronospora parasitica*. The Plant Cell 8:203-212.
Métraux, J.P., W. Burkhart, M. Moyer, S. Dincher, W. Middlesteadt, S. Williams, G. Payne, M. Carnes, and J. Ryals. 1989. Isolation of a complementary DNA encoding a chitinase with structural homology to a bifunctional lysozyme/chitinase. Proceedings of the National Academy of Sciences USA 86:896-900.
Métraux, J.P., H. Signer, J. Ryals, E. Ward, M. Wyss-Benz, J.Gaudin, K. Raschdorf, E. Schmid, W. Blum, and B.Inverardi. 1990. Increase in salicylic acid at the onset of systemic acquired resistance in cucumber. Science 250:1004-1006.
Métraux, J.P., L. Streit, and T. Staub. 1988. A pathogenesis-related protein in cucumber is a chitinase. Physiological and Molecular Plant Pathology 33:1-9.
Meuwly, P., W. Molders, A. Buchala, and J.-P. Métraux. 1995. Local and systemic biosynthesis of salicylic acid in infected cucumber plants. Plant Physiology 109:1107-1114.
Mölders, W., T. Buchala, and J.-P. Métraux. 1996. Transport of salicylic acid in tobacco necrosis virus-infected cucumber plants. Plant Physiology 112:787-792.
Molina, A., M.D. Hunt and J.A. Ryals. 1998. Impaired fungicide activity in plants blocked in disease resistance signal transduction. The Plant Cell 10:1903-1914.
Nair, M.G., G.R. Safir, and J.O. Siqueira. 1991. Isolation and identification of vesicular-arbuscular mycorrhiza-stimulatory compounds from clover (*Trifolium repens*) roots. Applied and Environmental Microbiology 57:434-439.
Neuenschwander, U., B.Vernooij, L. Friedrich, S. Uknes, H. Kessmann, and J. Ryals. 1995. Is hydrogen peroxide a second messenger of salicylic acid in systemic acquired resistance? The Plant Journal 8:227-233.
Nicholson, R.L., and R. Hammerschmidt. 1992. Phenolic compounds and their role disease resistance. Annual Review of Phytopathology 30:369-389.
Niki, T., I. Mitsuhara, S. Seo, N. Ohtsubo, and Y. Ohashi. 1998 Antagonistic effect of salicylic acid and jasmonic acid on the expression of pathogenesis-related (PR) protein genes in wounded mature tobacco leaves. Plant and Cell Physiology 39:500-507.
Noritake T., K. Kawakita, and N. Doke. 1996. Nitric oxide induces phytoalexin accumulation in potato tuber tissue. Plant and Cell Physiology 37:113-116.
Pallas, J.A., N.L. Paiva, C.J. Lamb, and R.A. Dixon. 1995. Tobacco plants epigenetically suppressed in phenylalanine ammonia lyase expression do not develop systemic acquired resistance in response to infection by tobacco mosaic virus. The Plant Journal. 10:281-293.
Pierpont, W.S. 1994. Salicylic acid and its derivatives in plants: Medicines, metabolism and messenger molecules. Advances in Botanical Research 20:164-235.
Pieterse, C.M.J., S.C.M. van Wees, J.A. van Pelt, M. Knoester, R. Laan, H. Gerrits, P.J. Weisbeek, and L.C. van Loon.1998. A novel signalling pathway controlling induced systemic resistance in *Arabidopsis*.The Plant Cell 10:1571-1580
Raskin, I. 1992. Role of salicylic acid in plants. Annual Review of Plant Physiology and Plant Molecular Biology. 43:439-463.
Rasmussen, J.B., R. Hammerschmidt, and M.N. Zook. 1991. Systemic induction of salicylic acid accumulation in cucumber after inoculation with *Pseudomonas syringae* pv. *syringae*. Plant Physiology 97:1342-1347.
Rasmussen, J.B., J.A. Smith, S. Williams, W. Burkhart, E. Ward, S.C. Somerville, J. Ryals, and R. Hammerschmidt. 1995. cDNA cloning and systemic expression of acidic peroxidases associated with systemic acquired resistance to disease in cucumber. Physiological and Molecular Plant Pathology 46:389-400.
Ribnicky, D.M., V. Shulaev, and I. Raskin. 1998. Intermediates of salicylic acid biosynthesis in tobacco. Plant Physiology 118:565-572.
Ryals, J., K. Weymann, K. Lawton, L. Friedrich, D. Ellis, H-Y. Steiner, J. Johnson, T. Delaney, J. Taco, P. Vos, and S. Uknes. 1997. The *Arabidopsis* NIM1 protein shows homology to the mammalian transcription factor inhibitor IkB. The Plant Cell 9:425-439.

Scutareanu, P., B. Drukker, J. Bruin, M.A. Posthumus, and M.W. Sabelis. 1997. Volatiles from *Psylla*-infested pear trees and their possible involvement in attraction of anthocorid predators. Journal of Chemical Ecology 23:2241-2260

Shah, J., F. Tsui, and D.F. Klessig. 1997. Characterization of a salicylic acid-insensitive mutant (*sai1*) of *Arabidopsis thaliana*, identified in a selective screen utilizing the SA-inducible expression of the tms2 gene. Molecular Plant-Microbe Interactions 10:69-78.

Sharma, Y., J. Leon, I. Raskin, and K. R. Davis. 1996. Ozone-induced responses in *Arabidopsis thaliana*: The role of salicylic acid in the accumulation of defense-related transcripts and induced resistance. Proceedings of the National Academy of Sciences USA 93:5099-5104.

Shulaev, V.P., J. Leon, and I. Raskin. 1995. Is salicylic acid a translocated signal of systemic acquired resistance in tobacco? The Plant Cell 7:1691-1701.

Shulaev, V, P. Silverman, and I. Raskin. 1997. Airborne signalling by methyl salicylate in plant pathogen resistance. Nature 385:718-721.

Shirasu, K., H. Nakajima,V.K. Rajasekhar, R.A. Dixon, C. Lamb. 1997. Salicylic acid potentiates an agonist-dependent gain control that amplifies pathogen signals in the activation of defense mechanisms. The Plant Cell 9:261-270.

Siegrist, J., W. Jeblick, and H. Kauss. 1994. Defense responses in infected and elicited cucumber (*Cucumis sativus* L.) hypocotyl segments exhibiting acquired resistance. Plant Physiology 105:1365-1374.

Smith, J.A., D.W. Fulbright, and R. Hammerschmidt. 1991. Rapid induction of systemic resistance in cucumber by *Pseudomonas syringae* pv. *syringae*. Physiological and Molecular Plant Pathology. 38:223-235.

Smith-Becker, J., E. Marois, E.J. Huguet, S.L. Midland, J.J. Sims, and N.T. Keen. 1998. Accumulation of salicylic acid and 4-hydroxybenzoic acid in phloem fluids of cucumber during systemic acquired resistance is preceded by a transient increase in phenylalanine ammonia-lyase activity in petioles and stems. Plant Physiology 116:231-238.

Stout, M.J., K.V. Workman, R.M. Bostock, and S.S. Duffey. 1998. Stimulation and attenuation of induced resistance by elicitors and inhibitors of chemical induction in tomato *(Lycopersicon esculentum*) foliage. Entomologia Experimentalis et Applicata 86:267-279.

Takahashi, H., Z. Chen, H. Du, Y. Liu, and D.F. Klessig. 1997. Development of necrosis and activation of disease resistance in transgenic tobacco plants with severely reduced catalase levels. The Plant Journal 11:993-1005.

Thomma, B.P.H.J., K. Eggermont, I.A.M.A. Penninckx, B. Mauch-Mani, R. Vogelsang, B.P.A. Cammue, and W.F. Broekaert. 1998. Distinct jasmonate-dependent and salicylate-dependent defense pathways in *Arabidopsis* are essential for resistance to distinct microbial pathogens. Proceeding s of the National Academy of Science USA 95:15107-15111.

Uknes, S., B. Mauch-Mani, M. Moyer, S. Potter, S. Williams, S., Dincher, D. Chandler, A. Slusarenko, E. Ward, and J. Ryals. 1992. Acquired resistance in *Arabidopsis*. The Plant Cell 4:645-656.

van Loon, L.C., P.A.H.M. Bakker, and C.M.J. Pieterse. 1998. Systemic resistance induced by rhizosphere bacteria. Annual Review of Phytopathology 36:453-483.

Vernooij, B., L. Friedrich, A. Morse, R. Reist, R. Kolditz-Jawhar, E. Ward, S. Uknes, H. Kessmann, and J. Ryals. 1994. Salicylic acid is not the translocated signal responsible for inducing systemic acquired resistance but is required in signal transduction. The Plant Cell 6:959-965.

Vijayan, P., J. Shockey, C.A. Levesque, R.J. Cook, and J. Browse. 1998. A role of jasmonate in pathogen defense of *Arabidopsis*. Proceedings of the National Academy USA 95:7209-7214.

Ward, E.R., S.J. Uknes, S.C. Williams, S.S. Dincher, D.L. Wiederhold, D.C. Alexander, P. Ahl-Goy, J.P. Métraux, and J.A. Ryals. 1991. Coordinate gene activity in response to agents that induce systemic acquired resistance. The Plant Cell 3:1085-1094.

Weymann, K., M. Hunt, S. Uknes, U. Neuenschwander, K. Lawton, H-Y Steiner, and J. Ryals. 1995. Suppression and restoration of lesion formation in *Arabidopsis* lsd mutants. The Plant Cell 7:2013-2022.

White, R.F. 1979. Acetylsalicylic acid (aspirin) induces resistance to tobacco mosaic virus in tobacco. Virology 99:410-412.

Yalpani, N., J. Leon, M.A. Lawton, and I. Raskin, 1993. Pathway of salicylic acid biosynthesis in healthy and virus-inoculated tobacco. Plant Physiology 103:315-321.

Yu, D.Q., Y.D. Liu, B.F. Fan, D.F. Klessig and Z.X. Chen. 1997. Is the high basal level of salicylic acid important for disease resistance in potato? Plant Physiology 115:343-349.

Zhou, N., T.L. Tootle, F. Tsui, D.F. Klessig, and J. Glazebrook. 1998. *PAD4* functions upstream from salicylic acid to control defense responses in *Arabidopsis*. The Plant Cell 10:1021-1030.

A Survey of Plant Defense Responses to Pathogens

Ray Hammerschmidt and Ralph L. Nicholson

Abstract

Plants are thought to resist pathogen infection through the use of a variety of constitutive and induced defenses. Among the pathogen-induced responses observed in resistant plants, the hypersensitive response, production of active oxygen species, phytoalexin accumulation, cell wall modifications, and accumulation of PR proteins have received attention as putative defenses. The possible role and evidence for those roles for each defense response is discussed along with a discussion of future research directions that will be most useful in evaluating the contribution of these responses to resistance.

Introduction

Plant resistance to pathogens is generally thought to require the presence of constitutive defenses and/or the activation of a variety of defense responses at the cellular level (Nicholson and Hammerschmidt 1992, Dixon et al. 1994, Hammerschmidt and Schultz 1996). There are a few good examples of pre-formed or constitutive defenses playing a decisive role in resistance; for example, we will later discuss a group of saponins called avenacins that confer resistance in oats and dictate host range to the fungal pathogen *Gaumanomyces graminis* (Osbourn et al. 1994, 1995, Bowyer et al. 1995). However, most resistance is thought to be expressed as multiple mechanisms which are elicited after infection (Hammerschmidt and Schultz 1996). This chapter will discuss several of the active responses that are thought to be involved in resistance with emphasis on production of phytoalexins, a classical defense response. Several recent reviews cover many of these defenses in more detail and can be consulted for further information (Nicholson and Hammerschmidt 1992, Cutt and Klessig 1993, Dixon et al. 1994, Goodman and Novacky 1994, Baker and Orlandi 1995, Kuć 1995).

The Hypersensitive Response

The terms "hypersensitive response" (HR) and "hypersensitivity" describe the localized and rapid death of one or a few host plant cells in response to invasion by an avirulent isolate of a pathogen (Goodman and Novacky 1994). The HR is characterized by a rapid loss of membrane integrity in the infected host cells and the accumulation of brown phenolic compound oxidation products (Goodman and Novacky 1994). Although the HR may be an effective defense against obligate parasites that require living host cells for nutrition, it is likely that this host response is only a single part of the defensive strategy of the plant because some host responses do not result in the HR (e.g., Hammerschmidt et al. 1985).

Time course studies suggest that the HR is coordinately expressed with other defenses (Goodman and Novacky 1994, Heath 1998a). Because of these correlations, it has been suggested that the hypersensitive response is a type of programmed cell death (see Gilchrist 1998 and Heath 1998a for recent reviews). In support of this idea, several plant mutants that express spontaneous necrotic lesion development that is similar to the HR (Dietrich et al. 1994, Greenberg et al. 1994) also express a number of genes and metabolites that are produced during active defense responses resistance (Greenberg et al. 1994).

Parallels have been drawn between the type of programmed cell death seen in the HR and the developmentally regulated apoptosis seen in animal cells (Gilchrist 1998). For example, Ryerson and Heath (1996) demonstrated that DNA laddering and a positive TUNEL reaction occurred during the HR of cowpea to *Uromyces vignae*. These authors also reported that the reaction in cowpea was not pathogen specific, as the same response was seen in reaction to treatment with KCN. *Arabidopsis* leaves and tobacco and soybean suspension cultured cells also expressed what appears to be a programmed cell death similar to mammalian apoptosis in response to treatments with avirulent pathogens or elicitors of the HR (Levine et al. 1996). The apoptosis-like cell death response also occurs in response to host selective toxins that cause plant cell death. Wang et al. (1996) showed that treatment of tomato tissue with AAL toxin that selectively kills cells of sensitive tomato cultivars also induces an apoptosis-like response.

Regardless of whether or not the hypersensitive response is a type of apoptosis or other form of programmed cell death, it is well-known through inhibitor studies that new transcription and translation is required for the development of the HR, as well as the expression of the putative defense responses that accompany host cell death (Goodman and Novacky 1994). What may be more important in terms of host plant defense is not if, or how the HR is programmed, but how (if at all) the HR directly contributes to host resistance and how the onset of the HR is related to the expression of the other putative defenses that will be discussed below.

Increases in Active Oxygen and Oxidative Enzyme Activity

The oxidation of phenolic compounds as cells undergo the HR suggests that there is an increase in phenol oxidizing enzymes and the production of active oxygen species such as hydrogen peroxide (H_2O_2), the hydroxyl radical ($OH^{\cdot}$) and the superoxide anion (O_2^-) (Goodman and Novacky 1994). Pathogen induced production of active oxygen species (often referred to as the oxidative burst) as part of active defense has received considerable attention in the last few years and has been recently reviewed (Sutherland 1991, Mehdy 1994, Baker and Orlandi 1995, Bolwell and Wojtaszek 1997). Therefore, we will only discuss this topic briefly.

An early event in the interaction between a pathogen and a host plant cell is the production of active oxygen species prior to the onset of the HR in what is called the oxidative burst (Baker and Orlandi 1995, Bolwell and Wojtaszek 1997). It has been suggested that hydrogen peroxide produced during the oxidative burst may function in the defense-signaling pathway which leads to the expression of the HR (Levine et al. 1994). However, it is not at all clear that the oxidative burst is required for the development of the HR for all plant-pathogen interactions. For example, Heath (1998b) has recently shown that the HR induced in cowpea by an avrirulent race of *U. vignae* is not elicited by the oxidative burst. These results and recent studies suggesting that along with active oxygen species, nitric oxide is required for HR expression (Delledonne et al. 1998, Durner et al. 1998) indicate that the specific involvement of active oxygen species in the induction of defense responses needs further clarification.

Regardless of the specific signaling role that active oxygen species may play in the HR, these molecules have the potential to directly contribute to defense. Each of these active oxygen species can be toxic (Baker and Orlandi 1995), and each may aid in reactions that strengthen the plant cell wall through crosslinking reactions (Brisson et al. 1994). Most studies on active oxygen have used cell suspension cultures (Baker and Orlandi 1995), but studies with leaves have shown that there can be induction of active oxygen in whole tissues after elicitor treatment (Lu and Higgins 1998) and in papilla that form as potential barriers at the point of attempted infection by fungal hyphae (Thordal-Christensen et al. 1997). These studies suggest that active oxygen could be produced in tissues as a direct defense when the pathogen attempts to invade host tissues.

Resistance expression is often accompanied by the activation and/or *de novo* synthesis of the phenol-oxidizing enzymes peroxidase and phenoloxidase and the lipid peroxidizing enzyme lipoxygenase (Goodman and Novacky 1994). Peroxidase activity often increases in response to infection (Rasmussen et al. 1995), and this enzyme may function in defense though production of antimicrobial quantities of hydrogen peroxide (Peng and Kuć 1992) as well as in more traditional cell wall lignification and crosslinking (Nicholson and

Hammerschmidt 1992). Phenoloxidases may contribute to defense through the production of toxic quinones (Appel 1993). Increase in activity (but apparently not synthesis) of phenoloxidase has been correlated with the onset of the hypersensitive response in soybean (Lazarovits and Ward 1982), and the interaction of phenoloxidase with endogenous phenols in the dying cell could be a major cause of the browning observed in hypersensitively responding cells. Another oxidase, lipoxygenase, may contribute to the hypersensitive response via disruption of cell membrane lipids (Goodman and Novacky 1994), to defense through the formation of toxic lipid oxidation products (Croft et al. 1993), as well as in the formation of lipid-derived signals (Farmer 1994).

Phytoalexins

Most flowering plant families respond to infection with the synthesis of low molecular weight antibiotics known as phytoalexins (Kuć 1995). The original concept of phytoalexins was of compounds that were specifically involved in host resistance. However, phytoalexins are now generally defined more broadly as compounds that are antimicrobial and induced after infection (Paxton 1981, Ebel 1986, van Etten et al. 1994). When considering phytoalexins, it must be remembered that although such compounds may represent a significant element of a plant's defense mechanisms, they rarely serve as the only element in a plant's defense arsenal (Nicholson and Hammerschmidt 1992, Grayer and Harborne 1994). In 1982, only 15 plant families were known to produce phytoalexins, and this did not include members of the Poaceae, unless one considers the agmatines and hordatines found in barley as active response compounds (Stoessl 1967). Now, species within more than 30 families are known to produce phytoalexins (Grayer and Harborne 1994) and this includes oat, rice, sugarcane, and sorghum which are members of the Poaceae (Lo et al. 1996).

Phytoalexins represent a diverse group of compounds from a number of different secondary metabolic pathways. For example, plants in the Solanaceae produce sesquiterpenoid phytoalexins via the mevalonate pathway while legumes produce isoflavonoid or pterocarpan derived phytoalexins that are synthesized using both the shikimate pathway and the acetate-malonate pathway (Kuć 1995). An excellent review of the overall synthetic profile for the involvement of phenylpropanoids and the routes of the eventual conversion to a variety of phenolic phytoalexins is that of Dixon and Paiva (1995). Several examples of phytoalexins representing a diversity of biosynthetic origins are shown in Fig. 1. While the chemistry of phytoalexins differs from one plant family to another, it can be generalized that an incompatible interaction between a pathogen and host plant results in the rapid *de novo* synthesis and accumulation of phytoalexins (Kuć 1995).

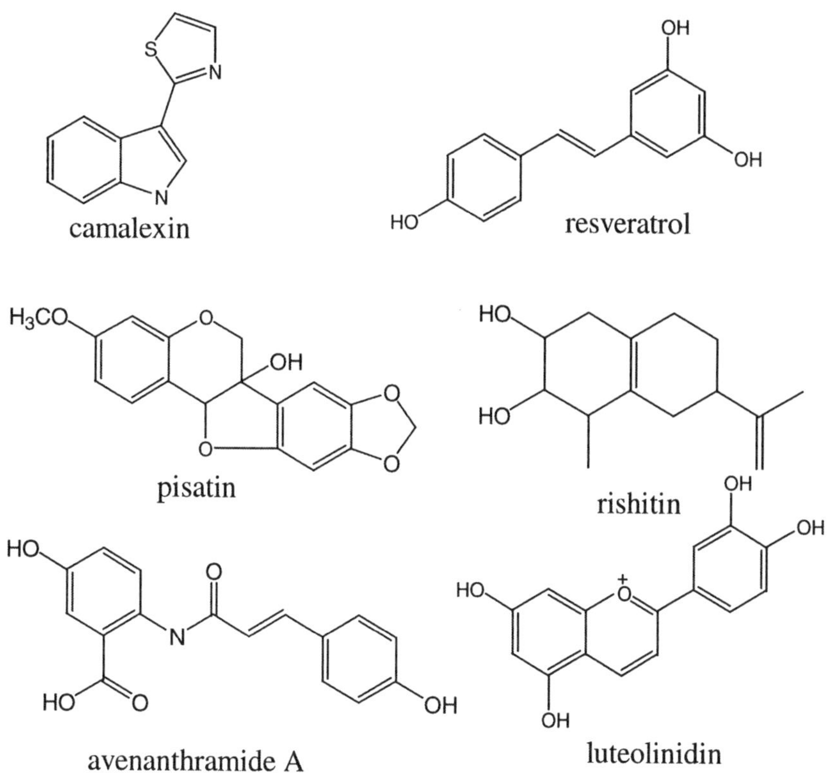

Figure 1: Structures of phytoalexins representing diverse biosynthetic origins.

Although many phytoalexins have been identified, it is interesting to note that relatively few studies have definitively confirmed that the inhibition of pathogen development is more than a correlative relationship with phytoalexin synthesis. A significant problem is that often phytoalexins can be shown to accumulate in areas of necrosis, especially necrosis associated with the hypersensitive response. However, few investigations have demonstrated that the timing, and indeed, the location of phytoalexin synthesis and accumulation is directly associated with specific sites of pathogen perturbation. Two systems that do address these problems involve the synthesis of very different classes of phytoalexins, the sesquiterpenoid phytoalexins synthesized in cotton in response to *Xanthomonas campestris* (Pierce et al. 1996), and the deoxyanthocyanidin phytoalexins synthesized in sorghum in response to infection by either pathogens or avirulent pathogens (Nicholson et al. 1987, Snyder and Nicholson 1990, Snyder et al. 1991). Genetic analysis has also shown co-segregation of resistance with pathogen-induced accumulation of avenanthramide (Fig. 1) phytoalexins in oat lines resistant to the crown rust pathogen, *Puccinia coronata* (Mayama et al. 1995a,b).

Concomitant with the synthesis and accumulation of phytoalexins is the induction of new transcription and translation of genes that are involved in the biosynthesis of these compounds. For example, infection of legumes with incompatible pathogens, or treatment of tissues with molecules known to elicit phytoalexin synthesis, will result in the rapid transcription and translation of genes; these genes are coordinately regulated in the phenylpropanoid and flavonoid pathways (e.g., phenylalanine ammonia-lyase and chalcone synthase) (Dixon and Paiva 1995), while in solanaceous plants, genes that regulate terpene synthesis (e.g., HMGR CoA reductase and sesquiterpene cyclases) are induced (Kuć 1995).

Unfortunately, most of the evidence for phytoalexins in defense is correlative and definitive evidence has not always been readily forthcoming to support a role for these compounds in defense. However, there are several examples where a case can be made for a specific, and primary, role in defense. *In situ* analysis of phytoalexin concentrations in relation to fungal or bacterial pathogen development has also been used to support a role for phytoalexins in resistance. Nicholson and co-workers have carried out a series of microspectrophotometric measurements of the accumulation of deoxyanthocyanidin (e.g., luteolinidin, Fig. 1), phytoalexins in the epidermal cells of sorghum leaves undergoing a resistant reaction to incompatible fungi (Snyder and Nicholson 1990, Snyder et al. 1991). These compounds absorb visible light (red to orange in color), and thus are easily observed in fresh tissue. They found that as the pathogen attempts to infect, the host cell begins to produce small inclusion bodies that appear to be the site of synthesis of the phytoalexins. The inclusions gradually migrate to the site of the infecting hyphae and, as they migrate, change in color from clear, to red-orange (indicating that synthesis of deoxyanthocyanidins has occurred). Microspectrophotometric analysis of these inclusion-containing cells indicated that the concentration of phytoalexin in these cells exceeds what is needed to kill the pathogen *in vitro* (Snyder et al. 1991). This response does not occur in susceptible interactions, and thus is a specific response of resistant host cells to infection.

Microspectrofluorometric analysis has demonstrated that the accumulation of the fluorescent phytoalexins of oats (avananthramides) and cotton (aromatic sesquiterpenes such as lacinoline c) correlate very well with the cessation of pathogen growth and development in resistant cultivars (Mayama and Tani 1982, Essenberg et al. 1992, Pierce et al. 1996). However, since most phytoalexins do not have color or distinctive fluorescence properties, localization/quantitation studies as described above are not possible.

Mutants deficient in phytoalexin production provide another strategy to study the role of these compounds in plant defense. *Arabidopsis* produces the indolic phytoalexin, camalexin, (Tsuji et al. 1992, Fig. 1) which is typical of the indole derived phytoalexins of the Brassicaceae (Hammerschmidt et al. 1993). The accumulation of camalexin after inoculation with incompatible bacterial and fungal pathogens suggests that it may play a role in resistance (Tsuji et al. 1992, Zook and Hammerschmidt 1997).

To further test the role of camalexin in defense, *Arabidopis* mutants that are either reduced or totally deficient in the production of camalexin have been isolated (Glazebrook and Ausubel 1994, Glazebrook et al. 1997). Analysis of these mutants indicated that some were regulatory mutants (Glazebrook et al. 1997), and thus might be less useful in studies on the role camalexin because other defense-associated processes might be effected. The regulatory role of at least one mutant *(pad4)* has been confirmed (Zhou et al. 1998).

Biosynthetic studies (Tsuji et al. 1993, Zook and Hammerschmidt 1997, Zook 1998) indicate that camalexin synthesis may be relatively simple with the last step possibly involving the condensation of cysteine with indole-3-carboxaldehyde to form camalexin. Cysteine has been shown to be a precursor (Zook and Hammerschmidt 1997), and recent time course studies suggest that indole-3-carboxaldehyde accumulation in infected tissue follows proper precursor-product kinetics (Kagan and Hammerschmidt, unpublished results). Thus, it is possible that only a few steps may be involved in camalexin biosynthesis. Understanding the biosynthetic pathway should lead to isolation of specific biosynthetic genes that could be potential targets for mutations that would effectively block camalexin biosynthesis. Such mutants would be invaluable for testing the specific contribution of camalexin in pathogen resistance.

Two problems that seem to be consistently not addressed are associated with phytoalexins and the constitutively produced phytoanticipins. First is the question of the mechanism of action of phytoalexins, that is, what accounts for their individual properties of toxicity. Second, there has been little correlative work that addresses possible relationships between structure and activity. Grayer and Harborne (1994) pointed out that an important feature of toxicity is that of the relative hydrophilicity or lipophilicity of individual compounds. For example, fungitoxic triterpenoid saponins are relatively hydrophilic because they are glycosylated and it is the carbohydrate moiety itself which accounts for toxicity. An excellent example of this relationship is that of the constitutively produced oat saponin avenacin (Fig. 2). Deglycosylation of the molecule by the enzyme avenacinase allows for the expression of pathogenicity by the fungus *Gaumanomyces graminis* (Osbourn et al. 1994, 1995, Bowyer et al. 1995). In effect, deglycosylation renders the molecule more lipophilic. In direct contrast to this is the phytoalexin pisatin (Fig. 1) produced by pea in response to fungal infection. The pathogen *Nectria haematococca* produces the enzyme pisatin demethylase that detoxifies the phytoalexin by demethylation, a process that

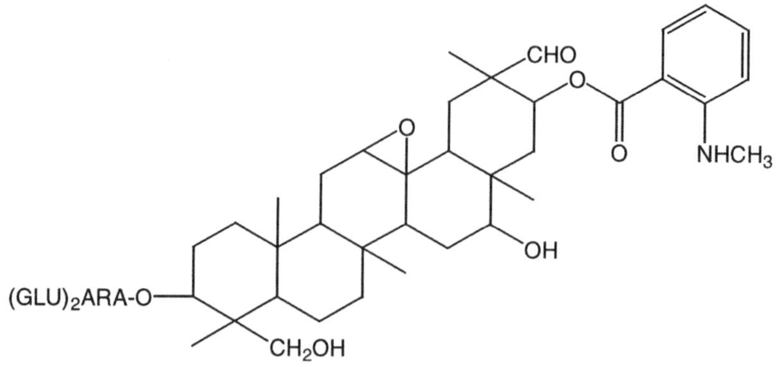

Figure 2: The structure of Avenacin A.

renders the compound more hydrophilic (Matthews and van Etten 1983, van Etten et al. 1989). Another example of the importance of relative hydrophilicity/ lipophilicity to phytoalexin toxicity is evident in the phytoalexins produced by sorghum. For example, five 3-deoxyanthocyanidins are known to be produced in most sorghum lines and all function as phytoalexins (e.g., luteolinidin, Fig. 1). Importantly, each of the phytoalexins has been shown to be fungitoxic to both pathogens and non-pathogens (Nicholson et al. 1987, Hipskind et al. 1990, Aida et al. 1996, Lo et al. 1996). Sorghum cultivars known as tan lines fail to produce any of the deoxyanthocyanidins and as expected are very susceptible to the anthracnose pathogen *Colletotrichum sublineolum.* It is interesting, however, that some sorghum inbred lines that produce only two of the phytoalexins (apigeninidin and the caffeic acid ester of arabinosyl 5-O-apigeninidin) exhibit greater susceptibility to the fungus although these compounds are themselves quite fungitoxic. Closer examination of the phytoalexin complex reveals two very important features for consideration of toxicity. First, luteolinidin and 5-methoxyluteolinidin are the most fungitoxic of the phytoalexins (Lo et al. 1996) although the apigeninidin derivatives are also toxic. Consistent with the work on the pea phytoalexin pisatin, results have shown that the methoxylated derivatives of apigeninidin and luteolinidin are significantly more toxic than the nonmethoxylated compounds (Aida et al. 1996, Lo et al. 1996). Another important factor is the timing of synthesis. For example, Wharton and Nicholson (unpublished data) have found that synthesis of apigeninidin and its derivatives lag considerably behind that of the other deoxyanthocyanidins. Thus, an important consideration concerning the efficacy of phytoalexins concerns not only their specific levels of toxicity but also the timing of their synthesis. It is also known that other sorghum cultivars exist that synthesize only apigeninidin derivatives, and these cultivars are considerably more susceptible to *C. sublineolum*, provided that the timing of phytoalexin synthesis compared to the timing of attempted infection is somewhat delayed (Lo and Nicholson unpublished data).

Finally, not all novel secondary products that are synthesized in host tissue during the expression of resistance are phytoalexins or phytoalexin-like compounds. For example, maize does not synthesize phytoalexins, rather, it up-regulates the synthesis of normal phenylpropanoids as a response to infection. Some of these compounds are converted to caffeic acid esters, which are then assumed to serve as a source of potentially oxidizable ortho-diphenols and are associated with the phenomenon of tissue necrosis and browning. Other phenylpropanoids are bound as weak esters to cell wall carbohydrates, and yet others are converted to a unique stress lignin high in syringaldehyde residues (Lyons et al. 1993). Each of these phenomena is thought to represent a collective of events that serve to inhibit the growth and development of a pathogen. It is interesting that at the termination of this series of biochemical events in resistant plants, anthocyanin synthesis often occurs (Hammerschmidt and Nicholson 1977). Recently, it was shown by plasma desorption mass spectrometry that the anthocyanin pigment synthesized is cyanidin 3-dimalonyl glucoside (Fig. 3) (Hipskind et al. 1996).

Cyanidin-3-dimalonylglucoside accumulates in living cells that surround the necrotic cells of the infection site. Because the compound is a zwitterionic anthocyanin, it is believed that it functions as a protectant of the host tissue. For example, co-pigmentation occurs when compounds such as free phenols and organic acids react with the positively charged anthocyanin flavilium cations to form molecular complexes. Thus, this anthocyanin pigment may serve to increase the capacity of living host cells to sequester toxic intermediary metabolites.

Cell Wall Modifications

The first barrier that most pathogens encounter is the cell wall (Aist 1983, Schäfer 1994), a structure that most pathogens are able to pass through without apparent difficulty (Schäfer 1994). However, plant cells may respond quickly to infection by modifying cell walls in such as way that the walls become more effective barriers to pathogen ingress into and through tissues. These changes may increase the mechanical strength of the cell walls (Ride 1978), decreasing the susceptibility of the wall to cell wall degrading enzymes (Ride 1980, Stermer and Hammerschmidt 1987), and possibly act as diffusion barriers that block the flow of nutrients to the pathogen or toxins to the host cell (Ride 1978).

Lignification and similar phenolic compound deposition is one type of cell wall modification that has been correlated with resistance (Nicholson and Hammerschmidt 1992). During the early stages of infection into a resistant plant, the host cell in contact with the pathogen often deposits lignin or some type of

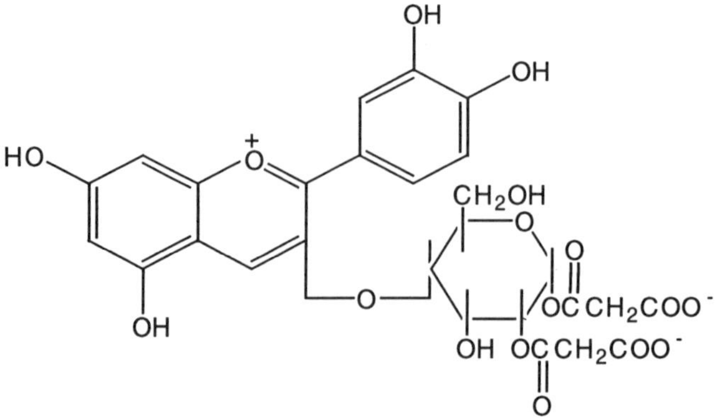

Figure 3: Cyanidin-3-dimalonylglucoside.

phenolic materials in its cell wall at the point of fungal invasion (e.g., Hammerschmidt et al. 1985). These deposits of lignin are often highly localized and appear to block the progression of the hyphae (Stein et al. 1993). If lignification occurs after cell wall penetration, the entire cell may lignify (Hammerschmidt et al. 1985) thus potentially trapping the pathogen in a lignified chamber. Another way that lignification may aid in resistance is by lignification of the pathogen cell walls. This has been observed for hyphae of *Colletotrichum orbiculare* in vitro (Hammerschmidt and Kuć 1982) and in cucumber leaf tissue (Stein et al. 1993). Thus, lignification may function to contain the pathogen in one place until other defenses (e.g., accumulation of phytoalexins or hydrolytic enzymes) are induced (Aist 1983). In maize, there is yet another series of events which take place where lignification of cell walls occurs as one of the final stages of the resistance response and follows esterification of various phenylpropanoids to cell wall components (Lyons et al. 1993).

In addition to lignin, other materials have been reported to accumulate in host cell walls in response to infection. These include the β -1,3-glucan callose (Aist 1983), hydroxyproline-rich glycoproteins (Hammerschmidt et al. 1984, Benhamou et al. 1991) and silicon oxides (Stein et al. 1993). Quite frequently, more than one type of cell wall modification may occur in a plant actively restricting infection. For example, the epidermal papilla-like structures that form in cucumber epidermal cells resisting fungal infection have been shown to contain both lignin and silicon (Stein et al. 1993). Either, both (or neither?) of these responses could contribute to the observed resistance. However, the relative contribution (if any) of each of these wall modifications to the defense response is not known.

Pathogenesis-Related Proteins

A group of proteins that are induced and accumulate locally and often systemically in plant tissues in response to infection are collectively known as the pathogenesis-related or PR proteins. The PR proteins are classified into several groups given the designations PR1 through PR11 (van Loon et al. 1994). Several very good reviews on PR proteins have been published (e.g., Cutt and Klessig 1992, van Loon 1997). As such, we will only present an overview of the evidence that supports a role for PR proteins in resistance.

Several lines of evidence have been used to support a role for PR proteins in resistance. These include the observations that the proteins are induced during the HR and appear systemically in plants expressing SAR (Cutt and Klessig 1992). Some of the PR proteins such as the β-1,3-glucanases (PR2) and chitinases (PR3) are antifungal *in vitro* (Cutt and Klessig 1992) and may also release elicitors from the cell wall of the pathogen (Yoshikawa et al. 1993). Transformation of plants with PR genes has also provided a means to test the role of these proteins in resistance. For example, the constitutive expression of the PR1 in transgenic tobacco or PR5 in transgenic potato increases resistance to oomycete pathogens (Alexander et al. 1993, Liu et al. 1994). Transformation of tobacco with a soybean β-1,3-glucanase (Yoshikawa et al. 1993) or a chitinase (Broglie et al. 1991) increases resistance to other fungi. Thus, there is some direct evidence for a role of PR proteins in defense.

Synthesis

There are two undeniable facts that have been demonstrated by studies on the interaction between plants and their pathogens: resistance exists and resistant plants quickly react to pathogen ingress with a number of different responses. Several of these responses have been discussed in this chapter, but the relationship between the resistant phenotype and the expression of these many putative defenses has generally not been proven and is based only on correlation.

Future studies on the nature of plant defense to pathogens will be more valuable if the focus is on the actual contribution that a putative defense makes to resistance. Most of the research has shown that, in infected resistant plants, there is an increase in one or many putative defenses at the site of infection at the time when pathogen development is being arrested (e.g., Snyder and Nicholson 1990, Stein et al. 1993, Thordal-Christensen et al. 1997). These spatial and temporal relationships are supported by *in vitro* assays that suggest that the response being studied can play some defensive role. For example, it is easy to visualize lignification in plant tissues and to show that a pathogen has progressed only up to the lignified cell wall. The role of lignified wall in defense can be supported by showing that the lignified host cell walls are much more resistant to the cell wall degrading enzymes that many pathogens produce. It is much more difficult to assess cause and effect by this approach. The same basic correlative arguments can be made for the role of any other defense. Transformation of plants with specific PR proteins has resulted in some plants becoming more resistant to fungal infections (e.g., Broglie et al. 1991). Similarly, transformation of tomato with the stilbene synthase that is responsible for resveratrol (Fig. 1) synthesis increases resistance of this plant to *Phytophthora infestans* (Thomzik et al. 1997). However, transformation of potato with the peroxidase that is associated with systemic acquired resistance in cucumber had no effect on resistance of potato to several pathogens (Ray et al. 1998). These studies are valuable because they demonstrate that increasing specific defenses can be directly involved in defense, but they are not able to provide definitive proof that the defense compound or proteins acts the same in the native host-pathogen interaction.

Future Directions

A possibly better approach than the ones described above would involve a genetic analysis of the putative defense response in association with the systematic elimination of defenses by either mutagenesis or, when practical, gene disruption. Using mutations has met limited success in understanding defense (e.g., Glazebrook et al. 1997), but this approach should be exploited more. Genes that are specific for the biosynthesis of several phytoalexins have been cloned (e.g., Wu et al. 1997, Back et al. 1998, Buell, this volume), and these genes provide tools that can be used to ask more specific questions about the relative contribution of a putative defense to resistance. These same genes also have potential use in engineering new forms of resistance (Dixon et al. 1996).

There is little doubt that plants can actively (and passively) defend themselves against pathogen attack, and there is a large literature that supports this. However, in order to truly understand the nature of host defense, future studies must concentrate of sorting out whether the host reactions that are studied are defenses or not. The tools to do just that are being developed, we only need to focus attention on evaluating pathogen defense more critically and precisely.

Acknowledgements

Support by the Michigan Agricultural Experiment Station and Purdue Agricultural Experiment Sation, and grants from NSF (IBN-9220912), USDA/NRICGP and USDA/CSREES to RH and an NSF grant (MCB-9603439) to RLN are gratefully acknowledged.

Literature Cited

Aida, Y., S. Tomogami, O. Kodama and T. Tsukinboshi. 1996. Synthesis of 7-methoxyapigeninidin and its fungicidal activity against *Gloeocercospora sorghi*. Bioscience Biotechnology and Biochemistry 60:1495-1496.

Aist, J.R. 1983. Structural responses as resistance mechanisms. Pages 33-70 *in* J. A. Bailey and B.J. Deverall, editors. The Dynamics of Host Defence. Academic Press, Sydney.

Alexander D., R.M. Goodman, M. Gut-Rella, C. Glascock, K. Weyman, L. Freiedrich, D. Maddox, P. Ahl-Goy, T. Luntz, E. Ward and J. Ryals. 1993. Increased tolerance to two oomycete pathogens in transgenic tobacco expressing pathogenesis-related protein 1a. Proceedings of the National Academy of Sciences USA 90:7327-7331.

Appel, H.M. 1993. Phenolics in ecological interactions: The importance of oxidation. Journal of Chemical Ecology 19:1521-1552.

Back, KW, S.L. He, K.U. Kim and D.H. Shin. 1998. Cloning and bacterial expression of sesquiterpenoid cyclase, a key enzyme for the synthesis of sesquiterpenoid phytoalexin capsidiol in UV-challenged leaves of *Capsicum annuum*. Plant and Cell Physiology 39:899-04.

Baker, C.J and E.W. Orlandi. 1995. Active oxygen in plant pathogenesis. Annual Review of Phytopathology 33:299-321.

Benhamou, N. D. Mazau, J. Grenier, and M.T. Esquerre-Tugaye.1991.Time-course study of the accumulation of hydroxyproline-rich glycoproteins in root cells of susceptible and resistant tomato plants infected by *Fusarium oxysporum* f.sp. *radicis-lycopersici.* Planta 184:196-208.

Bolwell, G.P. and P. Wojtaszek. 1997. Mechanisms for the generation of active oxygen species in plant defense - a broad perspective. Physiological and Molecular Plant Pathology 51:347-366.

Bowyer, P., B.R. Clarke, P. Lunness, M.J. Daniel, and A.E. Osbourn. 1995. Host range of a plant pathogenic fungus determined by a saponin detoxifying enzyme? Science 267:371-374.

Brisson, L.F., R. Tenhaken, and C. Lamb. 1994. Function of oxidative cross-linking of cell wall structural proteins in plant disease resistance. Plant Cell 6:1703-1712.

Broglie, K., I. Chet, M. Holliday, R. Cressman, P. Riddle, S. Knowlton, C.J. Mauvias, and R. Broglie. 1991. Transgenic plants with enhanced resistance to fungal pathogens. Science 254:1194-1197.

Croft, K.P.C., F. Juttner, and A.J. Slusarenko. 1993. Volatile products of the liopxygenase pathway evolved from *Phaseolus vulgaris* L. leaves inoculated with *Pseudomonas syringae* pv. *phaseolicola*. Plant Physiology 101:13-24.

Cutt, J.R., and D.F. Klessig. 1992. Pathogenesis-related proteins. Pages 181-216 *in* F. Meins and T. Boller, editors. Plant Gene Research. Springer-Verlag, New York.

Delledonne, M., Y.J. Xia, R.A. Dixon and C. Lamb. 1998. Nitric oxide functions as a signal in plant disease resistance? Nature 394:585-588.

Dietrich, R.A., T.P. Delaney, S.J. Uknes, E.R. Ward, J.A. Ryals, and J.L. Dangl. 1994. *Arabidopsis* mutants simulating disease resistance response. Cell 77:565-577.

Dixon RA., C.J. Lamb, S. Masoud, V.J.H. Sewalt and N.L. Paiva. 1996. Metabolic engineering: Prospects for crop improvement through the genetic manipulation of phenylpropanoid biosynthesis and defense responses - A review. Gene 179:61-71.

Dixon, R.A., M.J. Harrison, and C.J. Lamb. 1994. Early events in the activation of plant defense responses. Annual Review of Phytopathology 32:479-01.

Dixon, R.A., and N.L. Paiva. 1995. Stress-induced phenypropanoid metabolism. The Plant Cell 7:1085-1097.

Durner, J., D. Wendehenne, and D.F. Klessig. 1998. Defense gene induction in tobacco by nitric oxide, cyclic GMP, and cyclic ADP-ribose. Proceedings of the National Academy of Sciences USA 95:10328-10333.

Ebel, J. 1986. Phytoalexin synthesis: the biochemical analysis of the induction process. Annual Review of Phytopathology 24:235-264.

Essenberg, M., M.L. Pierce, E.C. Cover, B. Hamilton, P.E. Richardson, and V.E. Scholes. 1992. A method for determining phytoalexin concentrations in flourescent hypersenstively necrotic cells in cotton leaves. Physiological and Molecular Plant Pathology 41:101-109.

Farmer, E.E. 1994. Fatty acid signalling in plants and their associated microorganisms Plant Molecular Biology 26:1423-1437.

Glazebrook , J., and F.M. Ausubel . 1994. Isolation of phytoalexin-deficient mutants of *Arabidopsis thaliana* and characterization of their interactions with bacterial pathogens. Proceedings of the National Academy of Sciences USA 91:8955-8959.

Glazebrook J, M. Zook, F. Mert, I. Kagan, E.E. Rogers, I.R. Crute, E.R. Holub, R. Hammerschmidt and F. Ausubel. 1997. Phytoalexin-deficient mutants of *Arabidopsis* reveal that *PAD4* encodes a regulatory factor and that four PAD genes contribute to downy mildew resistance. Genetics 146:381-392.

Goodman, R.N., and A.J. Novacky. 1994. The Hypersensitive Reaction in Plants to Pathogens. APS Press, St. Paul, MN. pp. 244.

Grayer, R.J., and J.B. Harborne. 1994. A survey of antifungal compounds from plants, 1982-1993. Phytochemistry 37:19-42

Greenberg, J.T., A. Guo, D.F. Klessig, and F.M. Ausubel. 1994. Programmed cell death in plants: a pathogen triggered response activated coordinately with multiple defense functions. Cell 77:551-563.

Greenberg, J.T. 1997. Programmed cell death in plant-pathogen interactions. Annual Review of Plant Physiology and Plant Molecular Biology 48:525-545.

Gilchrist, D.G. 1998. Programmed cell death in plant disease: The purpose and promise of cellular suicide. Annual Review of Phytopathology 36:393-414.

Hammerschmidt, R. and J.C. Schultz. 1996. Multiple defenses and signals in plant defense against pathogens and herbivores. Recent Advances in Phytochemistry 30:121-154.

Hammerschmidt, R. A.M. Bonnen, G.C. Bergstrom, and K.K. Baker. 1985. Association of epidermal lignification with nonhost resistance of cucurbits to fungi. Canadian Journal of Botany 63:2393-2398.

Hammerschmidt, R.., and J. Kuć. 1982. Lignification as mechanism for induced systemic resistance in cucumber against *Colletotrichum lagenarium*. Physiological Plant Pathology 20:61-71.

Hammerschmidt, R., and R.L. Nicholson. 1977. Resistance of maize to anthracnose: Changes in host phenols and pigments. Phytopathology 67:251-258.

Hammerschmidt, R., J. Tsuji, M. Zook, and S. Somerville. 1993. A phytoalexin from *Arabidopsis thaliana* and its realtionship to other phytoalexins of crucifers. Pages 73-84 *in* K.R. Davis and R. Hammerschmidt, editors. *Arabidopsis thaliana* as a Model for Plant-Pathogen Interactions, APS Press, St. Paul, MN.

Hammerschmidt, R., D.T.A. Lamport and E.P Muldoon. 1984. Cell wall hydroxyproline enhancement and lignin deposition as an early event in the resistance of cucumber to *Cladosporium cucumerinum*. Physiological Plant Pathology. 24:43-47.

Heath, M.C. 1998a. Apoptosis, programmed cell death and the hypersensitive response. Plant Cell 104:117-124

Heath. M.C. 1998b. Involvement of reactive oxygen species in the response of resistant (hypersensitive) or susceptible cowpeas to the cowpea rust fungus. New Phytologist 138:251-263.

Hipskind, J.D., R. Hanau, B. Leite, and R.L. Nicholson. 1990. Phytoalexin accumulation in sorghum: identification of an apigeninidin acyl ester. Physiological and Molecular Plant Pathology 36:381-396.

Hipskind, J., K. Wood, and R.L. Nicholson. 1996. Localized stimulation of anthocyanin accumulation and delineation of pathogen ingress in maize genetically resistant to *Bipolaris maydis* race O. Physiological and Molecular Plant Pathology 49:247-256.

Kuć, J. 1995. Phytoalexins, stress metabolism and disease resistance in plants. Annual Review of Phytopathology. 33:275-297.

Lazarovits, G., and E.W.B. Ward. 1982. Polyphenoloxidase activity in soybean hypocotyls at sites inoculated with *Phytophthora megasperma* f. sp. *glycinea*. Physiological Plant Pathology 21:227-236.

Levine, A., R.I. Pennell, M.E. Alverez, R. Palmer, and C. Lamb. 1996. Calcium-mediated apoptosis in a plant hypersenstive disease resistance response. Current Biology 6:427-437.

Levine, A., R. Tenhaken, R.A. Dixon, and C. Lamb. 1994. H_2O_2 from the oxidative burst orchestrates the plant hypersensitive disease resistance response. Cell 79:583-593.
Liu, D., K.G. Raghothama, P.M. Hasegawa and R.A. Bressan. 1994. Osmotin overexpression in potato delays development of disease symptoms. Proceedings of the National Academy of Sciences USA 91:1888-92
Lo, S.-C., I. Weiergang, C. Bonham, J. Hipskind, K. Wood and R.L. Nicholson. 1996. Phytoalexin accumulation in sorghum: identification of a methyl ether of luteolinidin. Physiological and Molecular Plant Pathology 49:21-31.
Lu, H., and V.J. Higgins. 1998. Measurement of active oxygen species generated in planta in response to elicitor AVR9 of *Cladosporium fulvum*. Physiological and Molecular Plant Pathology 52:35-71.
Lyons, P.C., J. Hipskind, J.R. Vincent and R.L. Nicholson. 1993. Phenylpropanoid dissemination in maize resistant or susceptible to *Helminthosporium maydis*. Maydica 38:175-181.
Matthews, D.E., and H.D. van Etten. 1983. Detoxification of the phytoalexin pisatin by a fungal cytochrome P-450. Archives of Biochemistry and Biophysics 224:494-505.
Mayama S, and T. Tani. 1982. Microspectro-photometric analysis of the location of avenalumin accumulation in oat leaves in response to fungal infection with *Puccinia coronata*. Physiological Plant Pathology 21:141-49.
Mayama, S., A.P.A. Bordin, T. Morikawa, H. Tanpo, and H. Kato. 1995a. Association between avenalumin accumulation, infection hyphae length and infection type in oat crosses segregating for resistance to *Puccinia coronata* f.sp. *avenae* race 226. Physiological and Molecular Plant Pathology 46:255-262.
Mayama, S., A.P.A. Bordin, T. Morikawa, H. Tanpo, and H. Kato. 1995b. Association between avenalumin accumulation with co-segregation of victorin sensitivity and crown rust resistance carrying the Pc-2 gene. Physiological and Molecular Plant Pathology 46:263-274.
Mehdy, M.C. 1994. Active oxygen species in plant defense against pathogens. Plant Physiology 105:467-472.
Nicholson, R.L., S.S. Kollipara, J.R. Vincent, P.C. Lyons, and G. Cadena-Gomez. 1987. Phytoalexin synthesis by the sorghum mesocotyl in response to infection by pathogenic and nonpathogenic fungi. Proceedings of the National Academy of Science USA 84:5520-5524.
Nicholson, R.L., and R. Hammerschmidt. 1992. Phenolic compounds and their role in disease resistance. Annual Review of Phytopathology 30:369-389.
Osbourn, A., P. Bowyer, P. Lunness, B. Clarke, and M. Daniels. 1995. Fungal pathogens of oat roots and tomato leaves employ closely related enzymes to detoxify different host plant saponins. Molecular Plant-Microbe Interactions 8:971-978.
Osbourn, A. E., B.R. Clarke, P. Lunness, P.R. Scott, and M.J. Daniels. 1994. An oat species lacking avenacin is susceptible to infection by *Gaeumannomyces graminis* var. *tritici*. Physiological and Molecular Plant Pathology 45:457-467.
Paxton J.D. 1981. Phytoalexins - a working redefinition. Phytopathologische Zeitschrift 101:106-109.
Peng, M. and J. Kuć. 1992. Peroxidase-generated hydrogen peroxide as a source of antifungal activity in vitro and on tobacco leaf disks. Phytopathology 82:696-699.
Pierce, M.L., E.C. Cover, P.E. Richardson, V.E. Scholes and M. Essenberg. 1996. Adequacy of cellular phytoalexin concentrations in hypersensitively responding cotton leaves. Physiological and Molecular Plant Pathology 48:305-324.
Ray, H., D. Douches and R. Hammerschmidt. 1998. Transformation of potato with cucumber peroxidase: expression and disease response. Physiological and Molecular Plant Pathology 53:93-103.
Rasmussen, J.B., J.A. Smith, S. Williams, W. Burkhart, E. Ward, S.C. Somerville, J. Ryals and R. Hammerschmidt. 1995. cDNA cloning and systemic expression of acidic peroxidases associated with systemic acquired resistance to disease in cucumber. Physiological and Molecular Plant Pathology 46:389-400.
Ride, J.P. 1978. The role of cell wall alterations in resistance to fungi. Annals of Applied Biology 89:302-306.

Ride, J.P.1980. The effect of induced lignification on the resistance of wheat cell walls to fungal degradation. Physiological Plant Pathology 16:187-196.

Ryerson, D.E. and M.C. Heath. 1996. Cleavage of nuclear DNA into oligonucleosomal fragments during cell death induced by fungal infection or by abiotic treatments. The Plant Cell 8:393-402.

Schäfer, W. 1994. Molecular mechanisms of fungal pathogenicity in plants. Annual Review of Phytopathology 32:461-477.

Snyder, B.A., B. Leite, J. Hipskind, L.G. Butler and R.L. Nicholson. 1991. Accumulation of sorghum phytoalexins induced by *Colletotrichum graminicola* at the infection site. Physiological and Molecular Plant Pathology 39:463-470.

Snyder B.A., and R.L. Nicholson. 1990. Synthesis of phytoalexins in sorghum as a site specific response to fungal ingress. Science 248:1637-1639.

Stein, B.D., K. Klomparens and R. Hammerschmidt. 1993. Histochemistry and ultrastructure of the induced resistance response of cucumber plants to *Colletotrichum lagenarium*. Journal of Phytopathology 137:177-188.

Stermer, B.A. and R. Hammerschmidt. 1987. Association of heat shock induced resistance to disease with increased accumulation of insoluble extensin and ethylene synthesis. Physiological and Molecular Plant Pathology 31:453-461.

Stoessl, A. 1967. The antifungal factors in barley. IV. Isolation, structure and synthesis of the hordatines. Canadian Journal of Chemistry 45:1745-1760.

Sutherland, M.W. 1991. The generation of oxygen radicals during host responses to infection, Physiological and Molecular Plant Pathology 39:79-94.

Thomzik, J.E., K. Stenzel, R. Stocker, P.H. Schreier, R. Hain, and D.J. Stahl. 1997. Synthesis of a grapevine phytoalexin in transgenic tomatoes (*Lycopersicon esculentum* Mill.) conditions resistance against *Phytophthora infestans*. Physiological and Molecular Plant Pathology 51:265-278.

Thordal-Christensen, H., Z. Zhang, Y. Wei, and D.B. Collinge. 1997. Subcellular localization of H_2O_2 in plants. H_2O_2 accumulation in papilla and hypersensitive response during barley-powdery mildew interaction. Plant Journal 11:1187-1194.

Tsuji, J., M. Zook, R. Hammerschmidt, S. Somerville, and R. Last. 1993. Evidence that tryptophan is not a direct biosynthetic intermediate of camalexin in *Arabidopsis thaliana*. Physiological and Molecular Plant Pathology 43:221-29.

Tsuji, J., E.P. Jackson, D.A. Gage, R. Hammerschmidt and S.C. Somerville. 1992. Phytoalexin accumulation in *Arabidopsis thaliana* during the hypersensitive reaction to *Pseudomonas syringae* pv. *syringae*. Plant Physiology 98:1304-09.

van Loon, L.C. 1997. Induced resistance in plants and the role of pathogenesis – related proteins. European Journal of Plant Pathology 103:753-765.

van Loon, L.C., W.S. Pierpoint, T. Boller, and V. Conejero. 1994. Recommendations for naming plant pathogenesis-related proteins. Plant Molecular Biology Reporter 12:245-264.

van Etten, H., P. Matthews, K. Tegtmeier, M.F. Deitert and J.I. Stein. 1989. Phytoalexin detoxification: Importance for pathogenicity and practical implications. Annual Review of Phytopathology 27:143-64.

van Etten, H.D., J.W. Mansfield , J.A. Bailey and E. Farmer. 1994. Two classes of plant antibiotics: Phytoalexins versus "phytoanticipins". The Plant Cell 6:1191-92.

Wang, H., J. Li, R.M. Bostock, and D.G. Gilchrist. 1996. Apoptosis: A functional paradigm for programmed cell death induced by a host-selective phytotoxin and invoked during development. The Plant Cell 8:375-391.

Wu Q., C.A. Preisig, and H.D. van Etten. 1997. Isolation of cDNAs encoding (+)6a-hydroxymaackiain 3-*O*-methyltransferase, the terminal step for the synthesis of the phytoalexin pisatin in *Pisum sativum*. Plant Molecular Biology 35:551-60.

Yoshikawa M., M. Tsuda and Y. Takeuchi. 1993. Resistance to fungal disease in transgenic tobacco plants expressing the phytoalexin elicitor-releasing factor, β-1,3-glucanase, from soybean. Naturwissenschaften 80:417-20.

Zhou, N., T.L. Tootle, F. Tsui, D.F. Klessig and J. Glazebrook. 1998. *PAD4* functions upstream from salicylic acid to control defense responses in *Arabidopsis*. Plant Cell 10:1021-1030.

Zook, M., and R. Hammerschmidt. 1997. Origin of the thiazole ring of camalexin, a phytoalexin from *Arabidopsis thaliana*. Plant Physiology 113:463-468.

Zook, M. 1998. Biosynthesis of camalexin from tryptophan pathway intermediates in cell-suspension cultures of *Arabidopsis*. Plant Physiology 118:1389-1393.

Genes Involved in Plant-Pathogen Interactions

C. Robin Buell

Abstract

Plants can utilize an array of biochemical mechanisms to protect themselves against the viral, bacterial, fungal, and nematode pathogens that assault them in the phylloplane and rhizosphere. The response by the plant to a potential pathogen can be envisioned in three phases: first, the pathogen is recognized by the plant, second, the appropriate signal is transmitted to the host transcriptional and translational machinery, and third, the synthesis and/or release of molecules that impede pathogen growth and development. The central hypothesis governing specificity in disease resistance is the gene-for-gene model as proposed by H. H. Flor in the 1940s. This model proposes that the interaction between a single plant resistance gene product with its complementary avirulence gene product governs the outcome of the interaction and that the dominant alleles mediate incompatibility (resistance). In 1984, the first pathogen avirulence gene was cloned, providing molecular evidence to support the gene-for-gene model. Nearly a decade later, using genetic and molecular analyses, the first complementary plant resistance genes were cloned. Sequence data of these resistance genes has revealed a surprising conservation of sequence among the angiosperms, regardless of the host or pathogen taxonomic classification, suggesting conservation among plants in not only the recognition, but also the subsequent signaling mechanisms that lead to resistance. Indeed, several genes involved in signaling of pathogen defense responses have been shown to function in resistance to multiple pathogens. In addition, genes involved in the synthesis of antimicrobial factors have been able to provide enhanced resistance in heterologous systems, consistent with the hypothesis that the basic mechanism(s) by which pathogen ingress is arrested is conserved, to a large part, among plant species.

Abbreviations: 1-aminocyclopropane-1-carboxylic acid (ACC), Expressed Sequence Tag (EST), Leucine Rich Repeat (LRR), Leucine Zipper (LZ), Nucleotide Binding Site (NBS), Pathogenesis-Related (PR) protein, Polymerase Chain Reaction (PCR), Toll-IL-1R homology region (TIR), Transmembrane (TM)

Plant Disease Resistance Genes

Gene-for-Gene Model

In the 1940s, working with flax and the obligate flax rust pathogen, H. H. Flor hypothesized a genetic mechanism by which plants are resistant to pathogens (Flor 1971). This hypothesis, termed the gene-for-gene model, served as the paradigm in genetic breeding programs for the last 50 years. In this model, Flor proposed that a single gene in the host can confer resistance to the plant if the pathogen contains the complementary gene for avirulence. It is the presence of a dominant allele for resistance in the host and the complementary dominant allele for avirulence in the pathogen that dictates the outcome of the interaction. If the dominant allele is lacking in either the host or the pathogen, disease is the result.

Except for elegant genetic studies that confirmed the existence of single genes in the host and pathogen, no additional data supporting the model was generated until 1984. With the advent of molecular biology, Staskawicz et al. (1984) were able to isolate the first pathogen avirulence gene. Using a molecular approach, an avirulence gene from *Pseudomonas syringae* pv. *glycinea* race 6 that confers race-cultivar specific resistance on soybeans was cloned. This finding opened the flood gates and to date, over 30 avirulence genes have been isolated from bacterial pathogens alone (for review see Leach and White 1996). Characterization of these avirulence genes has yielded insight into the mechanism by which they confer avirulence. Although it had been postulated for several years, direct biochemical evidence that the avirulence gene product entered plant cells, where it potentially could interact with a cytoplasmically localized host factor, was difficult to obtain. Recently, new investigative tools have revealed that some avirulence gene products do enter the plant cell. Using reporter constructs, it was demonstrated that nuclear localization sequences on the *avrb6* and *pthA* genes of *Xanthomonas campestris,* as well as the homologous *avrBs3* gene from *X. c. vesicatoria*, are functional and result in the translocation of these avirulence proteins into the nucleus of host cells (Yang and Gabriel 1995, van den Ackerveken et al. 1996). Direct evidence that at least one avirulence gene product interacts with its complementary resistance gene product has been obtained (see below). Fungal avirulence gene products have also been studied and eight fungal avirulence genes have been cloned to date (reviewed in Laugé and De Wit 1998). Clearly, the retention of avirulence genes through evolution would not appear beneficial to the pathogen unless the avirulence gene serves additional functions in pathogen growth and reproduction. Indeed, several avirulence genes have been demonstrated to function in pathogenicity, revealing the basis for retention of the avirulence gene within bacterial and fungal populations (see Leach and White 1996, Laugé and De Wit 1998).

In contrast to the rapid isolation of avirulence genes upon the advent of molecular biology, the first plant disease resistance gene that conforms to the gene-for-gene model was not cloned until 1993. Using a positional cloning approach, the tomato *Pto* gene that confers resistance to the bacterial speck pathogen was isolated (Martin et al. 1993). *Pto* encodes a functional serine/threonine protein kinase, suggesting a role for *Pto* in a phosphorylation cascade (Loh and Martin 1995). Within the last six years, numerous disease resistance genes have been cloned, bringing the total number to 25 (Table 1). What has been surprising, is that with the exception of *Pto*, all of the other genes that conform to the gene-for-gene model (n=21) encode a protein with leucine-rich repeats (LRR). Additional motifs can be found in the LRR class of resistance genes, including leucine zippers (LZ), nucleotide binding sites (NBS), transmembrane domains (TM), and kinase domains. LRR-containing resistance genes have been isolated not only from monocots, but also from dicots, and encode resistance to viral, bacterial, fungal, nematode and aphid pests, suggesting that a common mechanism for recognition and signal transduction is conserved within the plant kingdom. As several articles on plant resistance genes have been published recently, the reader is referred elsewhere for in-depth discussion of resistance gene function (Bent 1996, Jones and Jones 1997).

In addition to the apparent conservation of sequence in most of the resistance genes cloned to date, additional insights into the gene-for-gene model have been made. First, a subset of the LRR resistance genes have similarity to two animal proteins, the Drosophila Toll and the mammalian interleukin-1 receptor, that function in insect antifungal defense and in innate immunity, respectively. This suggests a conservation in the mechanism(s) by which plants and animals mediate resistance (reviewed in Baker et al. 1997, Wilson et al. 1997). Second, in two separate examples, it appears that the same resistance gene encodes resistance against two separate pathogens; one with two distinct avirulence gene products and the other with two distinct pathogen origins. *RPM1* mediates resistance in *Arabidopsis* to *Pseudomonas* strains expressing either the *avrB* gene or the *avrRpm1* gene (Grant et al. 1995). These two avirulence genes share no significant similarity, suggesting either dual recognition capabilities are present in *RPM1* or another factor is involved in mediating resistance. Whereas *RPM1* confers resistance to two bacterial avirulence genes, the tomato *Mi* gene can confer resistance to not only three species of the nematode, *Meloidogyne*, but also to the potato aphid, *Macrosiphum euphorbiae* (Milligan et al. 1998, Rossi et al. 1998), extending the gene-for-gene model and the LRR-class of resistance genes to insects. Thus, in some cases, the "gene-for-gene" model should be restated as the "gene-for-genes" model.

The conservation in sequence motif among disease resistance genes suggests a common mechanism(s) for resistance. Some progress has been made in dissecting the molecular mechanism(s) by which these resistance genes 1) recognize the pathogen avirulence factor, and 2) initiate a signal transduction

Table 1. Disease resistance genes cloned to date.

Gene	Host	Pathogen	Class of R Gene[a]	Reference
Bacterial Resistance Genes				
Prf	Tomato	*Pseudomonas syringae* pv. *tomato*	LZ-NBS-LRR	Salmeron et al. 1996
Pto	Tomato	*P. s. tomato*	Protein kinase	Martin et al. 1993
RPM1 (*RPS3*)	*Arabidopsis*	*P. s. maculicola* *P. s. glycinea*	LZ-NBS-LRR	Grant et al. 1995
RPS2	*Arabidopsis*	*P. s. tomato*	LZ-NBS-LRR	Bent et al. 1994 Mindrinos et al. 1994
RPS5	*Arabidopsis*	*P. s. phaseolicola*	NBS-LRR	Warren et al. 1998
Xa1	Rice	*Xanthomonas oryzae* pv. *oryzae*	NBS-LRR	Yoshimura et al. 1998
Xa21	Rice	*X. o. oryzae*	LRR-kinase	Song et al. 1995
Fungal Resistance Genes				
Cf-2	Tomato	*Cladosporium fulvum*	LRR-TM	Dixon et al. 1996
Cf-4	Tomato	*C. fulvum*	LRR-TM	Thomas et al. 1997
Cf-5	Tomato	*C. fulvum*	LRR-TM	Dixon et al. 1998
Cf-9	Tomato	*C. fulvum*	LRR-TM	Jones et al. 1994
Hm1	Maize	*Cochliobolus carbonum*	Reductase	Johal and Briggs 1992
Hm2	Maize	*C. carbonum*	Reductase	Multani et al. 1998
I2C-1	Tomato	*Fusarium oxysporum* f. sp. *lycopersici*	NBS-LRR	Ori et al. 1997

L6	Flax	*Melampsora lini*	TIR-NBS-LRR	Lawrence et al. 1995
M	Flax	*M. lini*	TIR-NBS-LRR	Anderson et al. 1997
Mlo	Barley	*Erysiphe gramini* f. sp. *hordei*	Novel	Büschges et al. 1997
RPP1-WsA	*Arabidopsis*	*Peronospora parasitica*	TIR-NBS-LRR	Botella et al. 1998
RPP1-WsB	*Arabidopsis*	*P. parasitica*	TIR-NBS-LRR	Botella et al. 1998
RPP1-WsC	*Arabidopsis*	*P. parasitica*	TIR-NBS-LRR	Botella et al. 1998
RPP5	*Arabidopsis*	*P. parasitica*	TIR-NBS-LRR	Parker et al. 1997
RPP8	*Arabidopsis*	*P. parasitica*	LZ-NBS-LRR	McDowell et al. 1998
Viral Resistance Genes				
N	Tobacco	Tobacco Mosaic Virus	TIR-NBS-LRR	Whitham et al. 1994
Animal Pest Resistance Genes				
Hs1[pro-1]	Sugar beet	*Heterodera schachtii*	LRR-TM	Cai et al. 1997
Mi (*Meu*)	Tomato	*Meloidogyne* spp. *Macrosiphum euphorbiae*	LZ-NBS-LRR	Milligan et al. 1998 Rossi et al. 1998

[a] Leucine-Rich Repeat (LRR), Leucine Zipper (LZ), Nucleotide Binding Site (NBS), Toll-IL-1R homology region (TIR), Transmembrane (TM)

cascade that leads to incompatibility. As discussed above, several lines of evidence suggest that some of the avirulence gene products enter the plant cell, allowing for direct interactions with cytoplasmic localized plant proteins, and potentially the resistance gene product itself. Using the heterologous yeast two-hybrid system, direct interaction between the *Pto* resistance gene product and the *avrPto* gene product was demonstrated (Scofield et al. 1996, Tang et al. 1996), consistent with a model in which the resistance gene product and the avirulence gene product interact directly. Unlike the other resistance genes that conform to the gene-for-gene model, *Pto* encodes a protein kinase and not a LRR protein. Evidence for a direct interaction between a pathogen avirulence gene product and a host resistance protein of the LRR class has not been elucidated as of yet. This may be due to technical difficulties with heterologous expression of the LRR proteins, the possibility that the LRR resistance proteins do not interact directly with their complementary avirulence gene products, or the fact that additional components need to be present for a functional interaction between a LRR protein and the complementary avirulence protein.

Efforts focused on the identification of additional host proteins that interact with the resistance gene product have been productive. Using a yeast heterologous two-hybrid screen, Zhou et al. (1995) identified several proteins, termed Pto-interacting proteins (*Pti*), that interact with *Pto*. Characterization of *Pti1* revealed that it encodes a protein kinase, similar to *Pto*. *In vitro* studies revealed that *Pto* phosphorylates *Pti1*, suggesting *Pti1* acts downstream of *Pto* in a signaling cascade (Zhou et al. 1995). Other *Pti* genes include the *Pti4*, *Pti5*, and *Pti6* genes which encode proteins with sequence similarity to transcription factors (Zhou et al. 1997). However, *Pto* does not function alone and requires the action of the LRR protein, *Prf*. *Prf* was identified through positional cloning efforts to identify a closely linked but independent locus that is involved in *Pto*-mediated resistance (Salmeron et al. 1996). Thus, with bacterial speck of tomato, direct interaction between *Pto* and *avrPto* has been demonstrated and several components of the downstream signaling cascade have been identified.

Resistance Genes that Do Not Conform to the Gene-for-Gene Model

Although the gene-for-gene model governs a large number of plant-pathogen interactions, resistance can be mediated through alternative mechanisms, such as detoxification of pathogen toxins, insensitivity to host-selective toxins, induced resistance, escape from the pathogen, and tolerance. Three genes that confer alternative forms of resistance have been cloned. Indeed, the first resistance gene cloned was the maize *Hm1* gene that encodes a reductase that detoxifies the HC toxin produced by the maize fungal pathogen, *Cochliobolus carbonum* (Johal and Briggs 1992). HC toxin is an example of a host-specific toxin and is encoded by the *HTS1* gene of race 1 isolates of *C. carbonum* (Scott-Craig et al. 1992). The *HTS1* gene contains a 15.7 kb open reading frame that encodes a cyclic peptide synthetase responsible for HC toxin biosynthesis. The *Hm2* gene, which is highly similar to *Hm1,* was recently cloned from maize (Multani et al. 1998). The barley *Mlo* gene has also been cloned and encodes a protein with six transmembrane repeats (Büschges et al. 1997). Mutation of *Mlo* results in plants that produce spontaneous cell wall appositions and are resistant to most isolates of the powdery mildew pathogen, suggesting *Mlo* functions as a negative regulator of the defense response. Hybridization analyses, coupled with database searches, indicates *Mlo* orthologs are present in other plant species, including rice and *Arabidopsis*, suggesting *Mlo* may be a common mechanism in the regulation of the defense response (Büschges et al. 1997).

Implications of Cloned Resistance Genes

The cloning and characterization of resistance genes provides immediate and substantial insights into the biochemical and molecular basis for pathogen recognition and the subsequent resistance that results. In addition, these genes can be used to further improve resistance in an agricultural context. Several avenues to exploit these genes can be pursued. First, the cloned resistance genes can be used as molecular probes to identify homologous resistance genes within the same species and/or to identify orthologous sequences in related species. This approach has been extremely productive. Using the wealth of information present within the *Arabidopsis* Expressed Sequence Tag (EST) database, Botella et al. (1997) were able to establish the map position of 42 ESTs that share sequence similarity to resistance genes, using either polymerase chain reaction-mediated (PCR) amplification of the EST on yeast artificial chromosome clones or through genetic mapping. Not surprising, most of these candidate resistance genes mapped to locations that contain resistance genes identified through classical genetic mapping, providing candidate genes for these resistance loci. Also revealed in this report is that some of the resistance gene candidates appear to be clustered in the *Arabidopsis* genome, suggesting enrichment of resistance genes within the genome. In an alternative approach to identify resistance genes in *Arabidopsis,* use of PCR with degenerate primers, designed for conserved sequences within the LRR-class of resistance genes, resulted in the identification of additional candidate resistance genes. Some of these mapped to known disease resistance loci (Aarts et al. 1998a, Speulman et al. 1998). Detection of candidate resistance genes is not limited to *Arabidopsis*, as degenerate PCR using primers designed for conserved sequences within the LRR-class of resistance genes has yielded candidate resistance gene analogs from soybean and lettuce. Some of these have been mapped to loci of genetically defined resistance genes (Kanazin et al. 1996, Yu et al. 1996, Shen et al. 1998). High resolution mapping efforts and complementation experiments should validate this approach as an expedient method to identify disease resistance genes.

A second and more agriculturally important application of these sequences is to modify resistance in transgenic plants. Through overexpression of these resistance genes, enhanced resistance to pathogens has been obtained. Expression of the tomato *Pto* gene in tobacco resulted in transgenic tobacco with heightened resistance to an avirulent *P. s. tabaci* strain that expressed *avrPto* (Thilmony et al. 1995). Whereas Thilmony et al. (1995) used a tobacco line with a functional tobacco *Pto* homolog, Rommens et al. (1995) were able to demonstrate that the tomato *Pto* gene could provide effective resistance to *avrPto*-containing bacteria in *Nicotiana benthamiana*, a species that lacks a functional *Pto* homolog. Enhancement of resistance through expression of resistance genes is not limited to *Pto*-mediated resistance involving a protein kinase. In tomato, expression of the tobacco *N* gene, a LRR-encoding resistance gene, yielded plants that were resistant to tobacco mosaic virus (Whitham et al. 1996). Similarly, heterologous expression of the tomato *Cf-9* gene in tobacco and potato resulted in plants that generated a hypersensitive response following infiltration with the complementary Avr9 peptide (Hammond-Kosack et al. 1998). Thus, the possibility of creating transgenic plants with enhanced resistance is feasible and lengthy breeding programs can be accelerated through introduction of the resistance gene into a specific cultivar via biotechnology.

Genes Involved in Signal Transduction

One model in which the resistance response is initiated suggests that recognition of the pathogen by the host is mediated through the products of the avirulence and resistance genes. Subsequent to this recognition, a signal transduction cascade is initiated that leads to the synthesis of molecules that are involved in arresting pathogen growth and development. Molecules that have been implicated in signaling of defense responses include ethylene, salicylic acid, jasmonic acid, activated oxygen species, phosphorylation, ion fluxes, cell wall fragments, peptide hormones, and calcium, among others. The biochemical mechanism(s) by which these molecules elicit the defense response have been reviewed recently (Hammond-Kosack and Jones 1996, Wojtaszek 1997, see also Hammerschmidt and Smith-Becker, Staswick and Lehman, this volume). Implementation of genetic and molecular approaches to signaling mechanisms has not only yielded insight into this phase of the plant-pathogen interaction, but it has also provided novel molecular resources to modify disease resistance in transgenic plants. Select examples of signal transduction components that were identified using a genetic and/or molecular approach, and have provided new insight into signal transduction pathways, are discussed below.

Signaling in Induced Resistance

As discussed above, an alternative resistance mechanism is that of induced resistance. Induced resistance can be induced not only by necrotizing pathogens but also by several chemical inducers, providing a target for chemical strategies in disease control (Ryals et al. 1996). Several genes involved in induced resistance have been identified (Ryals et al. 1996). One gene of particular interest is the *NPR1* gene from *Arabidopsis* that functions in induced resistance. Mutation of the *NPR1* gene results in plants that are unable to accumulate mRNA for several pathogenesis-related (PR) proteins nor generate induced resistance in reponse to either chemical inducers of induced resistance or avirulent pathogens (Cao et al. 1994). Using a positional cloning approach, the *NPR1* gene was cloned and encodes a protein with ankyrin repeats (Cao et al. 1997). Not only was expression of *NPR1* in *npr1* mutants (non-expressor of PR genes) able to restore wild-type responsiveness to induced resistance inducers (Cao et al. 1997), but overexpression of *NPR1* in transgenic *Arabidopsis* resulted in resistance to both a virulent bacterial and fungal pathogen (Table 2, Cao et al. 1998). Thus, overexpression of the *NPR1* gene, a single component in the induced resistance signaling pathway, yielded resistance to multiple pathogens, providing a novel tool to engineer broad spectrum resistance in plants.

Modulation of the *Pto-Prf* Signaling Pathway

As discussed above, expression of the resistance gene, *Pto*, in tobacco resulted in the recognition of bacteria that expressed the complementary *avrPto* gene, suggesting that other components in the *Pto*-mediated resistance pathway are conserved among solanaceous plants (Rommens et al. 1995, Thilmony et al. 1995). Some components of the *Pto*-mediated resistance pathway have been identified and include the protein kinase-encoding *Pti1* gene (Zhou et al. 1995), the LRR-encoding *Prf* gene (Salmeron et al. 1996), and the transcription factor genes, *Pti4*, *Pti5*, and *Pti6* (Zhou et al. 1997). Overexpression of the *Pti1* gene in tobacco resulted in a more rapid hypersensitive response following challenge with *avrPto*-expressing bacteria (Table 2, Zhou et al. 1995), suggesting that enhanced resistance can also be provided through downstream components of the signaling pathway. Although its position in the *Pto*-mediated resistance pathway has not been confirmed, it has been suggested that *Prf* functions downstream of *Pto* (Olyroyd and Staskawicz 1998). Overexpression of the *Prf* gene in tomato lead to enhanced resistance to virulent pathogens, both bacterial and viral in origin (Table 2, Oldroyd and Staskawicz 1998). *Prf* overexpression was correlated with increased levels of salicylic acid, an inducer of disease resistance, and was correlated with increased mRNA accumulation of genes involved in defense. Thus, the mechanism of protection in *Prf*-overexpressing lines may be through activation of induced resistance.

Ethylene as a Modulator of Disease Symptoms

The simple plant hormone ethylene has been implicated in both plant defense responses and in symptom formation in the compatible interaction. Using a genetic and a molecular approach to bypass the artifacts which can be generated through application of ethylene-generating molecules and ethylene inhibitors, Lund et al. (1998) demonstrated that ethylene has a fundamental role in the development of symptoms in the compatible interactions of tomato with several pathogens. Overexpression in wild-type tomato of the 1-aminocyclopropane-1-carboxylic acid (ACC) deaminase gene, which results in inhibition of ethylene biosynthesis, yielded plants with reduced disease symptoms following challenge with *X. c. vesicatoria* (Table 2, Lund et al. 1998). Intriguingly, expression of the dominant negative *Arabidopsis etr1-1-Le-ETR3* construct that results in reduced ethylene perception also resulted in a reduction in symptoms in the compatible interaction. These data demonstrate that ethylene synthesis and perception is essential to the compatible interaction and that through modulation of ethylene synthesis and/or perception, plants can be modified for increased protection against pathogens.

Other Genes Involved in Signaling Events

Additional genes involved in signaling events are conserved among resistance pathways, such as the *Arabidopsis NDR1* and *EDS1* genes. *NDR1* was identified in a screen for mutants (*ndr1*: nonrace-specific disease resistance) that were deficient in resistance to avirulent bacterial and fungal pathogens, suggesting *NDR1* functions in a conserved signaling cascade (Century et al. 1995). *NDR1* encodes a small protein with two putative membrane-spanning domains and is upregulated in response to pathogen treatment (Century et al. 1997). Similar to *NDR1*, the *Arabidopsis EDS1* gene also functions in a conserved signal transduction cascade that leads to resistance (Parker et al. 1996). Mutation of *EDS1* results in plants that exhibit enhanced disease susceptibility (eds) to *Peronospora parasitica* and *Albugo candida*, but not *P. syringae* expressing the *avrB* gene. Subsequent analysis of the requirement for *NDR1* and *EDS1* in signal transduction events revealed that two signaling pathways are present in *Arabidopsis*. These pathways function without respect to the pathogen, be it fungal or bacterial, but instead are dictated by the resistance gene that governs the specificity of the interaction (Aarts et al. 1998b). Future experiments with overexpression of these two signaling components may reveal their potential in modifying resistance *in planta*, analogous to *NPR1* and *Prf*.

Table 2. Signaling and defense genes that have been demonstrated to enhance resistance in transgenic plants.

Gene	Function	Transgenic Plant	Pathogen Protection	Reference
Genes Involved in Signaling				
ACC deaminase [a]	Inhibits ethylene biosynthesis	Tomato	*Xanthomonas campestris* pv. *vesicatoria*	Lund et al. 1998
etr1-1-LeETR3 [b]	Dominant negative ethylene receptor	Tomato	*X. c. vesicatoria*	Lund et al. 1998
Glucose oxidase [c]	Hydrogen peroxide production	Potato	*Erwinia carotovora* subsp. *carotovora* *Phytophthora infestans*	Wu et al. 1995
NPR1	Induced resistance signaling pathway	*Arabidopsis*	*Pseudomonas syringae* pv. *maculicola* *Peronospora parasitica*	Cao et al. 1998
Prf [d]	*Prf-Pto*-mediated resistance	Tomato	*P. s. tomato* *X. c. vesicatoria* *Ralstonia solanacearum* Tobacco Mosaic Virus	Oldroyd and Staskawicz 1998
Pti1 [e]	*Prf-Pto*-mediated signaling cascade	Tomato	*P. s. tomato*	Zhou et al. 1995
Genes Involved in Arresting Pathogen Growth and Development				
Chitinase	Fungal cell wall degradation	Tobacco Tobacco Tobacco	*Rhizoctonia solani* *R. solani* *Cercospora nicotianae*, *R. solani*	Broglie et al. 1991 Zhu et al. 1994 Jach et al. 1995
Cystatin [f]	Protease inhibitor	*Arabidopsis*	*Heterodora schachtii* *Meloidogyne incognita*	Urwin et al. 1997

β-glucanase	Fungal cell wall degradation	Tobacco Tobacco	*C. nicotianae* *R. solani*	Zhu et al. 1994 Jach et al. 1995
Glucose oxidase [c]	Hydrogen peroxide production	Potato	*Erwinia carotovora* subsp. *carotovora.*, *P. infestans*	Wu et al. 1995
LTP2	Lipid transfer	*Arabidopsis* Tobacco	*P. s. tomato* *P. s. tabaci*	Molina and Garcia-Olmeda 1997
Osmotin [g]	Pathogenesis-related protein	Potato	*P. infestans*	Liu et al. 1994
PR-1a	Unknown	Tobacco	*P. tabacina* *P. parasitica* var. *nicotianae*	Alexander et al. 1993
Ribosome-inactivating protein	Inactivation of foreign ribosomes	Tobacco Tobacco	*R. solani* *R. solani*	Logemann et al. 1992 Jach et al. 1995
RS-AFP2	Cysteine-rich protein	Tobacco	*Alternaria longipes*	Terras et al. 1995
α-Thionin	Cysteine-rich protein	Tobacco	*P. s. syringae* *P. s. tabaci*	Carmona et al. 1993
Stilbene synthase	Phytoalexin synthesis	Tobacco	*Botrytis cinera*	Hain et al. 1993

[a] ACC deaminase converts ACC to α-ketobutyric acid, thereby inhibiting ethylene biosynthesis.

[b] *etr1-1-LeETR3* is a chimeric construct consisting of the *Arabidopsis etr1-1* dominant negative allele fused to the tomato *Le-ETR3* gene. Transgenic plants expressing this construct are blocked in ethylene perception.

[c] Hydrogen peroxide can function either as a direct antimicrobial compound or as an inducer of defense responses.

[d] The position of Prf in the Pto-mediated resistance pathway is unknown.

[e] *Pti1* overexpression only accelerated the hypersensitive response.

[f] A modified cystatin was used.

[g] Osmotin has antifungal activity and only delayed symptom expression.

Genes Involved in Arresting Pathogen Growth and Development

Following the recognition event between the host and the pathogen, changes in host gene expression occur which lead to the synthesis of factors that arrest pathogen growth and development, e.g., phytoalexins, or antimicrobial proteins. Various molecular approaches have been entailed to identify the defense genes that are upregulated in resistance, providing correlative data that they function in resistance. One assessment of the contribution of these defense genes to resistance is to overexpress them in transgenic plants and test for enhanced resistance to pathogens. This approach has been extremely successful in elucidating the contribution of individual defense genes in the resistance response. As indicated in Table 2, a diverse group of genes (n=16), when overexpressed in transgenic plants, can lead to increased resistance and/or tolerance to pathogens.

One obvious defense response to modify in transgenic plants is that of phytoalexin biosynthesis (Kuć 1995). In a clear demonstration that correlative data does reflect biological function, expression of the grapevine stilbene synthase gene that is responsible for the synthesis of stilbene phytoalexins resulted in elevated disease resistance in transgenic tobacco to the fungal pathogen, *Botrytis cinerea* (Hain et al. 1993).

In addition to subverting secondary metabolism within a plant to produce antimicrobial factors, expression of a single gene that encodes for either an antimicrobial polypeptide or enzyme has been effective in promoting increased resistance in transgenic plants. One obvious group of proteins to target for modified resistance are enzymes that directly affect pathogen growth and development, such as the hydrolytic enzymes, β-glucanase and chitinase, which degrade polysaccharide components in fungal cell walls (Bowles 1990, Tuzun and Bent, this volume). In 1991, Broglie et al. reported that expression of a bean chitinase in tobacco enhanced resistance to the soil-borne fungal pathogen, *Rhizoctonia solani*. Expression of chitinase did not alter resistance to the non-chitinous fungus, *Pythium aphanidermatum*, consistent with the function of chitinase in fungal cell wall degradation. Using a chitinase and a β-glucanase from barley, Jach et al. (1995) were able to demonstrate reduced disease severity in transgenic tobacco challenged with *R. solani*, confirming the earlier report by Broglie et al. (1991) that cell wall degrading enzymes can reduce disease. Expression of chitinase and β-glucanase also afforded protection in transgenic tobacco to *Cercospora nicotianae* (Zhu et al. 1994).

Other pathogenesis-related proteins in addition to chitinase and β-glucanase have been demonstrated to provide resistance in transgenic plants. Overexpression of *PR-1a* in tobacco provided enhanced tolerance to the blue mold pathogen, *Peronospora tabacina*, and the blank shank pathogen, *P. parasitica* var. *nicotianae,* but not to tobacco mosaic virus, potato virus Y, *C. nicotianae*, or *P. s. tabaci* (Alexander et al. 1993). Overexpression of the osmotin gene in transgenic potato, but not tobacco, delayed the onset of symptoms following challenge with *Phytophthora* species (Liu et al. 1994). These experiments demonstrate the validity in modifying resistance capabilities in transgenic plants through overexpression of genes intimately involved in the defense response, however, they highlight the subtlety by which resistance is affected at the plant-pathogen interface, a mechanism that probably is unique to each disease.

Other proteins with antimicrobial activity include cysteine-rich proteins (reviewed in Broekaert et al. 1995). Heterologous expression of the barley α-thionin gene in tobacco yielded plants with increased resistance to two bacterial pathogens, *P. s. tabaci* and *P. s. syringae* (Carmona et al. 1993). Similarly, expression of the radish *Rs-AFP2* gene, a small cysteine-rich defensin, enhanced resistance to *Alternaria longipes* in tobacco (Terras et al. 1995), implicating this class of proteins in resistance against both fungal and bacterial pathogens. Another class of antimicrobial proteins are the ribosome-inactivating proteins that are capable of inactivating foreign ribosomes (reviewed in Stirpe et al. 1992). Two reports have shown that expression of these proteins in transgenic tobacco provides protection against *R. solani* (Jach et al. 1995, Logemann et al. 1992). Successful modification of resistance is not limited to fungal and bacterial pathogens. Through overexpression of a modified rice cysteine protease inhibitor in *Arabidopsis*, a severe reduction in the female reproductive capability of two nematode species, *Heterodera schachtii* and *Meloidogyne incognita*, was observed (Urwin et al. 1997).

A group of proteins with diverse functions are the non-specific lipid transfer proteins. These proteins have been implicated in lipid transfer between membranes, but due to their extracellular localization, have also been implicated in cutin and wax formation as well as defense responses (for review see Kader 1997). Indeed, heterologous expression of the barley *LTP2* gene reduced the frequency and intensity of lesion formation in transgenic *Arabidopsis* following challenge with the virulent bacterial pathogen, *P. s. tomato* and in transgenic tobacco when challenged with *P. s. tabaci* (Molina and Garcia-Olmeda 1997).

In a novel application of molecular biology towards the creation of pathogen resistant plants, the glucose oxidase gene from *Aspergillus niger* was expressed in transgenic plants, resulting in plants with elevated levels of hydrogen peroxide (Wu et al. 1995). Surprisingly, these plants appeared normal and the tubers had increased resistance to the soft rot pathogen, *Erwinia carotovora* subsp. *carotovora*. In addition, protection was also provided against *P. infestans*, suggesting a more broad based protection provided by elevated hydrogen peroxide levels. Hydrogen peroxide can have a direct antimicrobial effect against both bacterial and fungal pathogens, but can also function in elicitation of the defense response, as reactive oxygen species function at the early stages of defense responses (Wojtaszek 1997).

Synthesis and Future Directions

Plant pathologists have made an overwhelming amount of progress in the last 50 years since Flor proposed the gene-for-gene model of resistance. Not only have the pathogen but also the host components of this model have been isolated and are being dissected for the mechanism of recognition (Table 1). These data have not only confirmed a model that was based solely on genetic data, but also provided new insights. The prevalence of the LRR-class of resistance gene among the cloned resistance genes (21 out of 25) suggests that the recognition of the pathogen and the subsequent signal transduction events are highly conserved among the angiosperms. The discovery that the *Mi* gene from tomato can confer resistance to both a nematode and an aphid is intriguing, as not only does this reveal a dual function for this resistance protein in tomato, but it also provides the first evidence that a LRR-resistance gene can confer resistance to insects. Whether *Mi* is representative of other insect resistance genes that behave in a gene-for-gene manner remains to be discovered, but as LRR-encoding proteins confer resistance to viruses, bacterial, fungi, nematodes, and insects, it may be that similar mechanisms function in gene-for-gene resistance to both insects and pathogens.

The fact that the major class of resistance gene, the LRR-class, not only encompasses resistance against all pathogen taxonomic classifications, but also is conserved throughout the angiosperms, has and will continue to provide invaluable molecular tools to more readily clone resistance genes. One aspect of agriculture that could be immensely impacted by these molecular tools is that of disease resistance in forest trees. Major disease resistance genes have been identified in trees and genetic mapping efforts have identified closely linked molecular markers, such as the gene for white pine blister rust resistance in sugar pine and the gene for fusiform rust resistance in loblolly pine (Devey et al. 1995, Wilcox et al. 1996). Thus, if the conserved nature of the LRR-class of resistance gene extends to gymnosperms, more rapid molecular approaches can be entailed to clone these genes and lengthy positional cloning approaches can be avoided.

Using molecular and biochemical approaches, extensive information is now available on signaling events in the defense response and a wealth of correlative data is available suggesting which molecules function in arresting pathogen growth and development. Using genetics and biotechnology, scientists have been able to dissect the precise contribution of individual genes in disease resistance. Advances have not been limited to discoveries in the laboratory, and application of these discoveries has resulted in the successful modification of resistance in transgenic plants (Table 2).

However, a new vista is present in biology which will impact the type and level of question that can be answered in plant pathology. This is the availability of the genomic sequence of both the host and the pathogen. This has already been exemplified by the availability of ESTs for both rice and *Arabidopsis*, a resource that greatly accelerates cloning genes. Currently, the *Arabidopsis* Genome Initiative has sequenced 56.6 Mbp of the 120 Mbp *Arabidopsis* genome (http://genome-www3.stanford.edu/cgi-bin/Webdriver?MIval=atdb_agi_total) and the entire sequence of the first plant should be available by the end of the year 2000 (Meinke et al. 1998). With time, the full genomic sequence of several plant pathogens will be available, providing complementary information. These resources will allow researchers to identify all the resistance genes within an organism, allowing inferences to made regarding the evolution of such genes. Further endeavors into functional genomics using DNA microarrays and DNA chips will provide researchers with the expression pattern of the entire genome through development and stress responses, such as pathogen challenge (Somerville et al. 1998). In addition, high through-put mutagenesis screens will provide the genetic resources to ascertain the function of every gene in an organism. Thus, we will progress from having correlative data of gene function to direct testing of the function of all of the genes within an organism, accelerating the rate at which the contribution of individual genes have in the process of disease resistance.

Acknowledgements

The critical reading and suggestive comments by Bud Damann, James Oard, Jackie Stephens, and Iain Wilson are greatly appreciated. Work in my laboratory is supported by funds from the Louisiana Agricultural Experiment Station (LAB03262) and the Louisiana Board of Regents (LEQSF 1997-00-RD-A-01).

Literature Cited

Aarts, M. G. M., B. L. Hekkert, E. B. Holub, J. L. Beynon, W. J. Stiekema, and A. Pereira. 1998a. Identification of R-gene homologous DNA fragments genetically linked to disease resistance loci in *Arabidopsis thaliana*. Molecular Plant-Microbe Interactions 11:251-258.

Aarts, N., M. Metz, E. Holub, B. J. Staskawicz, M. J. Daniels, and J. E. Parker. 1998b. Different requirements for *EDS1* and *NDR1* by disease resistance genes define at least two *R* gene-mediated signaling pathways in *Arabidopsis*. Proceedings of the National Academy of Sciences USA 95:10306-10311.

Alexander, D., R. M. Goodman, M. Gut-Rella, C. Glascock, K. Weymann, L. Friedrich, D. Maddox, P. Ahl-Goy, T. Luntz, E. Ward, and J. Ryals. 1993. Increased tolerance to two oomycete pathogens in transgenic tobacco expressing pathogenesis-related protein 1a. Proceedings of the National Academy of Sciences USA 90:7327-7331.

Anderson, P. A., G. J. Lawrence, B. C. Morrish, M. A. Ayliffe, E. J. Finnegan, and J. G. Ellis. 1997. Inactivation of the flax rust resistance gene *M* associated with loss of a repeated unit within the leucine-rich repeat coding region. Plant Cell 9:641-651.

Baker, B., P. Zambryski, B. Staskawicz, and S. P. Dinesh-Kumar. 1997. Signaling in plant-microbe interactions. Science 276:726-733.

Bent, A. F. 1996. Plant disease resistance genes: Function meets structure. Plant Cell 8:1757-1771.

Bent, A. F., B. N. Kunkel, D. Dahlbeck, K. L. Brown, R. Schmidt, J. Giraudat, J. Leung, and B. J. Staskawicz. 1994. *RSP2* of *Arabidopsis thaliana*: A leucine-rich repeat class of plant disease resistance genes. Science 265:1856-1860.

Botella, M. A., M. J. Coleman, D. E. Hughes, M. T. Nishimura, J. D. G. Jones, and S. C. Somerville. 1997. Map positions of 47 *Arabidopsis* sequences with sequence similiarity to disease resistance genes. Plant Journal 12:1197-1211.

Botella, M. A., J. E. Parker, L. N. Frost, P. D. Bittner-Eddy, J. L. Beynon, M. J. Daniels, E. B. Holub, and J. D. G. Jones. 1998. Three genes of the *Arabidopsis RPP1* complex resistance locus recognize distinct *Peronospora parasitica* avirulence determinants. Plant Cell 10:1847-1860.

Bowles, D. J. 1990. Defense-related proteins in higher plants. Annual Review of Biochemistry 59:873-907.

Broekaert, W. F., F. R. G. Terras, B. P. A. Cammue, and R. W. Osborn. 1995. Plant defensins: Novel antimicrobial peptides as components of the host defense system. Plant Physiology 108:1353-1358.

Broglie, K., I. Chet, M. Holliday, R. Cressman, P. Biddle, S. Knowlton, C. J. Mauvais, and R. Broglie. 1991. Transgenic plants with enhanced resistance to the fungal pathogen *Rhizoctonia solani*. Science 254:1194-1197.

Büschges, R., K. Hollricher, R. Panstruga, G. Simons, M. Wolter, A. Frijters, R. van Daelen, T. van der Lee, P. Diergaarde, J. Groenendijk, S. Töpsch, P. Vos, F. Salamini, and P. Schulze-Lefert. 1997. The barley *Mlo* gene: A novel control element of plant pathogen resistance. Cell 88:695-705.

Cai, D., M. Kleine, S. Kifle, H. Harloff, N. N. Sandal, K. A. Marcker, R. M. Klein-Lankhorst, E. M. J. Salentijn, W. Lange, W. J. Stiekema, U. Wyss, F. M. W. Grundler, and C. Jung. 1997. Positional cloning of a gene for nematode resistance in sugar beet. Science 275:832-834.

Cao, H., S. A. Bowling, A. S. Gordon, and X. Dong. 1994. Characterization of an *Arabidopsis* mutant that is nonresponsive to inducers of systemic acquired resistance. Plant Cell 6:1583-1592.

Cao, H., J. Glazebrook, J. D. Clarke, S. Volko, and X. Dong. 1997. The *Arabidopsis NPR1* gene that controls systemic acquired resistance encodes a novel protein containing ankyrin repeats. Cell 88:57-63.

Cao, H., X. Li, and X. Dong. 1998. Generation of broad-spectrum disease resistance by overexpression of an essential regulatory gene in systemic acquired resistance. Proceedings of the National Academy of Sciences USA 95:6531-6536.

Carmona, M. J., A. Molina, J. A. Fernandez, J. J. Lopez-Fando, and F. Garcia-Olmedo. 1993. Expression of the alpha-thionin gene from barley in tobacco confers enhanced resistance to bacterial pathogens. Plant Journal 3:457-462.

Century, K. S., E. B. Holub, and B. J. Staskawicz. 1995. *NDR1*, a locus of *Arabidopsis thaliana* that is required for disease resistance to both a bacterial and a fungal pathogen. Proceedings of the National Academy of Sciences USA 92:6597-6601.

Century, K. S., A. D. Shapiro, P. P. Repetti, D. Dahlbeck, E. Holub, and B. J. Staskawicz. 1997. *NDR1*, a pathogen-induced component required for *Arabidopsis* disease resistance. Science 278:1963-1965.

Devey, M. E., A. Delfino-Mix, B. B. Kinloch, and D. B. Neale. 1995. Random amplified polymorphic DNA markers tightly linked to a gene for resistance to white pine blister rust in sugar pine. Proceedings of the National Academy of Sciences USA 92:2066-2070.

Dixon, M. S., D. A. Jones, J. S. Keddie, C. M. Thomas, K. Harrison, and J. D. G. Jones. 1996. The tomato *Cf-2* disease resistance locus comprises two functional genes encoding leucine-rich repeat proteins. Cell 84:451-459.

Dixon, M. S., K. Hatzixanthis, D. A. Jones, K. Harrison, and J. D. G. Jones. 1998. The tomato *Cf-5* disease resistance gene and six homologs show pronounced allelic variation in leucine-rich repeat copy number. Plant Cell 10:1915-1926.

Flor, H. H. 1971. Current status of the gene-for-gene concept. Annual Review of Phytopathology 9:275-296

Grant, M. R., L. Godiard, E. Straube, T. Ashfield, J. Lewald, A. Sattler, R.W. Innes, and J. L. Dang. 1995. Structure of the *Arabidopsis RPM1* gene enabling dual specificity disease resistance. Science 269:843-846.

Hain, R., H. J. Reif, E. Krause, R. Langebartels, H. Kindl, B. Vornam, W. Wiese, E. Schmelzer, P. H. Schreier, R. H. Stocker, and K. Stenzel. 1993. Disease resistance results from foreign phytoalexin expression in a novel plant. Nature 361:153-156.

Hammond-Kosack, K. E., and J. D. G. Jones. 1996. Resistance gene-dependent plant defense responses. Plant Cell 8:1773-1791.

Hammond-Kosack K. E., S. Tang, K. Harrison, and J. D. G. Jones. 1998. The tomato *Cf-9* disease resistance gene functions in tobacco and potato to confer responsiveness to the fungal avirulence gene product Avr9. Plant Cell 10:1251-1266.

Jach, G., B. Gornhardt, J. Mundy, J. Logemann, E. Pinsdorf, R. Leah, J. Schell, and C. Maas. 1995. Enhanced quantitative resistance against fungal disease by combinatorial expression of different barley antifungal proteins in transgenic tobacco. Plant Journal 8:97-109.

Jones, D. A., C. M. Thomas, K. E. Hammond-Kosack. P. J. Balint-Kurti, and J. D. G. Jones. 1994. Isolation of the tomato *Cf-9* gene for resistance to *Cladosporium fulvum* by transposon tagging. Science 266:789-793.

Jones, D. A., and J. D. G. Jones. 1997. The role of leucine-rich proteins in plant defences. Advances in Botanical Research 24:89-167.

Johal, G. S., and S. P. Briggs. 1992. Reductase activity by the *HM1* disease resistance gene in maize. Science 258:985-987.

Kader, J. C. 1997. Lipid transfer protein: a puzzling family of plant proteins. Trends in Plant Science 2:66-70.

Kanazin, V., L. F. Marek, and R. C. Shoemaker. 1996. Resistance gene analogs are conserved and clustered in soybean. Proceedings of the National Academy of Sciences USA 93:11746-11750.

Kuć, J. 1995. Phytoalexins, stress metabolism, and disease resistance in plants. Annual Review of Phytopathology 33:275-297.

Laugé, R., and P. J. G. M. De Wit. 1998. Fungal avirulence genes: structure and possible functions. Fungal Genetics and Biology 24:285-297.

Lawrence, G. L., E. J. Finnegan, M. A. Ayliffe, and J. G. Ellis. 1995. The *L6* gene for flax rust resistance is related to the *Arabidopsis* bacterial resistance gene *RPS2* and the tobacco viral resistance gene *N*. Plant Cell 7:1195-1206.

Leach, J. E., and F. F. White. 1996. Bacterial avirulence genes. Annual Review of Phytopathology 34:153-179.

Liu, D., K. G. Raghothama, P. M. Hasegawa, and R. Bressan. 1994. Osmotin overexpression of potato delays development of disease symptoms. Proceedings of the National Academy of Sciences USA 91:1888-1892.

Logemann, J., G. Jach, H. Tommerup, J. Mundy, and J. Schell. 1992. Expression of a barley ribosome-inactivating protein leads to increased fungal protection in transgenic tobacco plants. Bio/Technology 10:305-308.

Loh, Y., and G. B. Martin. 1995. The *Pto* bacterial resistance gene and the *Fen* insecticide sensitivity gene encode functional protein kinases with serine/threonine specificity. Plant Physiology 108:1735-1739.

Lund, S. T., R. E. Stall, and H. J. Klee. 1998. Ethylene regulates the susceptible response to pathogen infection in tomato. Plant Cell 10:371-382.

Martin, G. B, S. H. Brommonschenkel, J. Chunwongse, A. Frary, M. A. Ganal, R. Spivey, T. Wu, E. D. Earle, and S. D. Tanksley. 1993. Map-based cloning of a protein kinase gene conferring disease resistance in tomato. Science 262:1432-1436.

McDowell, J. M., M. Dhandaydham, T. A. Long, M. G. M. Aarts, S. Goff, E. B. Holub, and J. L. Dangl. 1998. Intragenic recombination and diversifying selection contribute to the evolution of downy mildew resistance at the *RPP8* locus of *Arabidopsis*. Plant Cell 10:1861-1874.

Meinke, D. W., J. M. Cherry, C. Dean, S. D. Rounsley, and M. Koornneef. 1998. *Arabidopsis thaliana*: A model plant for genome analysis. Science 282:662-689.

Milligan, S. B., J. Bodeau, J. Yaghoobi, I. Kaloshian, P. Zabel, and V. M. Williamson. 1998. The root knot nematode resistance gene *Mi* from tomato is a member of the leucine zipper, nucleotide binding, leucine-rich repeat family of plant genes. Plant Cell 10:1307-1319.

Mindrinos, M., F. Katagiri, G. Yu, and F. M. Ausubel. 1994. The *A. thaliana* disease resistance gene *RPS2* encodes a protein containing a nucleotide-binding site and leucine-rich repeats. Cell 78:1089-1099.

Molina, A., and F. Garcia-Olmedo. 1997. Enhanced tolerance to bacterial pathogens caused by the transgenic expression of barley lipid transfer protein LTP2. Plant Journal 12:669-675.

Multani, D. S., R. B. Meeley, A. H. Paterson, J. Gray, S. P. Briggs, and G. S. Johal. 1998. Plant-pathogen microevolution: Molecular basis for the origin of a fungal disease in maize. Proceedings of the National Academy of Sciences USA 95:1686-1691.

Oldroyd, G. E. D., and B. J. Staskawicz. 1998. Genetically engineered broad-spectrum disease resistance in tomato. Proceedings of the National Academy of Sciences USA 95:10300-10305.

Ori, N., Y. Eshed, I. Paran, G. Presting, D. Aviv, S. Tanksley, D. Zamir, and R. Fluhr. 1997. The *I2C* family from the wilt disease resistance locus *I2* belongs to the nucleotide binding, leucine-rich repeat superfamily of plant resisitance genes. Plant Cell 9:521-532.

Parker, J. E., E. B. Holub, L. N. Frost, A. Falk, N. D. Gunn, and M. J. Daniels. 1996. Characterization of *eds1*, a mutation in *Arabidopsis* suppressing resistance to *Peronospora parasitica* specified by several different *RPP* genes. Plant Cell 8:2033-2046.

Parker, J. E., M. J. Coleman, V. Szabo, L. N. Frost, R. Schmidt, E. A. van der Biezen, T. Moores, C. Dean, M. J. Daniels, and J. D. G. Jones. 1997. The *Arabidopsis* downy mildew resistance gene *RPP5* shares similarity to the Toll and Interleukin-1 receptors with *N* and *L6*. Plant Cell 9: 879-894.

Rommens, C. M. T., J. M. Salmeron, G. E. D. Oldroyd, and B. J. Staskawicz. 1995. Intergeneric transfer and functional expression of the tomato disease resistance gene *Pto*. Plant Cell 7:1537-1544.

Rossi, M., F. L. Goggin, S. M. Milligan, I. Kaloshian, D. E. Ullman, and V. M. Williamson. 1998. The nematode resistance gene *Mi* of tomato confers resistance against the potato aphid. Proceedings of the National Academy of Sciences USA 95:9750-9754.

Ryals, J. A., U. H. Neuenschwander, M. G. Willits, A. Molina, H. Y. Steiner, and M. D. Hunt. 1996. Systemic acquired resistance. Plant Cell 8:1809-1819.

Salmeron, J. M., G. E. D. Oldroyd, C. M. T. Rommens, S. R. Scofield, H. Kim, D. T. Lavelle, D. Dahlbeck, and B. J. Staskawicz. 1996. Tomato *Prf* is a member of the leucine-rich repeat class of plant disease resistance genes and lies embedded within the *Pto* kinase gene cluster. Cell 86:123-133.

Scofield, S. R., C. M. Tobias, J. P. Rathjen, J. H. Chang, D. T. Lavelle, R. W. Michelmore, and B. J. Staskawicz. 1996. Molecular basis of gene-for-gene specificity in bacterial speck disease of tomato. Science 274:2063-2065.

Scott-Craig, J. S., D. G. Pannaccione, J. A. Pocard, and J. D. Walton. 1992. The cyclic peptide synthetase catalyzing HC-toxin production in the filamentous fungus *Cochliobolus carbonum* is encoded by a 15.7 kilobase open reading frame. Journal of Biological Chemistry 267:26044-26049.

Shen, K. A., B. C. Meyers, M. N. Islam-Faridi, D. B. Chin, D. M. Stelly, and R. W. Michelmore. 1998. Resistance gene candidates identified by PCR with degenerate oligonucleotide primers map to clusters of resistance genes in lettuce. Molecular Plant-Microbe Interactions 11:815-823.

Somerville, S. C., M. Nishimura, D. Hughes, I. Wilson, and J. Vogel. 1998. Alternate methods of gene discovery--The candidate EST approach and DNA microarrays. Pages 297-309 *in* F. L. Schiavo, R. L. Last, G. Morelli, and N. V. Raikhel, editors. Cellular Integration of Signalling Pathways in Plant Development. Springer-Verlag, Berlin.

Song, W., G. L. Wang, L. L. Chen, H. S. Kim, L. Y. Pi, T. Holsten, J. Gardner, B. Weng, W. X. Zhai, L. H. Zhu, C. Fauget, and P. Ronald. 1995. A receptor kinase-like protein encoded by the rice disease resistance gene, *Xa21*. Science 270:1804-1806.

Speulman, E., D. Bouchez, E.B. Holub, and J. L. Beynon. 1998. Disease resistance gene homologs correlate with disease resistance loci of *Arabidopsis thaliana*. Plant Journal 14:467-474.

Staskawicz, B. J., D. Dahlbeck, and N. T. Keen. 1984. Cloned avirulence gene of *Pseudomonas syringae* pv. *glycinea* determines race specific incompatibility on *Glycine max* (L.) Merr. Proceedings of the National Academy of Sciences USA 81:6024-6028.

Stirpe, F., L. Barbieri, L. G. Battelli, M. Soria, and D. A. Lappi. 1992. Ribosome-inactivating proteins from plants: present status and future prospects. Bio/Technology 10:405-412.

Tang, X., R. D. Frederick, J. Zhou, D. A. Halterman, Y. Jia, and G. B. Martin. 1996. Initiation of plant disease resistance by physical interaction of AvrPto and Pto kinase. Science 274:2060-2063.

Terras, F. R. G., K. Eggermont, V. Kovaleva, N. V. Raikhel, R. W. Osborn, A. Kester, S. B. Rees, S. Torrekens, F. van Leuven, J. Vanderlyden, B. P. A. Cammue, and W. F. Broekaert. 1995. Small cysteine-rich antifungal proteins from radish: Their role in host defense. Plant Cell 7:573-588.

Thilmony, R. L., Z. Chen, R. A. Bressan, and G. B. Martin. 1995. Expression of tomato *Pto* gene in tobacco enhances resistance to *Pseudomonas syringae* pv. *tabaci* expressing *avrPto*. Plant Cell 7:1529-1536.

Thomas C. M., D. A. Jones, M. Parniske, K. Harrison, P. J. Balint-Kutri, K. Hatzixanthis, and J. D. Jones. 1997. Characterization of the tomato *Cf-4* gene for resistance to *Cladosporium fulvum* identifies sequences that determine recognitional specificity in *Cf-4* and *Cf-9*. Plant Cell 9:2209-2224.

Urwin, P. E., C. J. Lilley, M. J. McPherson, and H. J. Atkinson. 1997. Resistance to both cyst and root-knot nematodes confered by transgenic *Arabidopsis* expressing a modified plant cystatin. Plant Journal 12:455-461.

van den Ackerveken, G.V., E. Marois, and U. Bonas. 1996. Recognition of the bacterial avirulence protein AvrBs3 occurs inside the host plant cell. Cell 87:1307-1316.

Warren, R. F., A. Henk, P. Mowery, E. Holub, and R. W. Innes. 1998. A mutation within the leucine-rich repeat domain of the *Arabidopsis* disease resistance gene *RPS5* partially suppresses multiple bacterial and downy mildew resistance genes. Plant Cell 10:1439-1452.

Whitham, S., S. P. Dinesh-Kumar, D. Choi, R. Hehl, C. Corr, and B. Baker. 1994. The product of the tobacco mosaic virus resistance gene *N*: Similarity to Toll and the Interleukin-1 receptor. Cell 78:1101-1115.

Whitham, S., S. McCormick, and B. Baker. 1996. The *N* gene of tobacco confers resistance to tobacco mosaic virus in transgenic tomato. Proceedings of the National Academy of Sciences USA 93:8776-8781.

Wilcox, P. L., H. V. Amerson, E. G. Kuhlman, B. H. Liu, D. M. O'Malley, and R. R. Sederoff. 1996. Detection of a major gene for resistance to fusiform rust disease in loblolly pine by genomic mapping. Proceedings of the National Academy of Sciences USA 93:3859-3864.

Wilson, I., J. Vogel, and S. Somerville. 1997. Signaling pathways: A common theme in plants and animals? Current Biology 7:R175-R178.

Wojtaszek, P. 1997. Oxidative burst: an early plant response to pathogen infection. Biochemistry Journal 322:681-692.

Wu, G., B. J. Shortt, E. B. Lawrence, E. B. Levine, K. C. Fitzsimmons, and D. M. Shah. 1995. Disease resistance conferred by expression of a gene encoding H_2O_2-generating glucose oxidase in transgenic potato plants. Plant Cell 7:1357-1368.

Yang, Y., and D. W. Gabriel. 1995. *Xanthomonas* avirulence/pathogenicity gene family encodes functional plant nuclear targeting signals. Molecular Plant-Microbe Interactions 8:627-631.

Yoshimura, S., U. Yamanouchi, Y. Katayose, S. Toki, Z. Wang, I. Kono, N. Kurata, M. Yano, N. Iwata, and T. Sasaki. 1998. Expression of *Xa1*, a bacterial blight-resistance gene in rice, is induced by bacterial inoculation. Proceedings of the National Academy of Sciences USA 95:1663-1668.

Yu, Y. G., G. R. Buss, and M. A. S. Maroof. 1996. Isolation of a superfamily of candidate disease-resistance genes in soybean based on a conserved nucleotide-binding site. Proceedings of the National Academy of Sciences USA 93:11751-11756.

Zhou, J., Y. Loh, R. A. Bressan, and G. B. Martin. 1995. The tomato gene *Pti1* encodes a serine/threonine kinase that is phosphorylated by Pto and is involved in the hypersensitive response. Cell 83:925-935.

Zhou, J., X. Tang, and G. B. Martin. 1997. The Pto kinase conferring resistance to tomato bacterial speck disease interacts with proteins that bind a *cis*-element of pathogenesis-related genes. EMBO 16:3207-3218.

Zhu, Q., E. A. Maher, S. Masoud, R. A. Dixon, and C. J. Lamb. 1994. Enhanced protection against fungal attack by constitutive co-expression of chitinase and glucanase genes in transgenic tobacco. Bio/Technology 12:807-812.

The Role of Hydrolytic Enzymes in Multigenic and Microbially-Induced Resistance in Plants

Sadik Tuzun and Elizabeth Bent

Abstract

Interactions between plants and associative rhizosphere microorganisms, which lead to the induction of plant growth and/ or disease resistance responses, are discussed, as well as the possibility that these mutualistic interactions may have co-evolved from originally antagonistic ones. Mechanisms by which organisms (plant growth-promoting rhizobacteria,or PGPR, in particular) may elicit multigenic plant defense responses are presented. Plant defense responses are multi-component, meaning there are multiple and apparently complimentary defense responses involved (e.g., production of phytoalexins, the formation or strengthening of physical barriers, or the production of pathogenesis-related (PR) proteins). The pattern of expression of hydrolytic enzymes and other PR proteins has been correlated with the ability of a variety of plants to mount an effective response against disease. The probable roles of specific hydrolase isozymes in the induction and maintenance of plant disease resistance are discussed.

Introduction

Plants and microorganisms have been in close contact ever since the origin of the first plant cell. It is thought that modern chloroplasts and mitochondria evolved from bacterial cells which were engulfed by some ancient eukaryotic precursor (Ozeki et al. 1989, Hill and Singh 1997). It has been theorized that plant-microbe symbioses such as mycorrhizae (Malloch 1987) or rhizobial nodules (Spaink 1995) may have arisen from antagonistic plant-microbe interactions, although alternative hypotheses have also been presented for rhizobial nodules (Provorov 1998). Mycorrhizal fungi and rhizobia both initially elicit defense responses from the plant, which the plant must attenuate for the infection to succeed (Duc et al. 1989, Gianninazzi-Pearson et al. 1995, Salzer et al. 1996), and under some circumstances, mycorrhizal fungi are deleterious to plant growth (Francis and Read 1995) or can become pathogenic (Beyrie et al. 1995).

Associative (free-living), rhizosphere organisms exist in a mutualistic relationship with plants also, albeit a more subtle one. Rhizosphere microbes are influenced by plants via root exudates (Curl and Truelove 1986, Stevenson et al. 1995); in fact, microbial activity and diversity is far greater in the vicinity of roots than in the bulk soil, a phenomenon known as the "rhizosphere effect" (Curl and Truelove 1986). Some rhizosphere organisms alter the rate or extent of plant growth, and sometimes simultaneously, the ability of the plant to defend itself against disease (Alstrom 1991, Wei et al. 1996, van Loon et al. 1998). A rhizobacterium which triggers a systemic defense response within the plant may enable the plant to protect its leaves from damage when pathogens (or herbivores, discussed in Zehnder et al., this volume) are present. Having intact leaves, the plant is able to produce more photosynthetic assimilate than it would otherwise, and release a greater quantity of these into the rhizosphere as root exudates. By protecting the plant, the rhizobacterium protects and even increases its food supply. By attracting and nourishing plant-immunizing rhizobacteria via root exudates, the plant increases its ability to defend itself against disease (or herbivory). A similar scenario involving plants and bacterial biocontrol agents was outlined by Ryan and Jagendorf (1995), and similar views have been advanced on the transition from antagonism to mutualism in plant-herbivore interactions (Paige 1992, Lennartson et al. 1997).

Might mutualistic plant-microbe interactions between associative soil microorganisms and plants, in which defense responses are induced, have also evolved from an originally antagonistic relationship? In this chapter, we will outline the interactions between associative soil bacteria and plants which result in the induction of defense responses, and describe the mechanisms which control these responses. We will also summarize the points that suggest that mutualistic interactions between plants and associative rhizobacteria may have evolved from antagonistic ones.

Simultaneous Plant Growth Promotion and Induction of Resistance

We will periodically refer to the process of induction of plant defense responses as "immunization", and call plants in which these responses have been evoked "immunized". Contrary to some reports (e.g., van Loon et al. 1998), simultaneous plant immunization and growth promotion may be mediated by pathogens, or products of pathogens, as well as by non-pathogenic rhizobacteria. That these two phenomena may occur simultaneously was first reported by Tuzun et al. (1986, 1992), in tobacco naturally or artificially immunized with the pathogen *Peronospora tabacina.* The simultaneous induction of resistance and growth promotion responses in tomato, tobacco and cucumber inoculated with harpin (a virulence factor purified from the pathogen *Erwinia amylovora),* was recently reported (Wei and Beer 1996). It has been suggested that growth promotion of plants is a side-effect of the induction of defense responses with a variety of inducing agents, including pathogens, abiotic elicitors, and nonpathogenic organisms (Tuzun and Kuć 1991). The phenomenon has been well documented for rhizobacteria (Tuzun and Kloepper 1995) and we will mainly limit our discussion to interactions between these organisms and plants.

Nomenclature of rhizosphere microorganisms. A spectrum of microbial activity exists within the rhizosphere, ranging from plant-beneficial to deleterious. Plant growth promoters include rhizobacteria (PGPR) (Suslow and Schroth 1982) and fungi (PGPF) (Hyakumachi 1984), while growth retarding organisms include "minor pathogens", also known as deleterious microorganisms (DRMO), a subset of which are rhizobacteria (DRB) (Suslow and Schroth 1982, Schippers et al. 1987). Many of the concepts that will be applied below to rhizobacteria may also be applied to fungi; the mechanisms through which rhizofungi may promote or deplete plant growth are similar in principle, if not in execution (Howell 1990, Meera et al. 1995, Shearer 1995, Bertagnolli et al. 1995).

Before continuing, it should be noted that classifications of organisms as growth-promoting or deleterious are operational, not absolute (Bent and Chanway 1999). Whether a rhizobacterium appears to be a PGPR, DRB or neither of these depends upon which plant species or variety is inoculated, and the environment within which the plant and microbe interact. Not all plants will respond similarly to one bacterial strain; an organism which promotes the growth of one plant may have no effect or inhibit the growth of another (Yuen and Schroth 1986, Chanway and Holl 1992). Likewise, the beneficial effects of a PGPR inoculant can be difficult to reproduce under field conditions, and may depend upon the inoculation method used (van Elsas and Heijnen 1990, Mazolla et al. 1995) as well as the soil environment the inoculant must survive in. Soil structure, texture, pH, temperature, moisture content, levels of available nutrients, the presence of pesticides or pollutants, other rhizosphere microorganisms, viruses or soil fauna are all factors which will affect the ability of rhizobacteria (PGPR and DRB) to produce the structures, signals or metabolites through which they interact with plants (Curl and Truelove 1986, Milus and Rothrock 1993, Forlani et al. 1995, Janzen and McGill 1995, van Elsas et al. 1997, Howarth et al. 1998, Bent and Chanway 1999).

How Rhizobacteria Affect Plant Growth

Before examining how associative rhizobacteria may affect plant defense responses, it is useful to summarize how these organisms can affect plant growth. Since the phenomena of growth promotion and immunization appear to be linked, at leased in some cases (e.g., Tuzun et al. 1986, Wei and Beer 1996), examination of one may provide insight into the other. Mechanisms of growth promotion and inhibition have been reviewed for associative PGPR (Weller 1988, Glick 1995) and DRB (Schippers et al. 1987, Nehl et al. 1996), respectively.

PGPR may improve plant growth indirectly, by suppressing the growth of plant-deleterious organisms (e.g., DRB, pathogens) via competition for limited resources, including nutrients, minerals or infection sites upon the root, direct predation (Stirling 1984) or the production of antibiotics or other toxic or inhibitory substances (Weller 1988, Glick 1995). DRB may use similar mechanisms to inhibit the growth of PGPR (Schippers et al. 1987, Nehl et al. 1996). PGPR may stimulate the germination or growth of plant-beneficial mycorrhizal fungi (Garbaye 1994); DRB may stimulate the germination or growth of fungal phytopathogens (Schippers et al. 1987, Nehl et al. 1996).

PGPR may also affect plant growth more directly by providing nutrients to plants (fixed nitrogen, phosphorus or other minerals) (Glick 1995), producing growth-promoting phytohormones (Frankenberger and Arshad 1995) or destroying growth-inhibitory hormones or their precursors (Glick 1995). Conversely, DRB may sequester nutrients from plants, or produce growth-inhibitory compounds such as HCN or other toxins (Alstrom and Burns 1989, Nehl et al. 1996). DRB may also produce too much of a good thing: growth-promoting hormones such as auxins, cytokinins or giberrellins can, when exogenously supplied in excess, inhibit plant growth or cause damaging malformations (Frankenberger and Arshad 1995, Nehl et al. 1996).

How Rhizobacteria may Induce Plant Defense Responses

There are several means by which rhizobacteria may induce a defense response in a plant. By "defense response", we mean a biochemical response to pathogens leading to a reduction in pathogenesis. Rhizobacterially-mediated induction of resistance against herbivores may also occur and is described in this volume by Zehnder et al.

Interestingly, there have been reports of DRB which apparently *increase the susceptibility* of a plant to disease (Alstrom 1991), but the majority of research has focussed on interactions in which plant resistance to disease is enhanced, and we will limit our discussion to these. Some ideas as to how PGPR may stimulate plant defense responses are presented below, in the hope that they will stimulate discussion, new ideas, and further research in this area.

Bacterial surface structures have been demonstrated to immunize plants. *Pseudomonas* siderophores or outer membrane lipopolysaccharides (LPS) have been demonstrated to induce disease resistance in radish (van Loon et al. 1998). These cannot be the only substances involved in PGPR-mediated induction of resistance, since some PGPR known to induce resistance are Gram positive or do not produce siderophores (Wei et al. 1996).

The promotion of oxidative damage, such as that occurring during the hypersensitive response (HR), is associated with, and may play a role in, the elicitation of defense responses (Strobel and Kuć 1995). The lesions formed by purified harpin on inoculated plants are microscopic (S. Beer, personal communication). Recently, Jetiyonan (1997) reported that, when injected into tobacco leaves, all the PGPR strains capable of inducing resistance in plants (*Bacillus pumilis* strains SE34 and SE49, and *Pseudomonas fluorescens* strain 89B-61) also elicited a confluent HR. The magnitude of the HR elicited by the *Bacillus* strains was similar to that elicited by *E. amylovora.* Interestingly, the only PGPR strain included in this study, which lacked the ability to induce resistance (*Enterobacter absuriae* strain JM22), also lacked the ability to elicit a confluent HR. These observations challenge the assumption that all immunizing non-pathogens are unable to elicit a HR (van Loon et al. 1998). Perhaps some immunizing PGPR produce compounds sufficiently like those of pathogens to elicit a defense response, and their ability to do so was acquired from pathogens via genetic exchange, or inherited from a phytopathogenic ancestor.

Elicitors which trigger a defense response may be produced by the partial hydrolysis of plant cell wall compounds by invading pathogens (Hahn 1996, Dorey et al. 1997). Non-pathogenic endophytes, including strains able to induce plant defense responses, may gain entry to roots via cell wall hydrolysis (Quadt-Hallman et al. 1997). It is conceivable that such hydrolysis could produce defense response elicitors similar or identical to those of incompatible pathogens, and result in a similar induction of a defense response. Weak immediate induction of tomato defense responses, followed by increased resistance to *Fusarium oxysporum* f. sp. *radicis-lycopersici* has been observed in response to infection by *P. fluorescens* strain 63-28, an endophytic rhizobacterium (M'Piga et al. 1997). The production of hydrolases by this organism was greatly diminished or nonexistent *in planta* after root colonization. It is not clear whether the generation of elicitors via hydrolases during initial root colonization events could have led to the observed responses, or whether other signals were generated by the bacterium to elicit these responses in the plant. Some endophytic PGPR do not induce plant defense responses, yet still appear to enter plant roots via cell wall hydrolysis (Quadt-Hallman et al. 1997). It is not known whether the cell wall fragments presumably produced by these strains are rapidly degraded, thus preventing the formation of elicitors, whether the activity of hydrolytic enzymes cease (M'Piga et al. 1997) or whether these fragments are simply not recognized as elicitors for other reasons.

Inorganic phosphate can stimulate plant defense responses (Reuveni et al. 1994, 1998). Rhizobacteria which solubilize phosphate often do not seem to stimulate phosphate uptake by the plant, or solubilize rock phosphate in doses large enough to induce a defense response (de Freitas et al. 1997). However, mycorrhizal fungi are known to improve the phosphate nutrition of their plant partners (Harley and Smith 1983), resulting in better growth and apparently strengthened defense responses (Lambais and Mehdy 1993, 1995). Mycorrhizae are able to increase the plant's supply of phosphate because they effectively increase the absorptive surface area of the plant root, through which phosphate is obtained (Harley and Smith 1983). Rhizobacteria might stimulate phosphate uptake (and perhaps affect plant defense responses) indirectly, through the production of phytohormones which stimulate the production of lateral roots (Christiansen-Weniger 1996, de Freitas et al. 1997), allowing the plant to extract phosphate from a larger volume of soil.

Cytokinin content increases during the induction of systemic resistance (Sarhan et al. 1991), and exogenously applied cytokinin was shown to protect tobacco leaves from oxidative damage analogous to that produced by necrotizing pathogens (Strobel and Kuć 1995). Cytokinins have been shown to control endogenous levels of salicylic and jasmonic acids in tobacco (Sano et al. 1996), and transgenic disease-resistant tomato plants have been produced from a susceptible tomato variety by the insertion of *Agrobacterium* genes elevating cytokinin production (Bettini et al. 1998). Cytokinin production by PGPR and other rhizosphere microorganisms has been well documented (reviewed in Frankenberger and Arshad 1995). Might exogenously supplied cytokinin, synthesized by PGPR, improve the ability of a plant to react to defense response elicitors, in addition to improving plant growth?

Non-Specific, Multigenic Resistance in Plants

Plant defense responses appear to be multi-component responses, involving the activation of multiple, coordinated and apparently complimentary defense responses, such as the production of phytoalexins or other antimicrobial compounds, the formation of physical barriers through deposition of lignin, callose, or the strengthening of barriers through increased cross-linking, the elicitation of the hypersensitive response, or the elevated expression of hydrolytic enzymes, such as chitinases and β–1,3-glucanases, and other pathogenesis-related proteins (e.g., Tuzun et al. 1989, Ye et al. 1992, Schneider and Ullrich 1994, Stermer 1995, Dubery and Slater 1997, Anfoka and Buchenauer 1997, Sticher et al. 1997, Xue et al. 1998). The induction of plant defense responses may involve several signal transduction cascades (Karban and Kuć, Staswick and Lehman, this volume). "Single gene" disease resistance depends on the possession of a single resistance ("R") gene by the plant, which must interact with an elicitor molecule ("avirulence" or *avr* gene product) present in the pathogen to trigger these defense responses. If a plant lacks the correct R gene to match at least one of the *avr* genes possessed by an invading pathogen, that plant will be unable to use its R genes to detect and stop the pathogen (Hammond-Kossack and Jones 1995, Hutcheson 1998, Buell, this volume).

Multigenic resistance, also known as "horizontal" resistance, refers to plant disease resistance generated via interactions between the products of many plant genes, not a single R gene (Nelson 1978, Simmonds 1991). Multigenic resistance is considered to be non-specific in that the plant and pathogen do not require matching R and *avr* genes for a timely plant defense response to occur; plants bred to have multigenic resistance tend to resist a greater variety of pathogens and pathogen races than those bred to have particular R genes (Simmonds 1991). Multigenic resistance against airborne fungal diseases, pathogenic soil fungi, bacteria, viruses, insects, nematodes, and combinations of these pests has been observed in a wide variety of crops, including cereals, tubers, legumes, vegetables, flowers, and trees (Simmonds 1991).

Hydrolytic Enzyme Isozymes and Induced Resistance

In this section, we will discuss the probable roles of hydrolytic enzyme isozymes in plant defense responses. While it is true that production of hydrolytic enzymes *alone* may not be sufficient for the protection of plants from disease (e.g., Dalisay and Kuć, 1995a,b), this does not mean that hydrolase isozymes are *not* involved in disease resistance, or that they do not play an important role in resistance to some pathogens. In the following sections we will briefly summarize evidence regarding the roles of specific isozymes of hydrolytic enzymes in plant defense responses, and the importance of the pattern of expression of these in triggering defense responses or increasing the effectiveness of these responses.

Antimicrobial effects of hydrolase isozymes. When applied to the tomato pathogen *Alternaria solani in vitro*, tomato chitinase and β–1,3-glucanase isozymes had a concentration-dependent antifungal effect, and acted in a synergistic manner (Lawrence et al. 1996). Isozymes inhibitory to *A. solani* were not inhibitory to another tomato pathogen, *Cladosporium fulvum* (Lawrence et al. 1996), suggesting that the expression of a given isozyme, or even several isozymes, by a plant may not guarantee resistance to all pathogens, or all pathogens of a given type (e.g., fungal pathogens). Other studies support this idea (Ye et al. 1990, Lusso 1995, Masoud et al. 1996). However, it should be noted that transgenic plants which express a single fungal chitinase gene have recently demonstrated resistance to multiple pathogenic fungi (Lorito et al. 1998).

Synergistic antifungal effects of chitinases and β–1,3-glucanases *in vitro* have also been reported for enzymes purified from tobacco (Sela-Buurlage et al. 1993), pea (Mauch et al. 1988) and the tropical forage plant *Stylosanthes guianensis* (Brown and Davis 1992). The events involved in the hydrolysis of fungal hyphae *in vitro* by a bean chitinase have been clearly established (Behnamou et al. 1993a). A bifunctional chitinase/lysozyme isozyme highly purified from cabbage was demonstrated to be bactericidal to *Xanthomonas campestris* pv. *campestris* (Dodson et al. 1993, Dodson and Tuzun, unpublished data).

Evidence for the antimicrobial role of hydrolases *in planta* has also accumulated. In transgenic plants constitutively expressing a chitinase gene, hyphal alterations observed in *Rhizoctonia solani* were correlated with extensive degradation of chitin (Benhamou et al. 1993b). This finding supports that of an earlier study (Benhamou et al. 1990) which demonstrated the antifungal effects of tomato chitinases *in planta*. The *in planta* antifungal activity of tobacco β–1,3-glucanases has also been recorded (Benhamou 1992).

Benhamou et al. (1990) demonstrated that chitinase activity occurred where fungal hyphae had been previously attacked by other hydrolytic enzymes, and suggested that these enzymes may be β–1,3 glucanases. The chitin layers of some fungal cell walls appear be buried in β-glucans, rendering the chitin inaccessible to chitinases unless there is prior hydrolysis with β–1,3-glucanases (Wessels and Marchant 1974, Behnamou et al. 1990). These observations may explain why β–1,3-glucanases and chitinases have synergistic antifungal effects *in vitro*.

Production of oligosaccharide elicitors by hydrolytic isozymes. Fragments of fungal (hepta-β-glucoside, oligochitin and oligochitosan) and plant (oligogalacturonide) cell wall polysaccharides have been purified and determined to be elicitors of plant defense responses (Hahn 1996). These fragments can be generated from fungal or plant cell walls via partial enzymatic hydrolysis (Boller 1995, Okinaka et al. 1995, Hahn 1996). Soybean β–1,3-glucanases (Keen and Yoshikawa 1983, Ham et al. 1991) and tomato chitinase and β–1,3- glucanase isozymes (Lawrence 1998) have clearly been demonstrated to generate elicitors from fungal pathogens.

Degradation of fungal cell walls *in planta* has been observed in tobacco, tomato and bean, and associated with the activity of hydrolytic enzymes, including chitinases and β–1,3-glucanases (Benhamou et al. 1990, 1993a, Benhamou 1992). Interestingly, elicitors produced by β–1,3-glucanase activity on fungal cell walls were hypothesized to have triggered subsequent defense responses, including elevated chitinase activity (Benhamou et al. 1990, 1993a). In transgenic canola constitutively expressing chitinase, hyphal alterations observed in *Rhizoctonia solani* were correlated with extensive degradation of chitin, and it is likely that oligochitin fragments were generated during this process (Benhamou et al. 1993b).

Constitutive expression of hydrolase isozymes in resistant plants. Specific isozymes of hydrolytic enzymes (chitinases and β–1,3-glucanases in particular) may be constitutively expressed in disease-resistant plants to a higher level than in susceptible plants. This has been observed in cabbage (Tuzun et al. 1997), tobacco (Broglie et al. 1992, Sela-Buurlage et al. 1993, Lusso 1995), tomato (Lawrence et al. 1996, Bettini et al. 1998), barley (Ignatius et al. 1994), grape (Busam et al. 1997) and potato (Wegener et al. 1996) .

Elevated expression of hydrolase isozymes after pathogen challenge. Increases in the activity or expression of hydrolase isozymes have been observed after pathogen challenge, and have been correlated with the expression of disease resistance. Inoculation of tobacco with *Peronospora tabacina* spores or tobacco mosaic virus resulted in the induction of systemic resistance and the accumulation of β–1,3-glucanase and chitinase isozymes (Tuzun et al. 1989, Ye et al. 1990, Pan et al. 1991, 1992). Similar results were observed for tobacco inoculated with viruses, PGPR or various chemical inducers (Maurhofer et al. 1994, Schneider and Ullrich 1994, Lusso and Kuć 1995). Increases in lysozyme activity correlated with the induction of resistance responses have also been reported (Schneider and Ullrich 1994). The cell wall of *P. tabacina* is primarily composed of β–1,3-linked glucans, meaning that accumulated chitinases would not be expected to act upon this pathogen. However, β–1,3-glucanases and chitinases may act in concert to degrade the hyphae of other fungi (Benhamou et al. 1990, 1993a). The expression of these isozymes may be coordinated, so that the induction of a defense response will result in (among other responses) the synthesis of a variety of hydrolytic enzymes capable of affecting a variety of pathogens (Tuzun et al. 1997).

Tomato plants immunized with β–amino butyric acid (BABA) accumulated β–1,3-glucanase and chitinase (Cohen et al. 1994), and increases in the levels of isozymes of these enzymes have been correlated with resistance in tomato to *A. solani* (Lawrence and Tuzun, unpublished data). Tomato chitinase accumulations in the vicinity of fungal cell death have been observed *in planta* (Benhamou et al. 1990). Enkerli et al. (1993) report correlations between increased tomato chitinase activity, but not β–1,3-glucanase activity, with induction of resistance. Correlations between induced resistance in tomato and increased production of various antifungal proteins or activity of peroxidases, but not of β–1,3-glucanases, have also been reported (Anfoka and Buchenauer 1997).

Increases in the level of chitinase and/or β–1,3-glucanase isozymes after pathogen challenge have also been reported in disease resistant barley (Ignatius et al. 1994) and pea (Vad et al. 1991), disease-resistant as well as immunized disease-susceptible wheat (Liao et al. 1994, Siefert et al. 1996), and immunized cotton (Dubery and Slater 1996), coffee (Guzzo and Martins 1996), grape (Busam et al. 1997), cucumber (Schneider and Ullrich 1994, Ju and Kuć 1995, Dalisay and Kuć 1995a,b), bean (Dann et al. 1996, Xue et al. 1998), pepper (Hwang et al. 1997), chestnut (Schafleitner and Wilhelm 1997),

Cotoneaster watereri (Mosch and Zeller 1996) and *S. guianensis* (Brown and Davis 1992). Kogel et al. (1994) reported that the onset of disease resistance in barley was correlated with increases in PR-1, peroxidase and chitinase proteins, but not β–1,3-glucanases. Chitosanases, chitinases and β–1,3-glucanases were observed to accumulate in infected spruce seedlings, indicating that the defense responses of gymnosperms may resemble those of angiosperms (Sharma et al. 1993).

The timing of hydrolase isozyme expression is important. Increases in the activity or expression of hydrolase isozymes after pathogen challenge tend to occur more rapidly and sometimes to a greater extent in resistant plants, or in susceptible plants in which systemic resistance has been induced, than in non-induced disease-susceptible plants. Lawrence et al. (1996) clearly demonstrated that disease-resistant tomato plants expressed a chitinase isozyme more rapidly and to a greater extent than susceptible plants, supporting earlier work in which more rapid accumulation of chitinases in resistant plants during incompatible tomato-pathogen interactions was observed (Benhamou et al. 1990). More rapid accumulations were observed for β–1,3-glucanase isozymes in immunized tobacco (Pan et al. 1991, Lusso and Kuć 1995), chitinases in disease-resistant pea (Vad et al. 1991) and immunized tobacco (Pan et al. 1992), and both chitinases and β–1,3-glucanases in disease-resistant sugar beet (Nielsen et al. 1994) and immunized coffee (Guzzon and Martins 1996).

Disease resistant transgenic plants expressing hydrolase isozymes. Transgenic plants, including tobacco (Broglie et al. 1992, Sela-Buurlage et al. 1993, Howie et al. 1994, Zhu et al. 1994, Lusso 1995), tomato (Bettini et al. 1998), canola (Benhamou et al. 1993b) and alfalfa (Masoud et al. 1996) that produce higher constitutive levels of chitinases and/or β–1,3-glucanases have been demonstrated to be more resistant to disease. Interestingly, transgenic tobacco expressing tobacco hornworm chitinase (normally expressed in this insect during moulting) had increased resistance to this insect (XiongFei et al. 1998).

In an analogous system, transgenic potatoes constitutively expressing an *Erwinia carotovora* pectate lyase, which generates defense response elicitors from plant cell wall pectin, were demonstrated to resist *E. carotovora* infection (Wegener et al. 1996). The pectate lyase was stored intracellularly in these plants, meaning the enzyme would not be released to generate elicitors unless the potato tuber were wounded and therefore likely to be infected by *E. carotovora.*

Probable Roles of Hydrolase Isozymes in Plant Defense

Constitutively produced hydrolytic isozynes. The constitutive production of specific hydrolase isozymes may give an advantage to the plant by 1) attacking the pathogen directly and so slowing the development of infection, and 2) by generating oligosaccharide defense response elicitors from the pathogen. The elicitors may initiate an HR or other plant defense responses, or perpetuate or amplify defense responses already initiated. We must note that we are not alone in suggesting that these events occur during plant defense responses (Pan et al. 1991, 1992, Benhamou et al. 1993a, Ignatius et al. 1994, Dann et al. 1996, Xue et al. 1998).

Disease-resistant plants are able to detect and respond to pathogens more quickly than susceptible plants (e.g., Pan et al. 1991, Lusso and Kuć, 1995, Lawrence et al. 1996, Tuzun et al. 1997). It seems likely that the greater constitutive production of specific hydrolytic enzyme isozymes observed in some resistant plants (reviewed above) would result in a greater probability of an encounter between these enzymes and a pathogen, and therefore more rapid detection of, and responses to, pathogenic invaders. It should be noted that encounters will only occur between enzymes which accumulate in, or are released via cell lysis into, the areas invaded by the pathogen.

Roles for intracellular and extracellular enzymes. Pan et al. (1991, 1992) noted that some tobacco chitinase and β–1,3-glucanase isozymes were located in the intercellular fluid (extracellular), while others were present within the vacuoles or other organelles of tobacco cells as well (intracellular). Accumulations in both intracellular and extracellular isozymes have been noted during the induction of defense responses (Neilsen et al. 1994), although sometimes extracellular enzymes accumulate while intracellular ones do not (Pan et al. 1992, Bettini et al. 1998). Benhamou et al. (1990, 1993a,b) note that accumulations of hydrolytic enzymes tend to occur in the immediate vicinity of an invading pathogen. If this is generally the case, it may be difficult to accurately interpret data from analyses that rely upon the extraction of enzymes from the intercellular or intracellular fluids of whole leaves or other plant parts.

The potential role of extracellular enzymes seems obvious: many invading pathogens would be expected to encounter extracellular hydrolases during the process of infection. The elevated, constitutive production of extracellular hydrolases may improve the ability of a plant to rapidly initiate defense responses, via the generation of defense response elicitors upon contact of an enzyme with the pathogen surface.

Intracellular hydrolases may also play a role in plant defense responses (Mauch and Staehelin 1989), since these enzymes are released upon cell lysis, during an HR or during the formation of plant wounds. The elevated production of intracellular hydrolases in disease-resistant plants could conceivably improve the ability of those plants to limit the spread of pathogens at the site of an HR, or of pathogens introduced to the plant via wounds (such as those produced by insects).

Hydrolytic isozymes rapidly accumulated after induction. The more rapid accumulation of specific hydrolytic enzyme isozymes in disease-resistant plants after the initiation of defense responses (reviewed earlier) could enable these plants to contain some pathogens more effectively than disease-susceptible plants. This may occur via 1) direct antagonism of the pathogen (i.e., cell wall hydrolysis) by these isozymes, and/or 2) more rapid production of oligosaccharide elicitors (i.e., cell wall fragments), which may induce other plant defense responses upon contact of these isozymes with pathogenic surfaces. It should be noted that the isozymes which accumulate more rapidly in resistant plants are not necessarily the same ones which may be constitutively expressed at a higher level (e.g., Ignatius et al. 1994).

Disease-susceptible plants which have been exposed to an inducing agent become more resistant to disease, and often also demonstrate patterns of hydrolase isozyme expression similar to those observed in plants with multigenic resistance to disease: the elevated constitutive expression, and/or upon pathogen challenge, more rapid production of hydrolase isozymes (Tuzun et al. 1989, 1997, Pan et al. 1991, 1992, Brown and Davis 1992, Cohen et al. 1994, Maurhofer et al. 1994, Schneider and Ullrich 1994, Lusso and Kuć 1995, Ju and Kuć 1995, Dubery and Slater 1996, Guzzo and Martins 1996, Dann et al. 1996, Lawrence et al. 1996, Seifert et al. 1996, Busam et al. 1997, Schafleiter and Wilhelm 1997).

Evolution of Mutualistic Interactions from Antagonistic Ones?

It is not in the best interest of a pathogen to render its food supply scarce. While this is an oversimplification, it is conceivable that during the course of co-evolution with plants, some pathogens would become gradually less antagonistic, to the point where the interacting microbe had an attenuated, neutral or even beneficial effect on its plant partner (Keeler 1985). The mutation of a single locus was sufficient to transform a fungal plant pathogen (*Colletotrichum magna*) into a non-pathogenic mutualist (Freeman and Rodriguez 1993). Interestingly, the non-pathogenic mutant in this study retained its ability to activate plant defense responses, and was demonstrated to induce resistance to *Fusarium oxysporum* in watermelon. It is conceivable that a similar process of conversion from pathogen to non-pathogen could occur in bacteria, producing non-pathogenic, resistance-inducing strains from pathogenic ones.

As has already been noted, associative rhizobacteria have a range of effects upon plant growth, ranging from beneficial to detrimental, and operate with a degree of host specificity. A particular rhizobacterial strain will not affect all plants in a similar manner. Rhizobacteria may provoke plant defense responses via compounds or mechanisms similar to those of pathogens. For example, endophytic rhizobacteria may enter the plant via cell wall hydrolysis, in a manner analogous to that used by phytopathogens, and may produce elicitors similar or identical to those generated by the invasion of a pathogenic, cell-wall degrading organism. Like pathogens, associative rhizobacteria induce the production of PR proteins in at least some instances (e.g., Zdor and Anderson 1992, Maurhofer et al. 1994) and can elicit the HR when injected into plant tissue (Jetiyanan 1997). Could the relationship between plant-immunizing rhizobacteria and plants have evolved from an antagonistic one? We hope that future investigations into the origin and evolution of associative plant-microbe interactions will shed light on this intriguing issue.

Synthesis

Rhizobacteria may induce defense responses in plants via production of LPS, siderophores, cytokinins, cellulases or other plant cell wall-degrading hydrolytic enzymes, or by stimulating the formation of lateral roots and increasing the phosphate uptake of the plant. Plants in which nonspecific, multigenic defense responses have been induced tend to express higher constitutive levels of specific isozymes of hydrolytic enzymes, and also tend to produce these more rapidly upon exposure to a pathogen. These enzymes not only may attack the pathogen directly, they may serve to trigger systemic defense responses by producing cell wall fragments, or elicitors, from contact with an invading pathogen. Symbiotic plant-microbe interactions may have arisen from originally antagonistic relationships, and it is possible that relationships between plants and associative rhizobacteria have arisen in a similar fashion.

Future Directions

1. Continued research into the fundamental mechanisms of induced resistance and the signal transduction pathways leading to the generation of a defense response is essential. Do extracellular and intracellular hydrolytic enzymes have different roles in disease resistance, and do intracellular hydrolases in particular affect the spread of pathogens introduced to the plant by wounding, as we have hypothesized? Are plant defense responses induced by non-pathogenic organisms truly distinct from those produced by pathogens, and if so, why is this? The signal transduction pathway(s) activated by disease resistance-inducing Gram positive PGPR have not been investigated. Are these the same as those reported for some Gram negative PGPR (reviewed in van Loon et al. 1998)? What plant receptors are involved in the perception of various inducing agents? Are these receptors highly conserved among plants, or do receptors vary widely between plant families?

2. When dealing with living organisms, one must ensure that the organisms remain viable and able to function in the range of environmental conditions into which they will be placed. In the case of plant-immunizing PGPR, the environmental conditions must also be such that 1) the microbe will produce the signaling compound(s) which induce defense responses in the plant,

2) these compounds are able to diffuse through soil air or water films to plant cells capable of perceiving them, and 3) if there is a threshold below which an effective response will not occur, the signaling compounds will be produced in sufficient quantity to activate an effective response. The variability observed in plant responses to PGPR might be due to the fact that at least two living organisms must interact in order for the desired response to be produced. These organisms do not normally co-exist in an orderly and predictable environment; each will be affected in numerous and largely unpredictable ways by environmental factors (including other living organisms), and may be thought of as existing as components in a complex web of positive and negative environmental feedback loops.

While PGPR have proven to be useful in specific applications, we believe the application of inducing chemicals (or compounds purified from PGPR or other organisms) will provide more consistent results and ultimately prove to be of greater use in large-scale agriculture, and we encourage applied research in this direction. Study of the mechanisms involved in PGPR-mediated induction of plant defenses may also elucidate how plant immunization and growth promotion may occur simultaneously, and how these phenomena are affected by environmental factors. Further study of the mechanisms by which PGPR affect plants may allow for the discovery of new elicitors, or means of delivering these to plants, in addition to examining the role soil microbes may play in the composition of plant communities.

3. The interactions between plant-immunizing PGPR or PGPF and their plant hosts are subtle: the plant provides food for these microorganisms in exchange for growth promotion/ immunization benefits conferred by these. How did this interaction evolve? Are modern plant-immunizing organisms the result of fortuitous mutagenesis or genetic exchanges among non-pathogenic soil microorganisms during the long course of co-evolution, or could some plant-immunizing rhizosphere organisms be the descendants of plant pathogens?

Acknowledgements

Anurag Agrawal provided advice which proved useful in the preparation of this manuscript, and we would like to thank him for it.

Literature Cited

Alstrom, S. 1991. Induction of disease resistance in common bean susceptible to halo blight bacterial pathogen after seed bacterization with rhizosphere pseudomonads. Journal of General and Applied Microbiology 37:495-501.

Alstrom, S. and R.G. Burns. 1989. Cyanide production by rhizobacteria as a possible mechanism of plant growth inhibition. Biology and Fertility of Soils 7:232-238.

Anfoka, G. and H. Buchenauer. 1997. Systemic acquired resistance in tomato against *Phytophthora infestans* by pre-inoculation with tobacco necrosis virus. Physiological and Molecular Plant Pathology 50:85-101.

Benhamou, N. 1992. Ultrastrucural detection of β–1,3-glucans in tobacco root tissues infected by *Phytophthora parasitica* var. *nicotianae* using a gold-complexed tobacco β–1,3-glucanase. Physiological and Molecular Plant Pathology 41:351-370.

Benhamou, N., M.H.A.J. Joosten and P.J.G.M. De Wit. 1990. Subcellular localization of chitinase and of its potential substrate in tomato root tissues infected by *Fusarium oxysporum* f. sp. *radicis-lycopersici.* Plant Physiology 92:1108-1120.

Benhamou, N., K. Broglie, R. Broglie and I. Chet. 1993a. Antifungal effect of bean endochitinase on *Rhizoctonia solani*: ultrastructural changes and cytochemical aspects of chitin breakdown. Canadian Journal of Microbiology 39:318-328.

Benhamou, N., K. Broglie, I. Chet and R. Broglie. 1993b. Cytology of infection of 35S-bean chitinase transgenic canola plants by *Rhizoctonia solani*, cytochemical aspects of chitin breakdown *in vivo.* Plant Journal 4:295-305.

Bent, E. and C.P. Chanway. 1999. The growth-promoting effects of an endophytic rhizobacterium on lodgepole pine are partially inhibited by the presence of other rhizobacteria. Canadian Journal of Microbiology 44:980-988.

Bertagnolli, B.L., F.K.D. Soglio and J.B Sinclair. 1995. Extracellular enzyme profiles of the fungal pathogen *Rhizoctonia solani* isolate 2B-12 and of two antagonists, *Bacillus megaterium* strain BB153-2-2 and *Trichoderma harzianum* isolate Th008.I. possible correlations with inhibition of growth and biocontrol. Physiological and Molecular Plant Pathology 48:145-160.

Bettini, P., E. Cosi, M.G. Pellegrini, L. Turbani, G.G. Vendramin and M. Buiatti. 1998. Modification of competence for *in vitro* response to *Fusarium oxysporum* in tomato cells. III. PR-protein gene expression and ethylene evolution in tomato cell lines transgenic for phytohormone-related bacterial genes. Theoretical and Applied Genetics 97:575-583.

Beyrie, H.F., S.E. Smith, R.L. Peterson, and C.M.M. Franco. 1995. Colonization of *Orchis morio* protocorms by a mycorrhizal fungus: effects of nitrogen nutrition and glyphosate in modifying the response. Canadian Journal of Botany 73:1128-1140.

Boller, T. 1995. Chemoperception of microbial signals in plant cells. Annual Review of Plant Physiology and Plant Molecular Biology 46:189-214.

Broglie, K., I. Chet, M. Holliday, R. Cressman, P. Biddle, S. Knowlton, C.J. Mauvais and R. Broglie. 1992. Transgenic plants with enhanced resistance to the fungal pathogen *Rhizoctonia solani.* Science 254:1194-1197.

Brown, A.E. and R.D. Davis. 1992. Chitinase activity in *Stylosanthes guianensis* systemically protected against *Colletotrichum gloeosporioides.* Journal of Phytopathology 136:247-256.

Busam, G., H-H. Kassemeyer, and U. Matern. 1997. Differential expression of chitinases in *Vitis vinifera* L. responding to systemic acquired resistance activators or fungal challenge. Plant Physiology 115:1029-1038.

Chanway, C.P. and F.B. Holl. 1992. Influence of soil biota on Douglas-fir (*Pseudotsuga menziezii*) seedling growth: the role of rhizosphere bacteria. Canadian Journal of Botany 70:1025-1031.

Christiansen-Weniger, C. 1996. Endophytic establishment of *Azorhizobium caulinodans* through auxin-induced tumors of rice (*Oryza sativa* L.). Biology and Fertility of Soils 21:293-302.

Cohen, Y., T. Niderman, E. Mösinger and R. Fluhr. 1994. β-aminobutyric acid induces the accumulation of pathogenesis–related proteins in tomato (*Lycopersicon esculentum* L.) plants and resistance to late blight infection caused by *Phytophthora infestans.* Plant Physiology 104:59-66.

Curl, E.A. and B. Truleove. 1986. The Rhizosphere. Springer-Verlag, New York.

Dalisay, R.F. and J.A. Kuć. 1995a. Persistence of reduced penetration by *Colletotrichum lagenarium* into cucumber leaves with induced systemic resistance and its relation to enhanced peroxidase and chitinase activities. Physiological and Molecular Plant Pathology 47:329-338.

Dalisay, R.F. and J. A. Kuć. 1995b. Persistence of induced resistance and enhanced peroxidase and chitinase activities in cucumber plants. Physiological and Molecular Plant Pathology 47:315-327.

Dann, E.K., P. Meuwly, J.-P. Métraux and B.J. Deverall. 1996. The effect of pathogen inoculation or chemical treatment on activities of chitinase and β–1,3-glucanase and accumulation of salicylic acid in leaves of green bean, *Phaseolus vulgaris* L. Physiological and Molecular Plant Pathology 49:307-319.

de Freitas, J.R., M.R Bannerjee and J.J. Germida. 1997. Phosphate-solubilizing rhizobacteria enhance the growth and yield but not phosphorus uptake of canola (*Brassica napus* L.). Biology and Fertility of Soils 24:358-364.

Dodson, K. M., J.J. Shaw and S. Tuzun. 1993. Purification of a chitinase/lysozyme isozyme (CHL2) that is constitutively expressed in cabbage varieties resistant to black rot. Abstract A28 Phytopathology 83:1335.

Dorey, S., F. Bailleul, M.-A. Pierrel, P. Saindrenan, B. Fritig and S. Kauffman. 1997. Spatial and temporal induction of cell death, defense genes, and accumulation of salicylic acid in tobacco leaves reacting hypersensitively to a fungal glycoprotein elicitor. Molecular Plant-Microbe Interactions 10:646-655.

Dubery, I.A. and V. Slater. 1997. Induced defence responses in cotton leaf disks by elicitors from *Verticillium dahliae*. Phytochemistry 44:1429-1434.

Duc, G., S. Trouvelot, V. Gianninazzi-Pearson and S. Gianinazzi. 1989. First report of non-mycorrhizal plant mutants (Myc-) obtained in *pea (Pisum sativum* L.) and fababean (*Vicia faba* L.). Plant Science 60:215-222.

Enkerli, J., U. Gisi and E. Mösinger. 1993. Systemic acquired resistance to *Phytophthora infestans* in tomato and the role of pathogenesis related proteins. Physiological and Molecular Plant Pathology 43:161-171.

Forlani, G.,M. Mantelli, M. Branzoni, E. Nielsen and F. Favilli. 1995. Differential sensitivity of plant-associated bacteria to sulfonylurea and imidazolinone herbicides. Plant and Soil 176:243-253.

Francis, R. and D.J. Read. 1995. Mutualism and antagonism in the mycorrhizal symbiosis, with special reference to impacts on plant community structure. Canadian Journal of Botany 73 (Suppl. 1) S1301-S1309.

Frankenberger, W.T. and M. Arshad. 1995. Phytohormones in soils. Marcel-Dekker, Inc., New York.

Freeman, S. and R.J. Rodriguez. 1993. Genetic conversion of a fungal plant pathogen to a nonpathogenic, endophytic mutualist. Science 260:75-78.

Garbaye, J. 1994. Tansley review no. 76: helper bacteria: a new dimension to the mycorrhizal symbiosis. New Phytologist 128:197-210.

Gianinazzi-Pearson, V., A. Goliotte, J. Lherminier, B. Tisserant, P. Franken, E. Dumas-Gaudot, M.-C. Lemoine, D. van Tuinen and S. Gianinazzi. 1995. Cellular and molecular approaches in the characterization of symbiotic events in functional arbuscular mycorrhizal associations. Canadian Journal of Botany 73 (Suppl. 1):S526-S532.

Glick, B.R. 1995. The enhancement of plant growth by free-living bacteria. Canadian Journal of Microbiology 41:109-117.

Guzzo, S.D. and E.M.F. Martins. 1996. Local and systemic induction of β–1,3-glucanase and chitinase in coffee leaves protected against *Hemileia vastatrix* by *Bacillus thuringiensis*. Journal of Phytopathology 144:449-454.

Hahn, M.G. 1996. Microbial elicitors and their receptors in plants. Annual Review of Phytopathology 34:387-412.

Ham, K.-S., S. Kauffman, P. Albetsheim and A.G. Darvill. 1991. A soybean pathogenesis-related protein with β–1,3-glucanase activity releases phytoalexin elicitor-active heat-stable fragments from fungal walls. Molecular Plant-Microbe Interactions 4:545-552.

Hammond-Kossack, K.E. and J.D.G. Jones. 1995. Plant disease resistance genes: unravelling how they work. Canadian Journal of Botany 73 (Suppl. 1) S495-S505.

Harley, J.L. and S.E. Smith. 1983. Mycorrhizal Symbiosis. Academic Press, New York.

Hill, K.A. and S.M. Singh. 1997. The evolution of species-type specificity in the global DNA sequence organization of mitochondrial genomes. Genome 40:342-356.

Howarth, W.R., L.F. Elliott and J.M. Lynch. 1998. Influence of soil quality on the function of inhibitory rhizobacteria. Letters in Applied Microbiology 26:87-92.

Howell, C.R. 1990. Fungi as biological control agents. Pages 257-286 *in* J.P. Nakas and C. Hagedorn, edtiors. Biotechnology of Plant-microbe Interactions. McGraw-Hill, Toronto ON.

Howie, W., L. Joe, E. Newbigin, T. Suslow and P. Dunsmuir. 1994. Transgenic tobacco plants which express the *chiA* gene from *Serratia marcesens* have enhanced tolerance to *Rhizoctonia solani*. Transgenic Research 3:90-98.

Hutcheson, S.W. 1998. Current concepts of active defense in plants. Annual Review of Phytopathology 36:59-90.

Hwang, S.K., J. Y. Sunwoo, Y.K. Kim and B.S. Kim. 1997. Accumulation of β–1,3-glucanase and chitinase isoforms, and salicylic acid in the DL-β-amino-n-butyric acid-induced resistance response of pepper stems to *Phytophthora capsici*. Physiological and Molecular Plant Pathology 51:305-322.

Hyakumachi, M. 1984. Plant growth promoting fungi from turfgrass rhizosphere with potentials for disease suppression. Soil Microorganisms 44:53-68.

Ignatius, S.M.J., R.K. Chopra and S. Muthukrishnan. 1994. Effects of fungal infection and wounding on the expression of chitinases and β–1,3-glucanases in near-isogenic lines of barley. Physiologia Plantarum 90:584-592.

Janzen, R.A. and W.B. McGill. 1995. Community-level interactions control proliferation of *Azospirillum brasilense* Cd in microcosms. Soil Biology and Biochemistry 27:189-196.

Jetiyanon, K. 1997. Interactions between PGPR and cucumber during induced systemic resistance: recognition and early host defense responses. Ph.D. dissertation, Auburn University, Auburn, AL.

Ju, C. and J. Kuć. 1995. Purification and characterization of an acidic β–1,3-glucanase from cucumber and its relationship to systemic disease resistance induced by *Colletotrichum lagenarum* and tobacco necrosis virus. Molecular Plant-Microbe Interactions 8:899-905.

Keeler, K.H. 1985. Cost:benefit models of mutualism. Pages 100-127 *in* D.H. Boucher, editor. The Biology of Mutualism. Oxford University Press, New York.

Keen, N.T. and M. Yoshikawa. 1983. β–1,3-endoclucanase from soybean release elicitor-active carbohydrates from fungus cell walls. Plant Physiology 71:460-465.

Kogel, K.-H., U. Beckhove, J. Dreschers, S. Münch and Y. Rommé. 1994. Acquired resistance in barley: the resistance mechanism induced by 2,6-dichloroisonicotinic acid is a phenocopy of a genetically based mechanism governing race-specific powdery mildew resistance. Plant Physiology 106:1269-1277.

Lambais, M.R. and M.C. Mehdy. 1993. Suppression of endochitinase, β–1,3-endoglucanase and chalcone isomerase expression in bean vesicular-arbuscular mycorrhizal roots under different soil phosphate conditions. Molecular Plant-Microbe Interactions 6:75-83.

Lambais, M.R. and M.C. Mehdy. 1995. Differential expression of defense-related genes in arbuscular mycorrhiza. Canadian Journal of Botany 73 (Suppl. 1):S533-S540.

Lawrence, C.B. 1998. Physiological and molecular aspects of horizontal resistance in tomato to *Alternaria solani*. Ph.D. dissertation, Auburn University, Auburn, AL.

Lawrence, C.B., M.H.A.J. Joosten and S. Tuzun. 1996. Differential induction of pathogenesis-related proteins in tomato by *Alternaria solani* and the association of a basic chitinase isozyme with resistance. Physiological and Molecular Plant Pathology 48:361-377.

Lennartsson, T., J. Tuomi and P. Nilsson. 1997. Evidence for an evolutionary history of overcompensation in the grassland biennial *Gentianella campestris* (Gentianaceae). American Naturalist 149:1147-1155.

Liao, Y.C., F. Krutzaler, R. Fischer, J.-J. Reisner and R. Tiburzy. 1994. Characterization of a wheat class Ib chitinase gene differentially induced in isogenic lines by infection with *Puccinia graminis*. Plant Science 103:177-187.

Lorito, M., S.L. Woo, I.G. Fernandez, G. Colucci, G.E. Harman, J.A. Pintor-Toro, E. Filippone, S. Muccifora, C.B. Lawrence, A. Zonia, S. Tuzun and F. Scala. 1998. Genes from mycoparasitic fungi as a source for improving plant resistance to fungal pathogens. Proceedings of the National Academy of Sciences USA. 95:7860-7865.

Lusso, M.F.G. 1995. Association of β–1,3-glucanase, ribonuclease and protease with systemic induced resistance of tobacco to fungal and viral pathogens. Ph.D. thesis, University of Kentucky, Lexington, KY.

Lusso, M. and J. Kuć. 1995. Evidence for transcriptional regulation of β–1,3-glucanase as it relates to induced systemic resistance of tobacco to blue mold. Molecular Plant-Microbe Interactions 8:473-475.

Malloch, D. 1987. The evolution of mycorrhizae. Canadian Journal of Plant Pathology 9:398-402.

Masoud, S.A., Q. Zhu, C. Lamb and R.A. Dixon. 1996. Constitutive expression of an inducible beta-1,3 glucanase in alfalfa reduces diease severity caused by the oomycete pathogen *Phytophtora megasperma* f. sp. *medicaginis*, but does not reduce disease severity of chitin-containing fungi. Transgenic Research 5:313-323.

Mauch, F. and L.A. Staehelin. 1989. Functional implications of the subcellular localization of ethylene-induced chitinase and β–1,3-glucanase in bean leaves. Plant Cell 1:447-457.

Mauch, F., B. Mauch-Mani and T. Boller. 1988. Antifungal hydrolases in pea tissue. II. Inhibition of fungal growth by combinations of chitinases and β–1,3-glucanases. Plant Physiology 88:936-942.

Maurhofer, M., C. Hase, P. Meuwly, J.-P. Métraux and G. Défago. 1994. Induction of systemic resistance of tobacco to tobacco necrosis virus by the root-colonizing *Pseudomonas fluorescens* strain CHA0L Influence of the *gacA* gene and of pyoverdine production. Phytopathology 84:139-146.

Mazzola, M., Stahlman, P.W. and J. E. Leach. 1995. Application method affects the distribution and efficacy of rhizobacteria suppressive of downy brome (*Bromus tectorum*). Soil Biology and Biochemistry 27:1271-1278.

Meera, M.S., Shivanna, M.B., Kageyama, K. and M. Hyakumachi. 1995. Persistence of induced systemic resistance in cucumber in relation oto root colonization by plant growth promoting fungal isolates. Crop Protection 14:123-130.

Milus, E.A., and C.S. Rothrock. 1993. Rhizosphere colonization of wheat by selected soil bacteria over diverse environments. Canadian Journal of Microbiology 39:335-341.

Mosch, J. and W. Zeller. 1996. Further studies on plant extracts with a resistance induction effect against *Erwinia amylovora.* Acta Horticulturae 411:361-366.

M'Piga, P., R.R. Bélanger, T.C. Paulitz and N. Benhamou. 1997. Increased resistance to *Fusarium oxysporum* f. sp. *radicis-lycopersici* in tomato plants treated with the endophytic bacterium *Pseudomonas fluorescens* strain 63-28. Physiological and Molecular Plant Pathology 50:301-320.

Nehl, D.B., Allen, S.J. and J.F. Brown. 1996. Deleterious rhizosphere bacteria: an integrating perspective. Applied Soil Ecology 5:1-20.

Nelson, R.R. 1978. Genetics of horizontal resistance to plant diseases. Annual Review of Phytopathology 16:359-378.

Nielsen, K.K., K. Bojsen, D.B. Collinge and J.D. Mikkelsen. 1994. Induced resistance in sugar beet against *Cercospora beticola*: induction by dichloroisonicotinic acid is independent of chitinase and β–1,3-glucanase transcript accumulaion. Physiological and Molecular Plant Pathology 45:89-99.

Okinaka, Y., K. Mimori, K. Takeo, S. Kitamura and Y. Takeuchi. 1995. A structural model for the mechanisms of elicitor release from fungal cell walls by plant β–1,3-endoglucanase. Plant Physiology 109:839-845.

Ozeki, H., K. Umesono, K. Inokuchi, T. Kohchi and K. Ohyama. 1989. The chloroplast genome of plants: a unique origin. Genome 31:169-174.

Paige, K.N. 1992. Overcompensation in response to mammalian herbivory: from mutualistic to antagonistic interactions. Ecology 773:2076-2085.

Pan, S.Q., X.S. Ye and J. Kuć. 1991. Association of β–1,3-glucanase activity and isoform pattern with systemic resistance to blue mould in tobacco induced by stem injection with *Pernospora tabacina* or leaf inoculation with tobacco mosaic virus. Physiological and Molecular Plant Pathology 39:25-39.

Pan, S.Q., X.S. Ye and J. Kuć. 1992. Induction of chitinases in tobacco plants systemically protected against blue mold by *Peronospora tabacina* or tobacco mosaic virus. Phytopathology 82:119-123.

Provorov, N.A. 1988. Coevolution of rhizobia with legumes: facts and hypotheses. Symbiosis 24:337-368.

Quadt-Hallmann, A., N. Benhamou and J.W. Kloepper. 1997. Bacterial endophytes in cotton: mechanisms of entering the plant. Canadian Journal of Microbiology 43:577-582.

Reuveni, M., Agapov, V. and R. Reuveni. 1994. Induced systemic protection to powdery mildew in cucumber by phosphate and potassium fertilizers: effects of inoculum concentration and post-inoculation treatment. Canadian Journal of Plant Pathology 17:247-251.

Reuveni, M., D. Oppenheim and R. Reuveni. 1998. Integrated control of powdery mildew on apple trees by foliar sprays of mono-potassium phosphate fertilizer and sterol inhibiting fungicides. Crop Protection 17:563-568.

Ryan, C.A. and A. Jagendorf. 1995. Self defense by plants. Proceedings of the National Academy of Sciences of the USA. 92:4075.

Salzer, P., G. Hebe, A. Reith, B. Zitterell-Haid, H. Stramsky, K. Gaschler and A. Hager. 1996. Rapid reactions of spruce cells to elicitors released from the ectomycorrhizal fungus *Hebeloma crustiniliforme*, and inactivation of these elicitors by extracellular spruce cell enzymes. Planta 198:118-126.

Sano, H., S. Seo, N. Koizumi, T. Niki, H. Iwamura and Y. Ohashi. 1996. Regulation by cytokinins of endogenous levels of jasmonic and salicyclic acids in mechanically wounded tobacco plants. Plant Cell Physiology 37:762-769.

Sarhan, A.R.T., Z. Kiraly, I. Sziraki and V. Smedegaard-Petersen. 1991. Increased levels of cytokinins in barley leaves having the systemic acquired resistance to *Bipolaris sorkiniana* (Sacc.) Shoemaker. Journal of Phytopathology 131:101-108.

Schafleitner, R. and E. Wilhelm. 1997. Effect of virulent and hypovirulent *Cryphonectria parasitica* (Murr.) Barr on the intercellular pathogen related proteins and on total protein pattern of chestnut (*Castanea stativa* Mill.). Physiological and Molecular Plant Pathology 51:323-332.

Schippers, B., A.W. Bakker and P.A.H.M. Bakker. 1987. Interactions of deleterious and beneficial rhizosphere microorganisms and the effect of cropping practices. Annual Review of Phytopathology 25:339-358.

Schneider, S. and W.R. Ullrich. 1994. Differential induction of resistance and enhanced enzyme activities in cucumber and tobacco caused by treatment with various abiotic and biotic inducers. Physiological and Molecular Plant Pathology 45:291-304.

Sela-Buurlage, M.B., A.S. Ponstein, S.A. Bres-Vloemans, L.S. Melchers, P.M.J. van den Elzen and B.J.C. Cornelissen. 1993. Only specific tobacco chitinases and β–1,3-glucanases exhibit antifungal activity. Plant Physiology 101:857-863.

Sharma, P., D. Børja, P. Stougaard, and A. Lönneborg. 1993. PR-protiens accumulating in spruce roots infected with a pathogenic *Pythium* sp. isolate include chitinases, chitosanases and β–1,3-glucanases. Physiological and Molecular Plant Pathology 43:57-67.

Shearer, C.A. 1995. Fungal competition. Canadian Journal of Botany 73 (Suppl. 1):S1259-S1264.

Siefert, F., M. Thalmair, C. Langebartels, H. Sandermann and K. Grossmann. 1996. Epoxiconazole-induced stimulation of the antifungal hydrolases chitinase and β–1,3-glucanase in wheat. Plant Growth Regulation 20:279-286

Simmonds, N.W. 1991. Genetics of horizontal resistance to diseases of crops. Biological Reviews 66:189-241

Spaink, H.P. 1995. The molecular basis of infection and nodulation by rhizobia: the ins and outs of sympathogenesis. Annual Review of Phytopathology 33:345-368.

Stermer, B.A. 1995. Molecular regulation of systemic induced resistance. Pages 111-140 *in* R. Hammerschmidt and J. Kuć, editors. Induced Resistance to Disease in Plants. Kluwer Academic Publishers, Boston.

Stevenson, P.C., D.E. Padgham and M.P Haware. 1995. Root exudates associated with the resistance of four chickpea cultivars (*Cicer arietinum*) to two races of *Fusarium oxysporum* f.sp. *ciceri*. Plant Pathology 44:686-694.

Sticher, L., B. Mauch-Mani, and J.P. Métraux. 1997. Systemic acquired resistance. Annual Review of Phytopathology 35:235-270.
Stirling, G.R. 1984. Biological control of *Meloidogyne javanica* with *Bacillus penetrans*. Phytopathology 74:55-60.
Strobel, N.E. and J.A. Kuć. 1995. Chemical and biological inducers of systemic resistance to pathogens protect cucumber and tobacco plants from damage caused by paraquat and cupric chloride. Phytopathology 85:1306-1310.
Suslow, T.V. and M.N. Schroth. 1982. Role of deleterious rhizobacteria as minor pathogens in reducing crop growth. Phytopathology 72:111-115.
Tuzun, S., P. A. Gay, C.B. Lawrence, T.L. Robertson and R.J. Sayler. 1997. Biotechnological applications of inheritable and inducible resistance to diseases in plants. Pages 25-40 *in* P.M. Gresshoff, editor. Technology Transfer of Plant Biotechnology. CRC Press Inc., New York.
Tuzun, S., J. Juarez, W.C. Nesmith and J. Kuć. 1992. Induction of systemic resistance in tobacco against metalaxyl-tolerant strains of *Peronospora tabacina* and the natural ocurrence of the phenomenon in Mexico. Phytopathology 82:425-429.
Tuzun, S. and J. Kloepper. 1995. Practical application and implementation of induced resistance. Pages 152-168 *in* R. Hammerschmidt and J. Kuć, editors. Induced Resistance to Disease in Plants: Developments in Plant Pathology vol. 4. Kluwer Academic Publishers, Boston.
Tuzun, S. and J. Kuć. 1991. Plant immunization: an alternative to pesticides for control of plant diseases in the greenhouse and field. Pages 30-40 *in* The Biological Control of Plant Diseases. FFTC Book Series No. 42. Food and Fertilizer Technology Centre of the Asian and Pacific Region, Taipei, Taiwan.
Tuzun, S., W. Nesmith., R.S. Ferriss, and J. Kuć. 1986. Effects of stem injections with *Peronospora tabacina* on growth of tobacco and protection against blue mold in the field. Phytopathology 76:938-941.
Tuzun, S., M.N. Rao, U. Vogeli, S.L. Schardl, and J. Kuć. 1989. Induced systemic resistance to blue mold: early induction and accumulation of β–1,3-glucanases, chitinases and other pathogenesis-related protiens (b-protiens) in immunized tobacco. Phytopathology 79:979-983.
Vad, K., J.D. Mikkelsen and D.B. Collinge. 1991. Induction, purification and characterization of chitinase isolated from pea leaves inoculated with *Ascochyta pisi*. Planta 184:24-29.
van Elsas, J.D. and C.E. Heijnen. 1990. Methods for the introduction of bacteria into soil: a review. Biology and Fertility of Soils 10:127-133.
van Elsas, J.D., Trevors, J.T. and E.M.H. Wellington, editors. 1997. Modern Soil Microbiology. Marcel Dekker Inc., New York.
van Loon, L.C., Bakker, P.A.H.M., and C.M.J. Pieterse. 1998. Systemic resistance induced by rhizosphere bacteria. Annual Review of Phytopathology 36:453-483.
van Peer, R., G.J. Niemann, and B. Schippers. 1991. Induced resistance and phytoalexin accumulation in biological control of *Fusarium* wilt of carnation by *Pseudomonas* sp. Strain WCS417r. Phytopathology 81:728-734.
Wegener, C., S. Bartling, O. Olsen, J. Weber and D. von Wertstein. 1996. Pectate lyase in transgenic potatoes confers pre-activation of defence against *Erwinia carotovora*. Physiological and Molecular Plant Pathology 49:359-376.
Wei, G., Kloepper, J.W. and S. Tuzun. 1996. Induced systemic resistance to cucumber diseases and increased plant growth by plant growth-promoting rhizobacteria under field conditions. Phytopathology 86:221-224.
Wei, Z-M., and S.V. Beer. 1996. Harpin from *Erwinia amylovora* induces plant resistance. Acta Horticulturae 411:223-225.
Weller, D.M. 1988. Biological control of soilborne plant pathogens in the rhizosphere with bacteria. Annual Review of Phytopathology 26:379-407.
Wessels, J.G.H and R. Marchant. 1974. Enzymic degradation of septa in hyphal wall preparations from a monokaryon and a dikaryon of *Schizophyllum commune*. Journal of General Microbiology 83:359-368.

XiongFei, D., B. Gopalakrishnan, L.B. Johnson, F.F. White, W. XiaoRong, T.D. Morgan, K.J. Kramer, S. Muthukrishnan. 1998. Insect resistance of transgenic tobacco expressing an insect chitinase gene. Transgenic Research 7:77-84.

Xue, L., P.M. Charest, and S.H. Jabaji-Hare. 1998. Systemic induction of peroxidases, 1,3-β-glucanases, chitinases and resistance in bean plants by binucleate *Rhizoctonia* species. Phytopathology 88:359-365.

Ye, X.S., U. Järlfors, S. Tuzun, S.Q. Pan and J. Kuć. 1992. Biochemical changes in cell walls and cellular responses of tobacco leaves related to systemic resistance to blue mold (*Peronospora tabacina*) induced by tobacco mosaic virus. Canadian Journal of Botany 70:49-57.

Ye, X.S., S.Q. Pan and J. Kuć. 1990. Association of pathogenesis-related proteins and activities of peroxidase, β–1,3-glucanase and chitinase with systemic induced resistance to blue mould of tobacco but not to systemic tobacco mosaic virus. Physiological and Molecular Plant Pathology 36:523-531.

Yuen, G.Y., and M.N. Schroth. 1986. Interactions of *Pseudomonas fluorescens* strain E6 with ornamental plants and its effect on the composition of root-colonizing microflora. Phytopathology 76:176-180.

Zdor, R. E. and A.J. Anderson. 1992. Influence of root colonizing bacteria on the defense responses of bean. Plant and Soil 140:99-107.

Zhu, Q., E.A. Maher, S. Masoud, R.A. Dixon and C.J. Lamb. 1994. Enhanced protection against fungal attack by constitutive co-expression of chitinase and glucanase genes in transgenic tobacco. Bio/Technology 12:807-813.

Jasmonic Acid-Signaled Responses in Plants

Paul E. Staswick and Casey C. Lehman

Abstract

Jasmonic acid (JA) is representative of an important family of plant signaling molecules derived from fatty acid peroxidation. JA accumulates constitutively in plants, but also in response to a variety of stress stimuli. It has many of the attributes of traditionally recognized plant hormones: low molecular weight, translocatable, endogenous compounds that evoke a number of physiological and gene regulatory responses at low concentration. Despite its fairly recent discovery, a substantial amount of information on JA activity in plants has accumulated. Exciting new research demonstrates that JA has a direct role in plant defense against both insects and microorganisms. JA is also notable for its similarity with animal hormones of the eicosanoid family, which are involved in inflammatory stress response.

Introduction

The methyl ester of jasmonic acid (MeJA) was first identified in 1962 as an essential oil isolated from extracts of the jasmine plant, *Jasminum grandiflorum*. Reports suggesting that JA might be involved in regulating plant growth and development first appeared in 1980, however, little attention was paid to JA as a *bona fide* signaling molecule in plants for the next 10 years. Then Farmer and Ryan (1990) made the remarkable discovery that MeJA that had volatilized from *Artemesia* (sagebrush) could trigger defense gene expression in adjacent tomato plants. This finding was itself the trigger for an awakening to the importance of JA in plant biology, particularly in plant defensive responses. Many additional discoveries since then have firmly established that JA is indeed a key signal intermediate in plant biology.

Investigators have used a variety of experimental approaches to study JA function in plants. These include the analysis of: 1) its biosynthesis, abundance and distribution, 2) effects of JA on plant responses and gene expression, 3) inhibitors of JA biosynthesis or response, 4) mutants affecting JA biosynthesis or signaling, and 5) plants genetically engineered in JA production. Substantial evidence now points to a crucial role for JA in signaling plant defense responses to both insects and microorganisms. The literature relating to JA is extensive and this chapter is not intended to be a comprehensive review. Rather, we will focus on recent findings, particularly those relating to defense signaling in plants. Greater details and references for much of the earlier work can be found in several previous reviews (Vick and Zimmerman 1987, Anderson 1989, Koda 1992, Sembdner and Parthier 1993, Reinbothe et al. 1994, Staswick 1995, Creelman and Mullet 1997, Wasternak and Parthier 1997, Farmer et al. 1998).

Fatty Acid Oxidation Yields a Variety of Signaling Molecules in Plants

Vick and Zimmerman (1987) established the essential features of the JA biosynthetic pathway, which was recently reviewed in detail by Mueller (1997). Briefly, the fatty acid α-linolenate (18:3) is oxygenated by lipoxygenase (LOX), followed by conversion of the resulting hydroperoxy- to 12,13-epoxy-linolenic acid by allene oxide synthase (AOS). Cyclization via allene oxide cyclase yields 12-oxo-phytodienoic acid (OPDA) with its five-membered ketone ring that is characteristic of the JA family. Reduction and successive β-oxidations yield (3R,7S)-JA, which can then isomerize to the trans configuration (3R,7R)-JA, the predominant species at thermodynamic equilibrium. Overall, the synthetic pathway has notable similarity to that for animal eicosanoids derived from arachidonate, a fatty acid lacking in plants. The OPDA intermediate in particular is structurally similar to prostaglandins, which are important regulatory and stress-signaling hormones in animals (Fig. 1).

As for some other plant hormones, chloroplasts are involved in JA biosynthesis. Early biochemical evidence indicated that allene oxide synthase activity is associated with chloroplast membranes. The subsequent molecular cloning and sequencing of the AOS gene from flax (Song et al. 1993) (and later in *Arabidopsis*) confirmed the presence of an organellar transit peptide, which indeed targets the flax enzyme to chloroplasts in transgenic potato plants (Harms et al. 1995). The initial oxidation of 18:3 also involves a chloroplast-localized lipoxygenase in *Arabidopsis*. Down regulation of this gene eliminates wound-induced accumulation of JA (Creelman and Mullet 1997), although normal JA levels still occur in unwounded leaves. This suggests that another LOX enzyme may function in constitutive JA synthesis. The compartmentation or differential regulation of redundant biosynthetic enzymes could uniquely modulate JA synthesis in response to a variety of cues. In contrast with earlier pathway enzymes OPDA reductase is present in the cytosol, indicating OPDA is

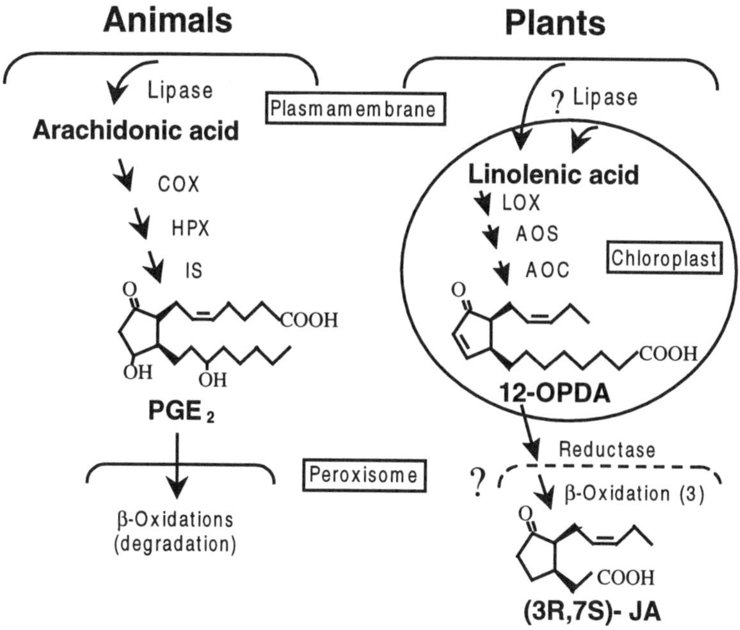

Figure 1: Comparison of plant and animal fatty acid oxidation pathways involved in signaling. Question marks indicate uncertainties about the source of linolenic acid and the residence of β-oxidases for JA synthesis. COX, cyclo-oxygenase; HPX, hydroperoxidase; IS, isomerase; LOX, lipoxygenase; AOS, allene oxide synthase; AOC; allene oxide cyclase; PGE_2, prostaglandin E_2; OPDA, oxo-phytodienoic acid; JA, jasmonic acid. (COX and HPX form the complex PGG/H synthase.)

transported out of chloroplasts for continued synthesis of JA. The residence of other enzymes is not known, although allene oxide cyclase (AOC) is most likely chloroplastic and β-oxidations might occur in peroxisomes, as is the case in animals.

It is now clear that JA is only one of several related signaling molecules derived from fatty acid peroxidation in plants. This is perhaps not surprising as a host of important signals are derived from fatty acid oxidation in animals as well. For simplicity the term JA will be used throughout this discussion, although we should be aware that other JA family members are also likely to play important roles in the response pathways discussed. Precursors of JA, including 18:3 and OPDA, exhibit biological activity in several test systems. While in some cases this could result from conversion of the precursor to JA, the efficacy of OPDA analogues that are unable to undergo β-oxidation to JA was demonstrated in the stimulation of alkaloid production in *Eschsholtzia californica* cell suspension cultures (Blechert et al. 1995). Furthermore, in some responses OPDA is more potent than JA (Gundlach and Zenk 1998) or has more rapid reaction kinetics (Weiler et al. 1993). Interestingly, in animals, reduction and β-oxidation of prostaglandins leads to their inactivation and degradation, although there is no evidence that β-oxidation initiates turnover in the plant JA pathway. Based on all of this information it has been suggested that OPDA may be a primary octadecanoid signal in plants (Mueller 1997). OPDA and JA might also carry out somewhat different signaling repertoires in plants (Laudert and Weiler 1998), although this is not the case in all response pathways. As plants contain a wide variety of structural analogues of JA, including amino acid (usually hydrophobic) conjugates and glucosyl esters (Sembdner and Parthier 1993) the possible variations in regulatory patterns are large. For most JA analogues however, biological activity is still an open question.

Linoleic acid (18:2) can also be converted to the biologically active 16-dihydro-12-OPDA (DH-OPDA) in some plants. Whether DH-OPDA undergoes reduction and oxidation to 9,10-dihydro-JA (DH-JA) in significant quantity *in planta* is still unclear (Blechert et al. 1995, Gundlach and Zenk 1998). The dihydro pathway does appear less consequential, based both on the lack of detectable DH-JA and the lack of response to applied DH-JA in many plant species. Limited evidence indicates DH derivatives may only be active in some JA responses, suggesting further that JA signaling pathways can discriminate among different oxylipin signals.

Weber et al. (1997) recently applied the term 'oxylipin signature' to describe unique profiles of JA family members that are obtained from their simultaneous extraction and detection by gas chromatography. This technique led to the discovery of a new biologically active family member, dinor-oxo-phytodienoic acid (DN-OPDA), derived from an alternate synthesis from hexadecatrienoic acid (16:3). It is not yet known if DN-OPDA is itself an important plant signal or if it is further metabolized to JA in plants. Oxylipin signatures will undoubtedly yield further insights about the important quantitative and qualitative changes in oxylipin profiles that precede plant responses. It will not be surprising if additional family members are found; some perhaps unique to particular species or to specific JA responses.

Structural analogues of JA have facilitated the identification of functional groups necessary for activity in plants. The cyclic five member ring, with or without a double bond, is crucial. Replacement of the 6-oxo group with a hydroxyl (cucurbic acid) or hydrogen eliminates induction of phytoalexin synthesis in *Eschsholtzia californica* suspension cells (Blechert et al. 1995) and tendril coiling in *Bryonia dioca* (Weiler et al. 1993), although cucurbic acid has growth inhibiting activity in some test systems. The acetyl and pentenyl side chains of JA are also important, although their length may not be. Stereochemistry around the two chiral centers also influences activity. The R configuration at C-3 is essential and apparently occurs exclusively in plants. The importance of the C-7 configuration varies with the response tested, although the trans-isomer is generally less active. Interestingly, structural requirements can vary between species and even among signaling pathways within a species. For example, 9,10-DH-JA was effective in the *E. californica* system but not in tendril coiling. Further study of the activity of JA analogues and isomers will likely reveal that the diversity of family members present in plants provides them with a much richer signaling repertoire than could be accomplished with a single molecule.

Changes in JA Quantity Are Associated With Response to Several Stimuli

JA is ubiquitous and undoubtedly a fundamental signaling molecule in all plants. It has been identified in over 160 diverse plant families (Sembdner and Parthier 1993), mostly angiosperms, but some gymnosperms, algae, and *Equisetum* are also included. Both the acid and methylester are present in plant tissues, although in some cases only one or the other has been reported. Documented as a fragrance component in only a handful of species, MeJA is likely a constituent of many other flowers and fruits. JA is also distributed throughout plant vegetative tissues, varying widely in quantity with tissue and developmental stage, as is true for other important plant signaling molecules. Reproductive and young vegetative tissues generally contain the highest levels, which can be more than 1 μg/g fresh weight.

JA level is increased by several stimuli, including mechanical wounding and water deficit (Creelman and Mullet 1997), herbivory (McCloud and Baldwin 1997), microbial cell wall elicitors (Blechert et al. 1995), and the plant signaling peptide systemin (see Bergey et al. 1996). The increase in JA also correlates well with increased transcript levels for several JA-inducible genes that are associated with response to these kinds of stimuli. On the other hand, inhibitors of JA accumulation also block the activation of JA-inducible genes by wounding and elicitor treatment, further supporting the role of JA in gene expression.

JA action can also be more subtle than a simple quantitative change, as is commonly detected in whole plant organs. In some cases the wound-stimulated elevation of JA is rapid, but transient, suggesting modulation of JA level may play an inducing role, but continued high levels are not always required for sustained activity. JA levels are also constitutively high in some young tissues without inducing defense pathways and elevated levels of JA in transgenic potato did not lead to *PinII* induction (Harms et al. 1995). In these cases JA might be inactivated or sequestered in some way that does not stimulate a response. Interestingly, wounding the transgenic potato leaves that overaccumulate JA raised JA levels even higher and led to an induction of *PinII*. This suggests that a modulation of JA level, not concentration *per se*, is important in some signaling paths (Harms et al. 1995). Finally, JA-inducible genes or responses may be stimulated without a detectable change in endogenous JA, as in the case of soybean *Vsp* (see Creelman and Mullet 1997) and PR-like genes induced in rice upon fungal infection (Schweizer et al. 1997). In these cases undetected minor or localized changes in JA, altered JA sensitivity, unknown octadecanoids, or entirely different signals might be involved.

An early model for herbivore/wound-stimulated induction of the protective leaf proteinase inhibitors linked signaling by JA and the peptide hormone systemin (see Bergey et al. 1996). The release of 18:3 substrate from membranes was proposed as a regulatory point for JA biosynthesis. Analogous to arachidonate liberation from plasmamembrane in animals (Smith 1989), it was suggested the free 18:3 level is controlled by phospholipase activity. Unesterified 18:3 (and 18:2) does increase after wounding in a manner consistent with the rise in JA level (Conconi et al. 1996), and phospholipase activity also rises following elicitor treatment (Chandra et al. 1996). On the other hand, many plant tissues appear to have sufficient unesterified 18:3 to account for the increased JA stimulated by wounding (Mueller 1997). Additional research is needed to clarify the role of phospholipase in JA signaling. It is also not known if 18:3 for JA synthesis originates from chloroplast and/or plasma membranes.

AOS is another important control point in the synthesis of JA. The high JA level attained in potato plants overexpressing the flax AOS gene (Harms et al. 1995) supports the earlier biochemical evidence that this step is rate limiting. This also confirms that available substrate is not limiting to high constitutive JA levels, although it might be in a rapid response to wounding. Interestingly, JA is a positive feedback regulator of its own biosynthetic pathway, stimulating expression of genes for both lipoxygenase and AOS. Thus, depletion of OPDA pools by conversion to JA could stimulate replenishment of OPDA. This would be important if both JA and OPDA are regulatory molecules with unique roles, each requiring maintenance of effective cellular concentrations.

Salicylic acid (SA) is an important signaling intermediate in pathogen defense, but interestingly, SA prevents accumulation of JA in response to wounding. An intriguing recent finding is that in *Arabidopsis* SA actually stimulates a modest increase in AOS transcript and protein, as well as a rise in OPDA level, without a concomitant increase in JA (Laudert and Weiler 1998). This led to the suggestion that OPDA is sequestered in chloroplasts and its release to the cytosol for reduction is another regulatory point for JA biosynthesis (Fig. 1). On the other hand, SA reportedly has the opposite effect in response to wounding in flax, where it inhibits accumulation of AOS transcripts (Harms et al. 1998). SA also inhibits tomato proteinase inhibitor synthesis at an unknown site downstream of octadecanoid signaling (Doares et al. 1995). Already well established as a regulator in plant defense pathways not involving JA, it appears that endogenous SA may also influence some octadecanoid signaling pathways.

Many of the Genes Induced by JA are Implicated in Defense

Considerable evidence implicates JA in gene regulation and many of these genes are involved in plant defense. Recent lists of these genes, and the corresponding primary references, are presented by Reinbothe et al. (1994), Creelman and Mullet (1997) and Wasternak and Parthier (1997) or they are mentioned elsewhere in this chapter. Gene products at least tentatively associated with defense response include various proteinase inhibitors, the systemic wound signal systemin, lipoxygenase and AOS enzymes for JA synthesis, ribosome-inactivating protein possibly regulating translation, phenylpropanoid pathway enzymes for phytoalexin and lignin production, various cell wall proteins, and the antifungal proteins thionin, osmotin and defensin. A number of secondary metabolites associated with defense are also JA-induced, implying many additional genes. Gene products not directly implicated in defense include vegetative and seed storage proteins, proteins involved in photosynthesis, and enzymes for amino acid biosynthesis. JA-stimulated genes are also associated with response to abiotic stresses such as UV light, dessication, salt and osmotic stress, but a direct demonstration of JA's role in protecting against these factors is presently lacking.

Among the genes that can be induced locally in response to pathogen attack are those for pathogenesis-related (PR) proteins and a variety of enzymes involved in phytoalexin biosynthesis. Early studies demonstrated that JA induces genes encoding key enzymes of the phenylpropanoid phytoalexin biosynthetic pathway, such as phenylalanine ammonia lyase and chalcone synthase in soybean and parsley. Subsequently, the accumulation of a wide variety of phytoalexins was documented in suspension cell cultures from 36 different plant species following stimulation by JA or OPDA. These include various alkaloids, flavonoids, terpenoids and anthraquinones (Blechert et al. 1995). Both free α-linolenic acid and JA levels rise rapidly, although transiently, following fungal elicitor treatment of cell cultures, further implicating JA in phytoalexin production. On the other hand, JA's role in phytoalexin synthesis is not universal. In potato tubers JA did not induce sesquiterpenoid phytoalexins that are accumulated in response to fungal elicitor, but JA did stimulate the wound-induced accumulation of steroid-glycoalkaloid (Choi et al. 1994).

The role of JA in PR protein induction is less clear, apparently varying with the species. In rice, locally induced resistance by inoculation with rice blast fungus does not stimulate JA accumulation. And although JA induces several PR proteins in common with those induced by the fungus, JA alone did not confer local resistance (Schweizer et al. 1997). Jasmonate probably also locally regulates a number of other genes that are associated with pathogen response, but further study is needed.

Promoter analysis has identified *cis* elements necessary for JA response in several genes. The palindromic G-box motif (CACGTG) is essential in soybean

VspA and potato *PinII* and is found in several other JA-induced genes. Analysis is complicated by the fact that the G-box also functions in response to other stimuli, such as light and abscisic acid (ABA), and is present in genes not responding to JA. Genes encoding nopaline synthase, lipoxygenase 1 and thionin use alternate palindromic sequences for JA response (Rouster et al. 1997, Vignutelli et al. 1998). As most genes induced by JA respond to various developmental and environmental cues, their overall regulatory patterns are undoubtedly influenced by an array of *cis* elements and their corresponding *trans* factors.

JA also represses both nuclear and chloroplast-encoded genes that are involved in photosynthesis. JA negatively affects the transcription of nuclear photosynthetic genes while the plastid-encoded *rbcl* transcript is alternately cleaved post-transcriptionally, leading to decreased translation (Reinbothe et al. 1994). Negative translational control is also exerted on nuclear-encoded photosynthetic genes, including *RbcS* and *Lhc*. Suppression of photosynthetic genes along with the accumulation of ribosome inactivating proteins by JA following pathogen infection may lead to cell damage or death, thereby limiting pathogen invasion (Reinbothe et al. 1994).

JA Signaling and Biosynthesis Mutants Provide Insight into JA Function

Mutants have provided important insights into what plant hormones do and how they work. Two classes of defects have been informative; those affecting hormone biosynthesis and those altering hormone response. Recent results have confirmed this approach is valid to study JA signaling as well. Not only do these mutants illuminate JA function, they also facilitate cloning of the respective loci by molecular genomic techniques, even when biochemical information about the gene's function is lacking.

Three loci affecting JA response have been reported in *Arabidopsis*. The first identified was *jar1,* isolated in a screen for decreased inhibition of seedling root growth in agar media containing MeJA (Staswick et al. 1992). A second locus, called *jin1,* was identified by a similar approach (Berger et al. 1996). The third locus (*coi1*) was isolated in a screen for insensitivity to the phytotoxin coronatine, produced by *Pseudomonas syringae* (Feys et al. 1994). Coronatine has structural similarity to JA, and consistent with the fact that coronatine is much more toxic to plants than is JA, the *coi1* mutant is more resistant to MeJA than are the other mutants. Another unique phenotype of *coi1* is male sterility due to a defect in late stages of pollen development and anther dehiscence. All three of these mutants are specifically insensitive to jasmonate, indicating they are not simply defective in general pathways influencing response to exogenous hormones, such as uptake or transport. The lesions also depress JA-inducible gene expression in leaves and seedlings, suggesting these mutants identify signaling pathway(s) acting throughout the plant.

The *Coi1* locus was recently cloned using a map-based strategy (Xie et al. 1998). The gene sequence revealed similarity to F-box proteins, some of which are involved in a ubiquitin-like E3 complex which targets the degradation of specific proteins. This result is intriguing. Over the past few years it has been established that a key facet of auxin signaling involves a ubiquitin-like pathway that includes TIR1, an F-box protein that is closely related to COI1 (Ruegger et al. 1998). How degradation of proteins mediates signaling in the auxin pathway is not yet clear but it could involve removal of negative pathway regulators. While it is not known if an analogous ubiquitination pathway operates in JA signaling, these results provide important avenues to explore.

Mutants affecting JA biosynthesis have also provided insight into JA function. An *Arabidopsis* mutant defective in three fatty acid desaturases (*fad3-2, fad7-2, fad8*) blocks the conversion of 18:2 to 18:3, resulting in no detectable JA accumulation (McConn and Browse 1996). Besides demonstrating that an 18:2 pathway is not a significant source for JA in *Arabidopsis*, this mutant is male sterile like *coi1*, confirming the role of octadecanoids in this aspect of development. The tomato mutant *def1,* which is blocked in the conversion of hydroperoxylinolenic acid to OPDA, does not respond to wounding (Howe et al.1996). This confirms that products generated after this point are essential in the systemin pathway for *PinII* induction and protection against insect larvae feeding (see Bergey et al. 1996). Manipulation of JA level by the genetic engineering of potato to over-accumulate AOS was discussed earlier.

Role of JA in Growth and Development

A wide variety of physiological responses to applied JA have been reported (see Sembdner and Parthier 1993) including stimulation of abscission, senescence, tuber formation, fruit ripening and pigment formation. Inhibitory responses have been noted for seed germination, callus growth, root growth, and photosynthesis, among others. While possibly informative, a response to applied JA (or lack thereof) is not necessarily indicative of the endogenous activity of a plant compound. Accumulating evidence from JA signaling and biosynthesis mutants and from transformed plants suggests that JA may not be critical for many aspects of plant growth and development.

The reported effects of exogenous JA on seed germination are contradictory. Inhibition was noted in *Brassica napus* and flax, stimulation was reported in apple, and no effect was evident in *Arabidopsis*. Furthermore, disruption of JA signal transduction in the *Arabidopsis* mutants did not affect germination, nor did the suppression of JA biosynthesis in *coi1*. These results suggest that a role for endogenous JA in germination, if any, is at least variable among species. JA's involvement in growth is also unclear. While exogenous JA inhibits root elongation at low concentration (10^{-8}M), *Arabidopsis* signaling mutants are not altered in root growth in the absence of applied JA. Growth of other plant organs also appears normal in the known biosynthesis and signaling mutants. The possible link between JA and senescence promotion was also questioned early in the study of JA because highest levels generally occur in young growing tissue and not in senescing tissue, as would be predicted. The more recently available mutants strengthen this argument as they appear to senesce normally, and are not obviously aberrant in most other aspects of development. Some of the reported responses to applied JA may be the result of stress signaling and not a reflection of the endogenous function of JA, although additional studies are needed.

This does not mean there are no developmental roles for JA. It is clearly essential for anther function in *Arabidopsis* and subsequent research may reveal other roles, some that are perhaps unique to certain species. For example, tendril coiling in *Bryonia* responds to very low concentrations of gaseous MeJA, which is consistent with hormone-like signaling and not simply a stress response. Furthermore, the evidence for multiple JA family members, some with differential response patterns, complicates the interpretation of results from JA signaling and biosynthesis mutants. Do JA and OPDA use the same signaling pathway or are alternate pathways involved? What about the role of products from the hexadecatrienoic acid pathway? Can these or additional unidentified signaling molecules compensate for a loss of JA signaling in the mutants presently available? These and other questions await further studies that may point to additional roles for oxylipin signals in plants.

JA Participates in Defense Against Insects

By far the best studied plant defense involving JA is the systemic induction of proteinase inhibitor genes in the Solanaceae. Details implicating JA in this important protection against insects can be found in previous reviews (see Bergey et al. 1996 and references therein). The model for this response includes the peptide hormone systemin as a wound-induced and systemically translocated signal that stimulates the release of linolenic acid from membranes, possibly via a phospholipase. The free 18:3 is then metabolized to JA, which activates *PinII* transcription through an unknown mechanism (Fig. 2). The wound-stimulated increase in endogenous JA is required for *PinII* gene expression and interestingly, JA also activates the prosystemin gene. ABA and ethylene can also positively influence *PinII*, while auxin inhibits induction by wounding (Peña-Cortés et al. 1995). The precise signaling role for these other hormones in *PinII* regulation is not clear, but ABA is apparently not a primary signal in this defense pathway (Birkenmeier and Ryan 1998). Although the function of proteinase inhibitors in protection against some insect pests is clear, the role of many other proteins that are also induced through the JA/systemin pathway in tomato is not yet known (Bergey et al. 1996).

Mechanical wounding has frequently been substituted for herbivory in studies of defensive responses. While this approach has greatly aided our understanding of wound response pathways, recent evidence indicates it may not faithfully mimic herbivory or other biotic assaults. Biotic stimuli generally

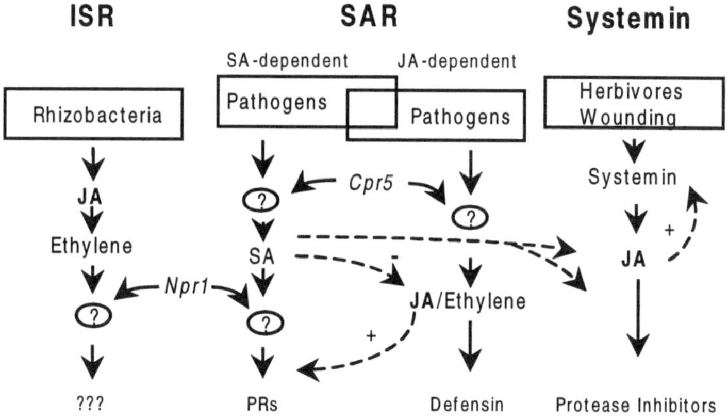

Figure 2: Major systemic defense signaling pathways in plants. Dashed arrows indicate possible interactions involving JA within or between pathways. Models are based primarily on results from only a few dicotyledonous species. Abbreviations are explained in the text.

induce higher levels of JA than mechanical wounding alone. In tobacco leaves this is partly due to local amplification effects caused by regurgitants from feeding larvae of *Manduca sexta* (McCloud and Baldwin 1997). Regurgitant from the same species also hastens the accumulation of *PinII* and other transcripts in potato compared with a mechanical stimulus alone (Korth and Dixon 1997). However, the resulting local elevation of JA in tobacco did not lead to a comparable systemic increase in JA or in the defense product nicotine. This suggests that regurgitant actually suppresses the systemic response in tobacco.

Insect saliva can also have a darker side for the producer. Volicitin is a fatty acid derivative present in caterpillar regurgitant that stimulates release of volatile terpenoids from corn leaves during caterpillar feeding (Alborn et al. 1997). These volatiles in turn attract parasitic wasps, whose offspring hatch from eggs laid in the caterpillar and eventually devour it. Because JA is known to stimulate the release of various volatiles from plant tissues, it has been suggested that volicitin might interact with the plant octadecanoid pathway to stimulate volatile terpenoid release. The benefit to the plant is clear, the value of volicitin to insects is not (see Felton and Eichenseer, this volume).

JA is also essential for protection against insects in *Arabidopsis*. The *fad* triple mutant is severely attacked by larvae of the fungal gnat *Bradysia impatiens,* a defect that is complimented by exogenous application of JA (McConn et al. 1997). The mechanism of resistance is not clear but may involve JA-stimulated production of toxic products, such as isothiocyanates, that are derived from glucosinolate metabolism (Farmer et al. 1998). Glucosinolate synthesis is influenced by JA in some plants in the Brassicaceae. The involvement of JA in insect defense in other plants is relatively unexplored, although several genes that respond to wounding and JA have been identified in other species, including woody plants. Wounding systemically induces PR proteins in rice through a JA-responsive pathway, but these proteins are apparently involved in defense against pathogens and not insects (Schweizer et al. 1998). It will be important to determine to what extent this pathway bears additional similarities with the systemin pathway or with the other pathogen response pathways discussed later.

Not all genes stimulated by JA are wound inducible. The small cysteine-rich antifungal peptide defensin found in *Arabidopsis* (*Pdf1.2*) and in radish is one example (Terras et al. 1998). In contrast with potato, induction of tomato papain inhibitor activity is weak and greatly delayed following wounding, although the gene is jasmonate responsive (Bolter et al. 1998). Recent evidence also suggests a JA-independent wound pathway in *Arabidopsis* (Titarenko et al. 1997). This route to gene induction appears to operate primarily near the wound site and on genes that are unresponsive to exogenous JA. Consistent with a JA-independent pathway, the genes remain wound-inducible in the JA signaling mutant *coi1*. This contrasts with the induction of another set of genes that respond systemically to wounding, are JA-responsive and are not expressed in *coi1*. Both pathways, however, appear to involve reversible protein phosphorylation (Rojo et al. 1998).

JA Participates in Defense Against Microorganisms

Evidence supporting a role for JA in plant defense against microorganisms was to some extent overshadowed by the exciting early work on the wound-induced proteinase inhibitors that are targeted toward insects. The apparent lack of a crucial role for JA in the well-documented SA-dependent systemic resistance pathway against microorganisms raised the question of whether jasmonate was primarily a "wound hormone" and not directly involved in defense pathways aimed at microorganisms. More recent work demonstrates conclusively that this is not the case.

Systemic Acquired Resistance. Pathogen attack can lead to systemic immunity against subsequent infection in healthy plant tissues. Systemic acquired resistance (SAR) was one of the first terms used to describe this phenomenon. SAR is characterized by the accumulation of SA and by the induction of PR proteins, including β-1-3-glucanases and chitinases that are capable of hydrolyzing fungal cell walls (Fig. 2) (reviewed in Ryals et al. 1996). Although exogenous SA induces PR gene expression and a pathogen-stimulated increase in SA is essential for establishing this resistance, SA is apparently not the systemic signal.

The finding that SA and related compounds inhibit JA production and block the wound induction of the JA-dependent systemin pathway in tomato cast doubt on a role for JA in SA-dependent SAR, even suggesting that wound- and pathogen-induced systemic signaling might be incompatible. *Arabidopsis* JA signaling mutants were also found competent for SA-dependent SAR, raising further doubt about JA's role in acquired resistance (Pieterse et al. 1998). Although there are scattered reports that JA induces some PR family members, it is generally assumed that octadecanoids and ethylene play only a minor role in SA-dependent SAR.

More recently, studies on *Arabidopsis* mutants that are defective in SA-dependent SAR revealed that these plants nevertheless display a systemic resistance response to some pathogens. In contrast to SA-dependent SAR, this response is ethylene and JA-dependent and SA-independent (Penninckx et al. 1996, Bowling et al. 1997). In *Arabidopsis* the biochemical marker for this JA-dependent pathway is the accumulation of the antifungal protein defensin (*Pdf1.2).* In contrast with PR proteins, defensin production is induced by either JA or ethylene, but not by SA, and it is not induced by pathogens in the JA signaling mutants *jar1* and *coi1* nor in ethylene signaling mutants. In addition to *Arabidopsis* the JA-dependent SAR pathway has been documented in radish (Terras et al. 1998) and possibly tobacco, where two groups of PR proteins respond differentially to SA and ethylene.

Interestingly, genetically engineered plants that lack SA accumulate more defensin in response to pathogen induction suggesting that while distinct, the JA- and SA-dependent pathways interact. One plausible mechanism for this interaction resides in the fact that SA blocks JA biosynthesis, and both JA and ethylene also appear to influence the accumulation of at least some PR proteins (see Fig. 2). However, the JA- and SA-dependent pathways are also linked by the *Cpr5* locus of *Arabidopsis*, which when mutated leads to constitutive expression of both the SA-dependent and the JA-dependent pathways. The biochemical function of CPR5 is unknown. Other antifungal proteins that appear to be induced along with defensin include thionins and hevein (Epple et al. 1995, Penninckx et al. 1996).

The extent to which the JA- and SA-dependent SAR pathways are co-induced, or whether they discriminate among some pathogens is unresolved. Limited evidence suggests the latter does occur. While both are induced by *Alternaria brassicicola* in *Arabidopsis* (Penninckx et al. 1996), thionin gene expression, but not PR production, is induced by *Fusarium oxysporum* (Epple et al. 1995). In radish, on the other hand, PR1 is accumulated only locally following induction by *Alternaria brassicicola,* but defensin accumulates both locally and systemically (Terras et al. 1998). The spectrum of resistance established by these pathways individually also differs. Both are involved in resistance to *Peronospora parasitica* Noco2 in *Arabidopsis* but only the JA-dependent pathway establishes resistance to *Pseudomonas syringae* pv. *maculicola* ES4326 (Bowling et al., 1997). Interaction between the JA-dependent and the systemin pathway is also possible because JA is involved in both. Although wounding does not stimulate defensin accumulation in *Arabidopsis*, both wounding and pathogens induce thionin genes, which are also JA inducible.

Induced Systemic Resistance. Recently a form of induced systemic resistance (ISR) thought to be distinct from salicylate-dependent SAR was described. Induction of resistance is not mediated by pathogens, but rather is established following the colonization of roots by nonpathogenic rhizobacteria (reviewed in van Loon et al. 1998, Zhender et al., this volume). Like SAR, ISR affords substantial protection against a broad range of pathogens, including a variety of fungi, bacteria and viruses, and occurs in a variety of plants. Organisms demonstrated to elicit ISR include sixteen strains of common soil bacteria, some of which also mediate SAR, belonging to the genera *Pseudomonas* and *Serratia*.

SA is not a required signalling intermediate in ISR. *Arabidopsis* mutants insensitive to JA or ethylene are not ISR-competent, suggesting these hormones are involved in signal transduction leading to ISR. Topical application of either hormone also induces ISR in the absence of rhizobacteria in wild type plants; however, commonly studied genes responding to JA or ethylene are not stimulated, distinguishing ISR from the JA-dependent SAR response. Presumably there are specific defense genes induced in the ISR pathway, but these have not yet been identified.

Although SAR and ISR utilize distinct signal transduction pathways, they require a common intermediate identified by the *Npr1* gene in *Arabidopsis* (Fig. 2). Mutation at this locus blocks PR protein accumulation and the establishment of SA-dependent SAR, and also prevents the establishment of ISR. These findings suggest there are components of each pathway that interact independently with *Npr1*.

Resistance to Opportunistic Soil Fungi. Recent reports also establish a role for JA in defense against opportunistic pathogens. A large number of *Pythium* species are widespread fungal soil inhabitants that are thought to frequently cause mild stunting of plants due to root damage or invasion of other tissues. More severe injury and death can also occur under certain conditions. The *jar1* mutant is susceptible to infection by *Pythium irregulare,* which leads to symptoms of wilting and tissue death, whereas wild type is largely asymptomatic (Staswick et al. 1998). Susceptibility in the mutant occurs not only in seedlings, a developmental stage that is typically more sensitive to *Pythium,* but also in mature plants. This indicates that the defense mechanism involving JA is required throughout the life of the plant. Similar results were found for the *Arabidopsis fad* mutant defective in octadecanoid synthesis (Vijayan et al. 1998). In this case susceptibility to *P. mastophorum* was significantly complemented by exogenous application of MeJA prior to inoculation. Tobacco transformed with a dominant *Arabidopsis* gene (*etr1)* conferring insensitivity to ethylene is also susceptible to *Pythium sylvaticum* (Knoester et al. 1998), suggesting that both JA and ethylene signaling are required for resistance to these oomycete fungi.

It is not clear if resistance to *Pythium* is a local or systemic response, nor whether the signaling path is one already described or illustrates a novel mechanism. Both JA and ethylene are implicated in both the CPR5- and the NPR1-dependent pathways. Several facts suggest the response to *Pythium* is distinct from ISR. First, ISR by definition can only be mediated by non pathogenic rhizobacteria. While *Pythium* generally causes only mild symptoms in wild type plants, it can cause serious infection in some situations. Second, *Pythium* stimulates defensin mRNA accumulation in *Arabidopsis* (Vijayan et al. 1998), which does not occur in ISR. Third, *Pythium* stimulates accumulation of basic PR1 proteins in wild type tobacco, but PR gene expression is not generally associated with ISR. Lack of PR induction in the *etr1* tobacco mutant was suggested as a possible explanation for susceptibility in the mutant, since these PR proteins are active against oomycete fungi (Knoester et al. 1998). On the other hand, PR1 is not induced in the defensin pathway in *Arabidopsis* (Penninckx et al. 1996), perhaps indicating a species difference.

As a weak pathogen with a broad host range, defense against *Pythium* may involve multiple pathways. Localized root damage could initiate wound-based defense responses involving JA. Consistent with this, *Pythium* induces two other genes in wild type *Arabidopsis* that are associated with JA and wound defense responses, those for lipoxygenase and chalcone synthase (Vijayan et al. 1998). On the other hand, the first symptom of infection in aerial plant parts is wilting, presumably due to loss of vascular function in root and/or crown tissues. JA has been implicated in drought response because some genes induced by water deficit are also JA-inducible, although a direct role is not known.

JA and Signaling Activity in Other Organisms

Signaling via JA in plants has interesting similarities with signaling paths in other organisms. The close relationship between the biosynthesis and structure of fatty acid-derived signaling molecules from plants and animals was noted earlier (for additional details see Bergey et al. 1996). JA is also used as a mating pheromone in oriental fruit moths, although it is not clear whether JA is synthesized by the moths or obtained from the fruit they consume (Baker et al. 1981)

A number of fungi, including some important plant pathogens, synthesize JA or related compounds (Weete and Weber 1980) but a role in these organisms is unknown. JA does affect the development of several fungal plant pathogens (e.g., Cohen et al. 1993). An obvious question is whether plant JA alters fungal development in a way that affords protection from potential pathogens. The relatively high doses that are necessary to induce developmental changes in fungi however, suggest this may not be biologically relevant in interactions with plants. Another question is whether plants can detect the JA present in certain fungi and interpret it as a signal to stimulate defensive responses. This also might seem unlikely, as the concentration a plant would encounter from a fungal source would probably be low relative to the endogenous levels in plants. However, in several cases plants respond to low levels of atmospheric MeJA, perceiving this as a signal above the background of endogenous JA. Further studies along this line might reveal exciting new roles for JA in interactions between plants and other organisms.

Synthesis – Applications of Jasmonic Acid

The practical application of what we have learned about JA will increase. MeJA shows promise as a post harvest treatment to extend the shelf life of fruits and vegetables (Buta and Moline 1998). MeJA applied as a vapor at low concentrations prior to cold storage significantly inhibits bacterial and fungal growth. This is apparently not a direct effect of MeJA, as very high concentrations are required to inhibit growth in the absence of plant material. Presumably, MeJA treatment induces defense gene product accumulation in the harvested tissue and these can protect fruits and vegetables during storage. The efficacy of MeJA as a vapor is particularly attractive because of the ease of application and the lack of harmful effects potentially caused by spraying or immersing produce in a liquid solution. An added benefit is that MeJA increases chlorophyll retention and decreases the discoloration of cut surfaces in some produce.

Could the application of JA to whole fields also be beneficial in protecting crops (Thaler 1999, this volume)? We still do not understand how induction of defense pathways by the exogenous application of JA relates to the induction of JA-dependent pathways by a specific pathogen. If the former signals a general stress response that activates multiple and perhaps conflicting pathways that are costly to plant productivity, then widespread spraying of JA would likely not be a desirable disease control strategy. On the other hand, careful control over the timing and amount of JA applied might afford increased protection that is not easily overcome by pathogens or pests. Individual disease response pathways might also be altered by genetically modifying plants in their JA metabolic or signaling pathways (Thaler et al. 1999, this volume). In this case, fine tuning the timing or magnitude of a response pathway while maintaining its inducible nature might afford increased protection that can survive in the environment. Clearly this would require a far more detailed understanding of JA function in signaling than is currently available.

Future Directions

In the past ten years JA has gone from a little regarded secondary metabolite to a widely recognized central player in plant defensive responses.

While our understanding of how plants use JA in signaling has advanced greatly, this wealth of information also reveals many new avenues to explore. For defense pathways that are JA-dependent, and even for those that are not, octadecanoids may help to coordinate activity among different signaling paths. Some of these interactions are just becoming evident, others will likely emerge in the future. New genes that are responsible for JA signal transduction will be isolated and will provide additional insights, not only on JA signaling but also about the interactions between hormone signaling pathways. Current models for plant defense responses are based primarily on work in a few dicotyledonous species. JA undoubtedly is universally important to plants, although the mechanism of its activity in other species probably differs, especially for monocots.

New roles for JA in plant development and abiotic stress response also likely await discovery, perhaps further tying defense to metabolism and growth. The link between JA signaling and ecological adaptation is also of great importance, while the more subtle aspects of JA signaling raise important questions. For example, why are some JA-responsive genes not induced in tissues with elevated JA? What are the signaling specificities of individual JA family members? How do they interact to control response pathways?

Literature Cited

Alborn, H.T., T.C.J. Turlings, T.H. Jones, G. Stenhagen, J.H. Loughrin and J.H. Tumlinson. 1997. An elicitor of plant volatiles from beet armyworm oral secretion. Science 276:945-949.

Anderson, J.M. 1989. Membrane derived fatty acids as precursors to second messengers. Pages 181-212 *in* W. Boss and D. Morre', editors. Second Messengers in Plant Growth and Development. Alan R. Liss, NY.

Baker,T.C., R. Nishida and W.L. Roelofs. 1981. Close-range attraction of female oriental fruit moths to herbal scent of male hairpencils. Science 214:1359-1361

Berger, S., E. Bell and J.E. Mullet. 1996. Two methyl jasmonate-insensitive mutants show altered expression of *Atvsp* in response to methyl jasmonate and wounding. Plant Physiology 111:525-531.

Bergey, D.R., G.A. Howe and C.A. Ryan. 1996. Polypeptide signaling for plant defensive genes exhibits analogies to defense signaling in animals. Proceedings of the National Academy of Sciences USA 93:12053-12058.

Birkenmeier, G.F. and C.A. Ryan. 1998. Wound signaling in tomato. Evidence that ABA is not a primary signal for defense gene activation. Plant Physiology 117:687-693.

Blechert, S., W. Brodschelm, S. Holder, L. Kammerer, T.M. Kutchan, M.J. Mueller, Z-H. Xia, and M.H. Zenk. 1995. The octadecanoic pathway: Signal molecules for the regulation of secondary pathways. Proceedings of the National Academy of Sciences USA 92:4099-4105.

Bolter, C.J., M. Latoszek-Green and M. Tenuta. 1998. Dependence of methyl jasmonate- and wound-induced cysteine proteinase inhibitor activity on nitrogen concentration. Journal of Plant Physiology 152:427-432.

Bowling, S.A., J.D. Clark, Y. Liu, D.F. Klessig and X. Dong. 1997. The *cpr5* mutant of *Arabidopsis* expresses both NPR1-dependent and NPR1-independent resistance. Plant Cell 9:1573-1584.

Buta, J.G. and H.E. Moline. 1998. Methyl jasmonate extends shelf life and reduces microbial contamination of fresh-cut celery and peppers. Journal of Agricultural and Food Chemistry 46:1253-1256.

Chandra, S., P.F. Heinstein and P.S. Low. 1996. Activation of phospholipase A analogues by plant defense elicitors. Plant Physiology 110:979-986.

Choi, D., R.M. Bostock, S. Avdiushko and D.F. Hildebrand. 1994. Lipid-derived signals that discriminate wound- and pathogen-responsive isoprenoid pathways in plants: Methyl jasmonate and the fungal elicitor arachidonic acid induce different 3-hydroxy-3-methylglutaryl-coenzyme A reductase genes and antimicrobial isoprenoids in *Solanum tuberosum* L. Proceedings of the National Academy of Sciences USA 91:2329-2333.

Cohen, Y., U. Gisi and T. Niderman. 1993. Local and systemic protection against *Phytophthora infestans* induced in potato and tomato plants by jasmonic acid and jasmonic methyl ester. Phytopathology 83:1054-1062.

Conconi, A., M. Miguel, J.A. Browse and C.A. Ryan. 1996. Intracellular levels of free linolenic and linoleic acids increase in tomato leaves in response to wounding. Plant Physiology111:797-803.

Creelman, R.A. and J.E. Mullet. 1997. Biosynthesis and action of jasmonates in plants. Annual Review of Plant Physiology and Plant Molecular Biology. 48:355-381.

Doares, S.H., J. Narvaez-Vasquez, A. Conconi and C.A. Ryan. 1995. Salicylic acid inhibits synthesis of proteinase inhibitors in tomato leaves induced by systemin and jasmonic acid. Plant Physiology 108:1741-1746.

Epple, P., K. Apel and H. Bohlman. 1995. An *Arabidopsis thaliana* thionin gene is inducible via a signal transduction pathway different from that for pathogenesis-related proteins. Plant Physiology 109:813-820.

Farmer, E. E. and C.A. Ryan. 1990. Interplant communication: Airborne methyl jasmonate induces synthesis of proteinase inhibitors in plant leaves. Proceedings of the National Academy of Sciences USA 87, 7713-7716.

Farmer, E.E., H. Weber and S. Vollenweider. 1998. Fatty acid signaling in *Arabidopsis*. Planta 206:167-174.

Feys, B.J.F., C.E. Benedetti, C.N. Penfold and J.G. Turner. 1994. *Arabidopsis* mutants selected for resistance to the phytotoxin coronatine are male sterile, insensitive to methyl jasmonate and resistant to a bacterial pathogen. Plant Cell 6:751-759.

Gundlach, H. and M.H. Zenk. 1998. Biological activity and biosynthesis of pentacyclic oxylipins: the linoleic acid pathway. Phytochemistry 47:527-537.

Harms, K., R. Atzorn, A. Brash, H. Kühn, C. Wasternak, L. Willmitzer and H. Peña-Cortés. 1995. Expression of a flax allene oxide synthase cDNA leads to increased endogenous jasmonic acid JA levels in transgenic potato plants but not to a corresponding activation of JA-responding genes. Plant Cell 7:1645-16544.

Harms, K., I. Ramirez and H. Peña-Cortés. 1998. Inhibition of wound-induced accumulation of allene oxide synthase transcripts in flax leaves by aspirin and salicylic acid. Plant Physiology 118:1057-1063.

Howe, G. A. J. Lightner, J. Browse and C.A. Ryan. 1996. An octadecanoid pathway mutant (JL5) of tomato is compromised in signaling for defense against insect attack. Plant Cell 8:2067-2077.

Knoester, M., L.C. van Loon, J. van den Heuvel, J. Hennig, J.F. Bol and H.J.M. Linthorst. 1998. Ethylene-insensitive tobacco lacks nonhost resistance against soil-borne fungi. Proceedings of the National Academy of Sciences USA 95:1933-1937.

Koda, Y. 1992. The role of jasmonic acid and related compounds in the regulation of plant development. International Review of Cytology 135:155-199

Korth, K.L. and R.A. Dixon. 1997. Evidence for chewing insect-specific molecular events distinct from a general wound response in leaves. Plant Physiology 115:1299-1305.

Laudert, D. and E.W. Weiler. 1998. Allene oxide synthase: a major control point for *Arabidopsis thaliana* octadecanoid signaling. Plant Journal 15:675-684

McCloud, E.S. and I.T. Baldwin. 1997. Herbivory and caterpillar regurgitant amplify the wound-induced increases in jasmonic acid but not nicotine in *Nicotiana sylvestris*. Planta 203:430-435.

McConn, M. and J. Browse. 1996. The critical requirement for linolenic acid is pollen development, not photosynthesis, in an *Arabidopsis* mutant. Plant Cell 8:403-416.
McConn, M., R.A. Creelman, E. Bell, J.E. Mullet and J. Browse. 1997. Jasmonate is essential for insect defense in *Arabidopsis*. Proceedings of the National Academy of Sciences USA 94:5473-5477.
Mueller, M.J. 1997. Enzymes involved in jasmonic acid biosynthesis. Physiologia Plantarum 100:653-663.
Peña-Cortés, H., J. Fisahn and L. Willmitzer. 1995. Signals involved in wound-induced proteinase inhibitor II gene expression in tomato and potato plants. Proceedings of the National Academy of Sciences USA 92:4106-4113.
Penninckx, I.A.M.A., K. Eggermont, F.R.G. Terras, B.P.H.J. Thomma, G.W. De Samblanx, A. Buchala, J.-P. Métraux, J.M. Manners. and W.F. Broekaert. 1996. Pathogen-induced systemic activation of a plant defensin gene in *Arabidopsis* follows a salicylic acid-independent pathway. Plant Cell 8:2309-2323.
Pieterse, C.M.J., S.C.M. van Wees, J.A. van Pelt, M. Knoester, R. Laan, H. Gerrits, P.J. Weisbeek and L.C. van Loon. 1998. A novel signaling pathway controlling induced systemic resistance in *Arabidopsis*. Plant Cell 10:1571-1580
Reinbothe, S., B. Mollenhauer and C. Reinbothe. 1994. JIPs and RIPs: The regulation of plant gene expression by jasmonates in response to environmental cues and pathogens. Plant Cell 6:1197-1209.
Rojo, E., E. Titarenko, J. León, S. Berger, G. Vancanneyt and J. Sánchez-Serrano. 1998. Reversible protein phosphorylation regulates jasmonic acid-dependent and -independent wound signal transduction pathways in *Arabidopsis thaliana*. Plant Journal 13:153-165.
Rouster, J., R. Leah, J. Mundy and V. Cameron-Mills. 1997. Identification of a methyl jasmonate-responsive region in the promoter of a lipoxygenase 1 gene expressed in barley. Plant Journal 11:513-523.
Ruegger, M., E. Dewey, W.M. Gray, L. Hobbie, J. Turner and M. Estelle. 1998. The TIR1 protein of *Arabidopsis* functions in auxin response and is related to human SKP1 and yeast Grr1p. Genes and Development 12:198-207.
Ryals, J.A., U.W. Neuenschwander, M.G. Willits, A. Molina, H.-Y. Steiner and M.D. Hunt. 1996. Systemic acquired resistance. Plant Cell 8:1809-1819.
Schweizer, P., A. Buchala, R. Dudler and J.-P. Métraux. 1998. Induced systemic resistance in wounded rice plants. Plant Journal 14:475-481.
Schweizer, P., A. Buchala, P. Silverman, M. Seskar, I. Raskin and J.-P. Métraux. 1997. Jasmonate-inducible genes are activated in rice by pathogen attack without a concomitant increase in endogenous jasmonic acid levels. Plant Physiology 114:79-88.
Sembdner, G. and B. Parthier. 1993. The biochemistry and the physiological and molecular actions of jasmonates. Annual Review of Plant Physiology and Plant Molecular Biology 44:569-589.
Smith, W.L. 1989. The eicosanoids and their biochemical mechanisms of action. Biochemical Journal 259:315-324.
Song, W.-C., C.D. Funk, and A.R. Brash. 1993. Molecular cloning of an allene oxide synthase: A cytochrome P450 specialized for metabolism of fatty acid hydroperoxides. Proceedings of the National Academy of Sciences USA 90:8519-8523.
Staswick, P.E. 1995. Jasmonate activity in plants. Pages 179-187 *in* P. Davies, editor. Plant Hormones: Physiology, Biochemistry and Molecular Biology, 2nd edition. Kluwer Academic Publishers, Dordrecht.
Staswick, P.E., W. Su and S.H. Howell. 1992. Methyl jasmonate inhibition of root growth and induction of a leaf protein are decreased in an *Arabidopsis thaliana* mutant. Proceedings of the National Academy of Sciences USA 89:6837-6840.
Staswick, P.E., G.Y. Yuen and C.C. Lehman. 1998. Jasmonate signaling mutants of *Arabidopsis* are susceptible to the soil fungus *Pythium irregulare*. Plant Journal 16:747-754.
Terras, F.R.G., I.A.M.A. Penninckx, I.J. Goderis and W.F. Broekaert. 1998. Evidence that the role of plant defensins in radish defense responses is independent of salicylic acid. Planta 206:117-124.

Thaler, J. S. 1999. Induced resistance in agricultural crops: effects of jasmonic acid on herbivory and yield in tomato plants. Environmental Entomology 28:30-37.

Thaler, J. S., A. L. Fidantsef, S. S. Duffey, and R. M. Bostock. 1999. Tradeoffs in plant defense against pathogens and herbivores: a field demonstration using chemical elicitors of induced resistance. Journal of Chemical Ecology, in press.

Titarenko, E., E. Rojo, J. León and J. Sánchez-Serrano. 1997. Jasmonic acid-dependent and -independent signaling pathways control wound-induced gene activation in *Arabidopsis thaliana.* Plant Physiology 115:817-826.

van Loon, L. C., P. A. H. M. Bakker, and C. M. J. Pieterse. 1998. Systemic resistance induced by rhizosphere bacteria. Annual Review of Phytopathology 36:453-483.

Vick, B.A. and D.C. Zimmerman. 1987. Oxidative systems for the modification of fatty acids. Pages 53-90 *in* P.K. Stumpf and E.E. Conn, editors. The Biochemistry of Plants. Academic Press, New York.

Vignutelli, A., C. Wasternack, K. Apel and H. Bohlman. 1998. Systemic and local induction of an *Arabidposis* thionin gene by wounding and pathogens. Plant Journal 14:285-295.

Vijayan, P., J. Shockey, C.A. Lévesque, R.J. Cook and J. Browse. 1998. A role for jasmonate in pathogen defense of *Arabidopsis*. Proceedings of the National Academy of Sciences USA 95:7209-7214.

Wasternak, C. and B. Parthier. 1997. Jasmonate-signaled plant gene expression. Trends in Plant Science 2:302-307.

Weber, H., B.A. Vivk and E.E. Farmer. 1997. Dinor-oxo-phytodienoic acid: A new hexadecanoid signal in the jasmonate family. Proceedings of the National Academy of Sciences USA 94:10473-10478.

Weete, J.D. and D.J. Weber. 1980. Lipid Biochemistry of Fungi and other Organisms. Plenum Press, New York.

Weiler, E.W., T. Albrecht, B. Groth, Z-Q. Xia, M. Luxem, H. Li, L. Andert. and P. Spengler. 1993. Evidence for the involvement of jasmonates and their octadecanoid precursors in the tendril coiling response of *Bryonia doica*. Phytochemistry 32:591-600.

Xie, D.-X., B.F. Feys, S. James, M. Nieto-Rostro and J.G. Turner. 1998. *COI1*: An *Arabidopsis* gene required for jasmonate-regulated defense and fertility. Science 280:1091-1094.

A Survey of Herbivore-Inducible Defensive Proteins and Phytochemicals

C. Peter Constabel

Abstract

The defense of plants against herbivory is often active and involves the rapid accumulation of proteins and phytochemicals which prevent or reduce further herbivore damage. In this chapter, herbivore- and wound-induced proteins and phytochemicals are surveyed. The diversity of wound-induced proteins is substantial; prominent among these are protease inhibitors, oxidative enzymes, cell wall proteins, and phenylpropanoid enzymes. Reports of induced phytochemicals are more restricted, but cover a variety of phenolic compounds and terpenoids, as well as some alkaloids and glucosinolates.

Introduction

Higher plants, being stationary organisms, have evolved the ability to persist under unfavorable and stressful conditions. Plants therefore possess biochemical defense mechanisms which can deter, poison, or starve herbivores that feed on them. Toxic proteins and metabolites are commonplace in the plant kingdom, and the diversity of phytochemicals found in the plant kingdom is due in part to the selective pressure of herbivory. The active nature of plant defense was emphasized by the discovery that wounding tomato leaves stimulates the rapid accumulation of proteinase inhibitor proteins (Green and Ryan 1972), indicating that plants may adaptively respond to herbivores with inducible defense mechanisms.

This chapter presents an overview of the range of proteins and phytochemicals induced by herbivory and wounding. Induced accumulation of a given metabolite following herbivory, while not proof of a defensive role, is a first clue. Here the emphasis is on describing the biochemical responses to herbivory rather than their specific effects on individuals (see also Agrawal, Zangerl, this volume) or populations (Underwood, this volume); the objective is to provide a starting point for more detailed analyses of how individual plant-herbivore interactions are mediated. Herbivory is a type of wounding, and for practical reasons it is often preferable to effect damage by mechanical means rather than using herbivores. However, plants are clearly capable of distinguishing pliers from live caterpillars, dramatically demonstrated by the specificity of volatiles released from plants (De Moraes et al. 1998), and by the ultimate fitness consequences of clipping versus real herbivore damage (Agrawal 1998). Thus, artificial wounding can only partially mimic herbivory. In tomato and some other species, the response to herbivores can also be artificially induced by jasmonic acid or other jasmonates (Sembdner and Parthier 1993, Weiler 1997, Thaler, this volume). These chemicals have been used as convenient tools for the study of defense responses ever since their implication in proteinase inhibitor induction in tomato (Farmer and Ryan 1990). However, the biological effects of jasmonates have been observed in areas beyond herbivore defense signaling. Jasmonates are known to trigger responses to other stresses, including pathogens, elicitors, drought, aluminum, and mechanical pressure (Gundlach et al. 1992, Sembdner and Parthier 1993, Weiler 1997, Xiang and Oliver 1998, Staswick and Lehman, this volume). Thus, a number of jasmonate-induced responses do not strictly represent herbivore defense, and plant products induced solely by jasmonates without other evidence of herbivore- or wound-induction are not included in this overview.

The induced defenses surveyed include both proteins and secondary plant metabolites, here referred to as phytochemicals. Phytochemicals are organic molecules of low molecular weight, which by definition are not required for normal physiological processes in growth and development, and which tend to be more species-specific in distribution than primary metabolites. Research on herbivore defense has often been aimed at phytochemicals rather than proteins, perhaps due to their intriguing structural diversity and generally higher biological activities (Duffey and Stout 1996). In the context of induction by herbivores, proteins must also be considered as important mediators of plant defense (Table 1).

Protein-Based Defenses

Proteins as defense mechanisms have an advantage over phytochemicals from a biotechnological viewpoint, in that each protein is encoded by a single gene which can be isolated and used to genetically engineer crops for enhanced pest resistance. This strategy is being actively pursued by the agricultural biotechnology industry, and is driving the search for novel and more potent defensive proteins in plants as well as other organisms (Boulter 1993, Estruch et al. 1997). Due to the complexity of the biosynthetic pathways of most other secondary plant metabolites, it is as yet impractical to genetically engineer most phytochemical defense mechanisms.

The power of molecular biology has led to many analyses of defense responses at the level of gene expression rather than direct assays of the encoded proteins. Since regulation of gene expression appears to be the primary mechanism for the induction of defense proteins in plants, induction demonstrated at the level of the mRNA is considered adequate for the purpose of this survey.

Protease Inhibitors

Protease Inhibitors (PIs) are proteins that tightly bind proteolytic enzymes and thereby inhibit their activity. PIs are undoubtedly among the most studied anti-herbivore proteins of plants, and have been attributed to a wide range of species (Ryan 1990, Richardson 1991). Well-protected tissues such as potato tubers or the seeds of soybean and other legumes have long been known to contain high constitutive levels of PIs (Richardson 1991). PIs accumulate in tomato leaves in response to herbivore attacks within hours of damage, the observation that first demonstrated the dynamic nature of plant defense against herbivores (Green and Ryan 1972). Since the induced PI accumulation occurs plant-wide, i.e., systemically, in response to local herbivory, this phenomenon stimulated much research on signal transduction in plants. This research ultimately led to the identification of the systemic wound signal systemin and

Table 1. Defense proteins induced by herbivory and wounding.

Serine Protease Inhibitors:
- potato inhibitor I
- potato inhibitor II
- Bowman-Birk
- Kunitz

Other Protease Inhibitors:
- cystatin (cysteine PI)
- cathepsin D inhibitor (aspartic PI)

Oxidative Enzymes:
- polyphenol oxidase
- peroxidase
- lipoxygenase
- ascorbate oxidase

Cell Wall Proteins:
- hydroxyproline-rich glycoprotein (HRGP)
- glycine-rich protein (GRP)
- proline-rich protein (PRP)

Phenylpropanoid and Related Enzymes:
- phenylalanine ammonia lyase (PAL)
- 4-cinnamic acid hydroxylase (4-CH)
- caffeic acid O-methyltransferase
- 4-coumarate CoA ligase (4CL)
- 3-deoxy-D-arabino-heptulosonate-7-phosphate (DAHP) synthase
- shikimate dehydrogenase

Lectins and carbohydrate-binding proteins:
- chitin-binding proteins
- *Brassica* lectin-like gene
- class I chitinase

other components of the signal transduction cascade in the tomato plant (Schaller and Ryan 1995). More detailed knowledge of the signaling pathway in turn led to the identification of new defense proteins in tomato (Constabel et al. 1995, Bergey et al. 1996).

The efficacy of specific PIs as a defense was unequivocally shown using PI-overexpressing transgenic plants (Hilder et al. 1987, Johnson et al. 1989). The metabolic consequences for insect herbivores ingesting a diet with high concentrations of PI are thought to include a lack of available amino acids, which may lead to oversecretion of trypsin and further loss of sulfur amino acids (Ryan 1990). Ingestion of large amounts of PIs may also have toxic and lethal effects (Duffey and Stout 1996). Furthermore, plant PIs are also known to inhibit microbial proteases, and since PIs may also be pathogen-induced they may play a role in defense against pathogens.

Mechanistically, PIs are classified as inhibitors of serine, cysteine, aspartic, or metallo- proteases (Ryan 1990). Within each class, a PI may belong to one of several protein families based on amino acid sequence. The great majority of known plant PIs are inhibitors of serine proteases, perhaps reflecting the predominance of this type of protease in herbivore digestive systems. However, PIs of cysteine proteases are also widely distributed; beetles use cysteine proteases in their digestive tracts. So far, there are fewer examples of aspartic and metalloprotease inhibitors from plants.

Serine protease inhibitors. At least eight distinct families of protease inhibitors of serine proteases have been identified (Ryan 1990, Richardson 1991), and wound- and herbivore-induced examples are known from the potato inhibitor I, potato inhibitor II, Bowman-Birk, and Kunitz families. Serine PIs often occur at high constitutive levels in the edible seeds of legumes and cereals, for example soybean, peanut, rice and wheat (Richardson 1991). Some PIs also possess α-amylase inhibitory activity; however, to date, none of these bifunctional inhibitor proteins has been shown to be herbivore-induced.

The potato Inhibitor I and Inhibitor II families are defined by proteins first isolated from potato tubers, which were later shown to be systemically wound-inducible in potato and tomato leaves (Ryan 1990). Inhibitor I in tomato and potato is a protein of approximately 8 kDa and is active against chymotrypsin. Wound-induced homologs are less common but are found in tobacco (Lindhorst et al. 1993) as well as in maize (Cordero et al. 1994). Potato and tomato Inhibitor II are slightly larger proteins (12 kDa) and have dual activity against both trypsin and chymotrypsin (Ryan 1990). Wound-induced homologs also occur in tobacco (Pearce et al. 1993). The anti-herbivore potential of tomato Inhibitor II was clearly demonstrated when *Manduca sexta* larvae showed reduced growth while feeding on transgenic tobacco overexpressing this gene (Johnson et al. 1989). By contrast, Inhibitor I-overexpressing plants showed no additional protection against this pest.

Bowman-Birk PIs are also relatively small proteins (8-9 kDa) commonly found in seeds of many legumes and cereals (Richardson 1991), but among leguminous plants only alfalfa shows wound-induction of this type of PI (Brown et al. 1985). More recently, Rohrmeier and Lehle (1993) reported the cloning of a systemically wound-inducible gene from maize with similarity to a Bowman-Birk PI. The efficacy of Bowman-Birk PIs in plant defense was shown by overexpressing the cowpea PI in transgenic tobacco, one of the first applications of genetic engineering for enhancing plant defense (Hilder et al. 1987). Transgenics suffered significantly less damage from *Heliothis virescens* than control plants, although later experiments using transgenic potato with the tomato moth, *Lacanobia oleracea* showed a negative effect on insect growth but no reduction in damage as a result of the transgene (Gatehouse et al. 1995).

The Kunitz-type PI was first characterized in soybean, but also occurs in other legumes and cereals (Richardson 1991). Wound-inducible homologs of Kunitz-type PIs are found in poplar (Bradshaw et al. 1989) and sweet potato (Yeh et al. 1997).

Cysteine protease inhibitors. As with the serine PIs, cereal and legume seeds often contain cysteine PIs, or cystatins (Richardson 1991). Rice seed and potato tuber cystatins have been have been particularly well characterized (Abe et al. 1992, Walsh and Strickland 1993). Cystatins are also constitutively expressed in other organs such as in leaves of *Sorghum bicolor* (Li et al. 1996), as well as in leaves, flowers and roots of *Brassica campestris* (Lim et al. 1996). Wound-inducibility of cystatin genes was demonstrated in leaves of potato and soybean (Hildmann et al. 1992, Bergey et al. 1996, Botella et al. 1996). These results suggest that cysteine PIs are components of the inducible defense arsenal of plants, and cysteine PIs were shown to have inhibitory activity against pest insects in artificial diet feeding assays using *Diabrotica* and *Tribolium* (Orr et al. 1994, Edmonds et al. 1996). In addition, transgenic potato and rice plants overexpressing rice and maize cystatin genes, respectively, showed an increased inhibitory activity against digestive proteases of coleopteran test insects (Benchekroun et al. 1995, Irie et al. 1996).

Inhibitors of other proteases. Inhibitors of aspartic proteases (Cathepsin D inhibitors) have been identified in potato tubers where they are expressed as storage proteins. These inhibitors were also shown to be wound-induced in leaves of tomato and potato (Hannapel 1993, Bergey et al. 1996). Cathepsin D inhibitors constitute a separate class of PIs, but show some sequence similarity to the Kunitz-type PIs discussed above. Several other classes of PIs are known from plants, many of them from seeds (Ryan 1990, Richardson 1991). Although they undoubtedly have defensive roles, to date they have not been shown to be induced by herbivore attack.

The widespread distribution of PIs in vulnerable storage tissues, their wound- and herbivore-inducibility in plants from diverse plant families, and the protective effects of overexpressing PIs in transgenic plants demonstrate the important role of PIs as defense proteins against herbivores. However, PIs are very specific for their protease targets, so that any one serine PI, for example, will inhibit only some of the serine proteases that may potentially be encountered (Ryan 1990). Therefore, individual PIs need to be tested against specific gut proteases and evaluated with specific insect pests. In addition, distinct proteases with different susceptibilities to any one PI are found in the gut of the same insects, and some insect pests actively alter their digestive protease profile in response to high concentrations of PIs in their diet. Jongsma et al. (1995) first demonstrated that *Spodoptera exigua* larvae feeding on PI II-overexpressing transgenic tobacco are able to adapt to this PI in the diet by producing new inhibitor-resistant trypsin activities. Broadway (1996) found similar changes in gut proteases using cabbage PIs and *Pieris rapae* and *P. napi*, and Bolter and Jongsma (1995) demonstrated adaptive shifts in digestive cysteine protease specificity in the Colorado potato beetle. These counter-adaptations have added a fascinating layer of complexity to the study of plant protease inhibitors and in our knowledge of co-evolution of plants with their insect herbivores.

Oxidative Enzymes

Plants contain a diverse set of stress-associated oxidative enzymes including phenol oxidases, peroxidases, and lipoxygenases (Butt 1980). The potential of oxidative enzymes as anti-nutritive defenses against insect herbivores has been recognized largely through the work of Duffey and coworkers (Felton et al. 1989, Duffey and Felton 1991, Duffey and Stout 1996). Support for a role of oxidative enzymes in anti-herbivore defense is provided by the induction of polyphenol oxidase and lipoxygenase in tomato by systemin, jasmonates, and the octadecanoid defense signaling pathway (Constabel et al. 1995, Thaler et al. 1996, Heitz et al. 1997). The anti-nutritive effects of all oxidative enzymes is based on their ability to destroy essential nutrients in insect diets. Importantly, the efficacy of oxidative enzymes as a defense generally depends on the chemical environment in the insect gut, such as the redox potential and the levels of anti-oxidants which rapidly reduce the reactive products. Likewise, the quality and quantity of protein in the diet strongly influence the effects of oxidative enzymes on the herbivore (Duffey and Stout 1996).

Polyphenol Oxidase. Polyphenol oxidases (PPOs) are enzymes which use molecular oxygen to catalyze the oxidation of monophenolic and *ortho*-diphenolic compounds. Surveys of PPO in plants have detected PPO activity in essentially all vascular plants examined, although at highly variable levels

(Sherman et al. 1991, Constabel and Ryan 1998). PPO is often expressed at higher levels in diseased and wounded tissues (Mayer and Harel 1979), although a recent survey detected wound-inducible PPO in leaves of only four of 20 species tested (Constabel and Ryan 1998). The availability of PPO cDNA probes has confirmed wound-induction at the level of PPO mRNA in potato, tomato, apple, and hybrid poplar (summarized in Constabel et al. 1996, Constabel and Christopher, unpublished data). In tomato, PPO is induced by caterpillar feeding, jasmonates, and mechanical damage, but not by mites or leafminers (Stout et al. 1994, Constabel et al. 1995, Thaler et al. 1996).

The products of the PPO-catalyzed reaction are highly reactive *ortho*-quinones which can spontaneously polymerize and cross-link with other biomolecules. This leads to the commonly observed browning phenomenon in damaged and diseased plant tissues, and PPO has in fact been hypothesized to have a role in wound healing and disease resistance (summarized in Constabel et al. 1996). During insect feeding and the resulting breakdown of compartmentalization, the enzymes gain access to their substrates, producing reactive *ortho*-quinones which readily alkylate electron-rich functions such as sulfhydryl and amine groups of proteins. These reactions result in protein modification, cross-linking, and precipitation, and thus prevent efficient digestion and assimilation of nitrogen by the insect, significantly impacting insect herbivores (Felton et al. 1992, Duffey and Stout 1996). For example, growth of the tomato fruit worm (*Helicoverpa zea*) and beet armyworm (*Spodoptera exigua*) was found to vary inversely with the quantity of PPO present in tomato tissue (Felton et al. 1989, Stout et al. 1996). In artificial diets, the addition of PPO and its substrate, chlorogenic acid, had severe impact on larval growth, but only if the oxidation occurred in the presence of the dietary protein (Felton et al. 1989, Duffey and Felton 1991).

Peroxidase. Peroxidases are widespread heme-containing enzymes which use H_2O_2 to oxidize a wide variety of biological substrates including phenolics, indole acetic acid, and ascorbate (Butt 1980). Peroxidase has been studied in various stress-related and developmental processes, which has resulted in the detection of several differentially-regulated isozymes. A primary role for peroxidase has been defined during lignification, where cell wall peroxidases are implicated in the generation of phenoxy radicals from hydroxycinnamyl alcohols which then polymerize nonenzymatically to form the lignin polymer (Douglas 1996). A similar role is proposed for suberization (Kollattukudy 1981). Furthermore, peroxidases are likely involved in the formation of cross-links between cell wall carbohydrate and proteins (Fry 1986, Cassab and Varner 1988).

Wounding commonly upregulates peroxidase, as documented for diverse plants including tomato, rice, peanut and bean (Breda et al. 1993, Felton et al. 1994b, Ito et al. 1994, Smith et al. 1994). Induction of peroxidase by insect herbivores has been observed in tomato leaves by Duffey and Felton (1991) and Stout et al. (1994, 1998). Like PPO, peroxidases are capable of oxidizing a variety of phenolics, and their presence in insect diets can lead to alkylation of dietary proteins as discussed above. A negative effect of peroxidase activity on insect growth was demonstrated by Duffey and Felton (1991), who manipulated the levels of H_2O_2 in insect diets with catalase. This experiment also underlined the dependence of peroxidase on a supply of H_2O_2 , and to date there is little evidence for adequate H_2O_2 levels in wounded plant tissue.

Based on its involvement in lignification and cross-linking of other cell wall components, peroxidase may also contribute to defense via cell wall reinforcement. This process is likely to increase leaf toughness and thus negatively impact insect performance (Coley 1983, see below). Robison and Raffa (1994) have observed an increase, both locally and systemically, in leaf toughness in poplar following damage by forest tent caterpillar.

Lipoxygenase. Lipoxygenases use molecular oxygen to oxygenate unsaturated fatty acids, such as linoleic and linolenic acid, producing fatty acid hydroperoxides (Gaillard and Chan 1980, Siedow 1991). These reactive and cytotoxic products do not accumulate *in planta* but are metabolized further to short chain alcohols and aldehydes, as well as cyclic compounds such as jasmonates or related oxylipins (Siedow 1991, Weiler 1997). Alternatively, fatty acid hydroperoxides undergo spontaneous decomposition to epoxides, and ketols as well as free radicals (Duffey and Felton 1991, Siedow 1991). Lipoxygenases are considered to be ubiquitous in plant tissues, yet are found at widely varying levels in different plant organs. Legume seeds and potato tubers show high levels of lipoxygenase activity (Gaillard and Chan 1980).

Lipoxygenases are primarily induced by pathogen attack, and to a lesser extent by wounding and herbivore damage (Siedow 1991). Wound-induction of lipoxygenase was demonstrated in leaves of *Arabidopsis*, potato, tomato, soybean and pea (Bell and Mullet 1991, 1993, Royo et al. 1996, Saravits and Siedow 1996, Heitz et al. 1997). In many cases, only some members of the lipoxygenase gene family are wound-induced while others are developmentally regulated, suggesting diverse roles for this enzyme. Induction by insect herbivory was observed in tomato (Stout et al. 1994), soybean (Felton et al. 1994b) and cotton (Bi et al. 1997).

Lipoxygenases have several potentially important roles in defense. A direct antinutritive effect on herbivores results from the destruction of polyunsaturated fatty acids, important nutrients for optimal growth of insects (Duffey and Stout 1996). Furthermore, the reaction of the lipoxygenase-generated fatty acid hydroperoxides (as well as other free radicals) with essential amino acids in dietary protein makes lipoxygenase an antinutritive defense analagous to PPO and peroxidase (Duffey and Felton 1991). These effects have been observed using insect bioassays and artificial diets; lipoxygenase plus linolenic acid resulted in reduced growth of *H. zea*, a chewing leaf eater, as well as *Nilaparvata lugens* and *Nephotettix cinciteps*, both sucking insects (Duffey and Felton 1991, Powell et al. 1993, Felton et al. 1994a). Lipoxygenase is also required for the synthesis of jasmonates (Staswick and Lehman, this volume). The central importance of lipoxygenase in signaling was demonstrated directly using transgenic *Arabidopsis* with reduced levels of one of the isozymes of lipoxygenase, LOX-2. These plants not only showed reduced levels of wound-induced jasmonic acid, but accumulated lower levels of wound-induced proteins (Bell et al. 1995).

Ascorbate oxidase. Ascorbic acid, or vitamin C, is a potent antioxidant and free radical scavenger. It is thus capable of reducing the PPO- and peroxidase-generated quinones, counteracting the alkylation of essential amino acids and preserving the nutritive value of the dietary protein (Duffey and Felton 1991, Duffey and Stout 1996). This has been demonstrated experimentally by adding ascorbate to artificial diets containing PPO plus phenolic substrates (Felton and Summers 1993). The level of ascorbic acid in a diet rich in oxidative enzymes is therefore of great importance in determining the degree of anti-nutritive effects and maintaining nutrition. Ascorbic acid levels are in turn very dependent on the enzyme ascorbate oxidase which converts it to its oxidized state as dihydroascorbic acid. The presence of this enzyme in insect diets would therefore enhance the effectiveness of other oxidative enzymes discussed above (Duffey and Felton 1991, Duffey and Stout 1996). Although this enzyme has not been widely investigated in the context of plant defense, induction by herbivory and wounding has been observed. Bi et al. (1997) measured significant increases in ascorbate oxidase in cotton following herbivory by *H. zea*, and soybean leaves accumulate increased levels of the enzyme after feeding by a phloem-feeding insect, *Spissistilus festinus* (Felton et al. 1994b).

Cell Wall Proteins

The cell wall of higher plants is a complex amalgam of carbohydrate, protein, and phenolics, all of which can be modified during stresses such as pest and pathogen attack (Bowles 1990, Carpita and Gibeaut 1993). Cell wall proteins are characterized by a highly repetitive structure. They are generally encoded by gene families, which are often differentially-regulated by both developmental and environmental cues (Ye and Varner 1991, Showalter 1993). Wounding of leaves or stems induces mRNA expression for several types of cell wall proteins including hydroxyproline-rich glycoproteins (HRGPs), glycine-rich proteins (GRPs), proline-rich proteins (PRPs), and arabinogalactan proteins (summarized in Showalter 1993). The best-characterized of the stress-induced proteins are the extensins, a type of hydroxyproline-rich glycoprotein (HRGP) characterized by the motif -Ser-$(Hyp)_4$-. Extensins are thought to provide intermolecular cross-links with cell wall components such as pectin, in addition to forming intra-molecular isodityrosine linkages (Sommer-Knudsen et al. 1998). Wound-induction of HRGPs mRNA has been demonstrated for many plants, including bean, potato, and tomato (Rumeau et al. 1990, Zhou et al. 1992, Wycoff et al. 1995). GRPs contain up to 70% glycine, predominantly in Gly-X repeating units where X is frequently Gly, and are induced by a variety of stresses including wounding. Their localization appears to correlate with vascular cells, in particular those that will become lignified (Showalter 1993). The proline-rich proteins (PRPs), also known as proline-rich HRGPs, have a high proline and hydroxyproline composition with Pro-Pro repeats contained within a variety of larger repeated motifs. There are a few reports of induction of PRPs by wounding (Showalter 1993).

Although changes in the cell wall are most often interpreted as preventing possible pathogen ingress (Bowles 1990), increasing cell wall toughness is also a potential impediment to insect chewing and other forms of herbivory. A major fuction of cell wall proteins is to cross-link other cell wall components and to provide a scaffold for other cross-linking molecules, and the induction of cell wall proteins is consistent with a role in defense via increasing wall toughness. An induced increase in the toughness of poplar leaves on plants which had been subjected to herbivory by forest tent caterpillars was observed by Robison and Raffa (1994). Similar results are reported by Raupp and Sadof (1991) in *Salix*. Increasing leaf toughness via stress-induced lignification and phenolic deposition has also been documented in plants defense (see below).

Enzymes of Phenylpropanoid Metabolism

Phenolics and phenylpropanoids form one of the largest classes of phytochemicals and their synthesis and accumulation constitute a common response to herbivory, wounding, and pathogen attack. Phenylpropanoids are derived from phenylalanine, which is successively deaminated, hydroxylated and methoxylated at one or more positions to give rise to several closely related hydroxycinnamic acids. Products of wound-induced phenylpropanoid metabolism are commonly lignin or lignin-like polymers important for cell wall reinforcement but may include other phenolics (see below). Several phenylpropanoid enzymes are wound-inducible in different experimental systems; in most cases the products themselves were not characterized. Inducible phenylpropanoids are discussed in a later section.

Phenylalanine ammonia lyase (PAL) catalyzes the deamination of phenylalanine to cinnamic acid, the entry and key regulatory step into the phenylpropanoid pathway. PAL is inducible by wounding and herbivory, pathogens, and abiotic stresses including UV light (Hahlbrock and Scheel 1989). Induction occurs via transcription of one or several PAL genes; well-studied wound-induced systems showing PAL induction include bean hypocotyls

(Cramer et al. 1989), tobacco leaves (Fukasawa-Akada 1996), potato tubers and leaves (Rumeau et al. 1990, Joos and Hahlbrock 1992), and parsley (Lois and Hahlbrock 1992). In addition to PAL, several other core phenylpropanoid enzymes and their genes are known to be induced by wounding. The 4-cinnamic acid hydroxylase (4-CH) gene is wound-induced in *Arabidopsis* and pea (Frank et al. 1996, Mizutani et al. 1997), as is the caffeic acid O-methyltransferase gene in maize (Capellades et al. 1996). 4-coumarate CoA ligase, an enzyme that activates the hydroxycinnamic acids via CoA ester formation, was shown to be wound-inducible in tobacco, *Arabidopsis*, and bean (Smith et al. 1994, Ellard-Ivey and Douglas 1996, Lee and Douglas 1996). In addition, a number of enzymes of the shikimic acid pathway, required for phenylalanine biosynthesis, are also induced by wounding. These include 3-deoxy-D-arabino-heptulosonate-7-phosphate (DAHP) synthase in wounded potato tubers (Dyer et al. 1989) and shikimate dehydrogenase from bell peppers (Diaz and Merino 1998).

Lectins, Carbohydrate-Binding Proteins, and Chitinases

Lectins are proteins with the ability to strongly bind carbohydrates, and were originally defined as proteins which agglutinate certain human erythrocytes. Peumans and van Damme (1995) have extended this definition to include all proteins which contain at least one carbohydrate-binding domain, and this usage will be adopted here. Many lectins are toxic to both insects and vertebrates, although to date only a few examples of this heterogeneous group of proteins are known to be herbivore- or wound-induced.

Direct evidence for an anti-herbivore role of lectins has come from insect feeding studies using purified lectins in artificial diets. For example, both wheat germ agglutinin and snowdrop (*Galanthus nivalis*) lectins strongly inhibit growth of sucking insects *Nilaparvata lugens* and *Nephotettix cinciteps* (Powell et al. 1993). Wheat germ agglutinin also has strong inhibitory effects on the cowpea weevil (Murdock et al. 1990). Snowdrop lectin is also effective against phloem-feeding insects in transgenic tobacco (Hilder et al. 1995). While the mechanism for these anti-insect effects is not entirely clear, it is presumably due to their carbohydrate-binding abilities. It has been suggested that the toxicity of chitin-binding lectins is mediated by a disruption of the peritrophic membrane lining the insect gut (Raikhel et al. 1993).

Lectins are especially abundant in seeds of cereals and legumes, with well-known examples from wheat and garden beans (Raikhel et al. 1993). Although wheat germ agglutinin is induced by both pathogenic and saprophytic fungi (Cammue et al. 1990), induction by herbivory has not been reported. However, in *Brassica*, lectin-like genes were found to be induced by wounding (Taipalensuu et al. 1997). Molecular analyses of other wound-induced genes have also revealed sequence similarity with known lectins. For example, chitin-binding domains known from wheat germ agglutinin were found in wound-induced genes of unknown function from potato and tobacco, and in the wound-induced protein hevein in latex of the rubber tree *Hevea brasiliensis* (Stanford et al. 1989, Broekart et al. 1990, Ponstein et al. 1994). The chitin-binding domains are also present in Ac-AMPs, small anti-fungal proteins from amaranth, as well as in class I chitinases from a variety of different species (Collinge et al. 1993, Raikhel et al. 1993). These lectin-like proteins are all wound-induced, although none have been directly shown to possess carbohydrate-binding or anti-herbivore activity. Chitinases are widespread among plants, and have been studied extensively as pathogen defenses. Although they can be induced by many different stimuli, their role in plant metabolism is poorly understood (Brederode et al. 1991). In tobacco they are also classified as belonging to the PR-3 family of pathogenesis-related proteins. Plant chitinases have been shown to be antifungal both *in vitro* and *in vivo* (Collinge et al. 1993, Raikhel et al. 1993), but again no anti-herbivore activity has been reported. Recently, however, chitinases from non-plant sources were shown to have potent anti-insect effects. A purified chitinase from the soil bacterium *Streptomyces albidoflavus* was active against a range of herbivorous insects (Broadway et al. 1998), and a *Manduca sexta* chitinase overexpressed in transgenic tobacco reduced growth of *Heliothis virescens* feeding on these plants (Ding et al. 1998).

Other Enzymes and Proteins

Many wound- or herbivore-induced genes encode proteins for which an anti-herbivore role is not immediately apparent. For example, in potato and tomato leaves threonine deaminase mRNA is strongly induced by wounding (Hildmann et al. 1992, Bergey et al. 1996). Its role in defense is not known, but it is required for biosynthesis of branched chain amino acids, used as precursors for the synthesis of glucose esters which protect leaves by entrapping small insects (Walters and Stephens 1990). However, to date there are no reports of a wound-induction of these compounds. Leucine aminopeptidase is also highly induced in this system but its defensive significance is not clear (Pautot et al. 1993). Recent studies in tomato have led to the discovery of a whole set of proteins which are systemically upregulated by wounding. These include a carboxypeptidase, an aspartic protease, a cysteine protease, and a ubiquitin-like protein, all proteins with possible involvement in stress-related protein turnover (Bergey et al. 1996). Other systemically induced proteins in tomato leaves have homology to proteins with known signaling functions. These include prosystemin, calmodulin, nucleotide diphospate kinase and acyl CoA-binding proteins (Bergey et al. 1996). Induction of signaling proteins by wounding may serve to amplify the wound signal.

Several wound-induced proteins of unknown biochemical function have been identified. This group includes the pathogenesis-related, or PR proteins, which are induced by viral, fungal and bacterial infection, as well as other stresses such as UV, ethylene and wounding (Bol et al. 1990, Brederode et al. 1991). In tobacco leaves, five families of PR-proteins have been characterized, but wound-induction occurs with only with the basic, but not acidic, forms (Brederode et al. 1991). The PR-2 and PR-3 families encode chitinases and β-glucanases, respectively. The anti-herbivore potential of chitinases has been described earlier, and that of β-glucanases is unknown. The PR-5 family includes proteins with some sequence similarity to a previously identified bifunctional protease inhibitor / alpha amylase inhibitor (Bol et al. 1990) although this activity has not been directly tested on the wound-induced proteins. Functions for PR-1 and PR-4 proteins are not yet clear but may include antifungal activities. Inducible proteins similar to the PR proteins have also been identified in other species including monocots (Bowles 1990). Other highly inducible proteins of unknown function include the wound- and pathogen-induced PR 10a family of proteins, representatives of which have been found in both monocots and dicots (Matton and Brisson 1989, Warner et al. 1992).

Phytochemical Defenses

The diversity of phytochemicals found in the plant kingdom is enormous, and many thousands of structures are known (Luckner 1990, Dey and Harborne 1997). Although phytochemicals perform many other ecological functions besides defense, co-evolution with insects is likely to have contributed to this chemical diversity (Harborne 1993). The breadth of distribution of phytochemicals is highly variable; for example, chlorogenic acid and other caffeic acid derivatives are found throughout the plant kingdom, whereas the opium alkaloids are specific to poppy and a few closely-related genera (Dey and Harborne 1997).

Phytochemicals have been widely investigated as constitutive anti-herbivore defenses, but demonstrations of phytochemical synthesis and accumulation as a result of herbivory and wounding are surprisingly limited considering the enormous number of compounds known (Table 2). By contrast, there is a richer literature on pathogen-induced respones, including phytoalexins (Kuć 1995, Hammerschmidt and Nicholson, this volume). Phytoalexins are generally not induced by wounding, and in most plants herbivore and pathogen responses are distinct and are controlled by separate induction mechanisms (see Stout and Bostock, this volume). Here the focus will be on plants and phytochemicals with clear wound- and herbivore-induction patterns, organized by structural class. Induction is here defined as *de novo* synthesis, and toxins, such as HCN, which are released from preexisting chemicals during tissue damage are not included. A more complete overview of the diversity of phytochemicals is found in Luckner (1990) and Dey and Harborne (1997). Induced volatiles, though clearly an induced response with potential benefit to the plant via the attraction of predators and parasitoids, will not be considered as

Table 2. Phytochemicals induced by herbivory and wounding.

Phytochemical	Species
Phenolics:	
Phenolic acids and conjugates (chlorogenic acid, tyramine conjugates)	lettuce, tomato, potato
Phenolic glycosides (salicortin, tremulacin)	trembling aspen
Furanocoumarins (xanthotoxin, bergapten)	*Pastinaca sativa*
Coumarins (scopoletin, ayapin)	sunflower, tobacco
Stilbenes (pinosylvin, pinosylvin methylether)	*Pinus radiata, Eucalyptus*
Tannins (condensed, hydrolyzable)	oak, birch, willow
Lignin	various
Total phenolics	oak, birch, trembling aspen
Terpenoids:	
Monoterpenes, diterpene acids	*Abies grandis,* other conifers
Diterpene aldehydes (hemigossypolone)	cotton
Steroidal alkaloids (solanidine)	potato
Phytoecdysteroids (20-hydroxyecdysone)	spinach
Cucurbitacins	*Cucurbita maxima,* other cucurbits
Alkaloids:	
Nicotine	tobacco
Quinolizidine	*Lupinus* spp.
Tropane (hyoscamine, apoatropine)	*Atropa acuminata*
Hydroxamic acids (DIMBOA, DIBOA)	wheat
Other:	
Indole glucosinolates	*Brassica napus,* several other mustards

these are discussed elsewhere in this book (see Paré et al., Sabelis et al., this volume).

Phenolic and Phenylpropanoid Metabolites

Plant phenolics comprise a diverse group of phytochemicals, ranging from small phenolic acids to complex polymers such as tannins and lignin (Dey and Harborne 1997). Phenolics are characterized by aromatic rings bearing one or more hydroxyl groups, and the majority of known compounds are derived from the shikimic acid and phenylpropanoid pathways described earlier. Of the major classes of phytochemicals, phenolics are the most commonly induced by stress such as excessive light, UV, cold, nutrient deficiencies, and attacks by herbivores and pathogens (Dixon and Paiva 1995). To date, only a minority of specific phenolics have been shown to be induced by herbivores.

Low molecular weight phenolics: phenolic acid conjugates, coumarins, stilbenes and flavonoids. All higher plants contain phenolic acids such as ferulic, caffeic, and *p*-coumaric acids, as these are precursors of lignin; in addition, many species also accumulate phenolic acids as conjugates with other organic compounds. For example, chlorogenic acid is a widespread caffeic acid conjugate; together with related phenolic conjugates, its accumulation was found to be induced by wounding of potato tubers (Hahlbrock and Scheel 1989) and lettuce (Loiza-Verde et al. 1997). Chlorogenic acid alone has mild anti-insect effects, but is further activated by polyphenol oxidase. Other common wound-induced phenolic acid conjugates include feruloyl- and p-coumaroyl-tyramine which accumulate in tomato leaves and wound periderm of potato tubers (Negrel et al. 1996, Pearce et al. 1998). These chemicals are thought to defend against fungal invasion but may also increase cell wall toughness. Phenolics are commonly found as glycosides, for example tremulacin and salicortin which are abundant in bark and leaves of *Salix* and *Populus* species. Trembling aspen leaves contain high concentrations of these phenolic glycosides, and wounding causes a modest increase in levels (Clausen et al. 1989, Lindroth and Kinney 1998). Ingestion of these chemicals has negative effects on performance of the lepidopteran aspen pests *Choristoneura conflictana*, *Malacosoma disstria*, and *Papilio* species (Lindroth and Hwang 1996).

Coumarins and furanocoumarins are cyclic derivatives of *p*-coumaric acid characteristic of the Umbelliferae and Rutaceae. Furanocoumarins have been studied in great detail as mediators of co-evolution between specialist insects and the wild parsnip *Pastinaca sativa* which contains a mixture of xanthotoxin, bergapten, and other furanocoumarins (Berenbaum and Zangerl 1996). Although xanthotoxin and related furanocoumarins are constitutively present in leaves of the wild parsnip, they are also induced several-fold by real or simulated herbivory (Zangerl 1990, Zangerl, this volume). These compounds possess a bewildering spectrum of biological activities including DNA crosslinking and many cytotoxic effects, and they have been shown to deter feeding or reduce growth of several leaf-eating insects, acting both individually and synergistically (Berenbaum and Zangerl 1996). Other inducible coumarins include scopoletin and ayapin. These compounds are induced in sunflower by the sunflower beetle *Zygogramma excalmationis* to which they act as feeding deterrents (Olson and Roseland 1991).

Stilbenes, *p*-coumaric acid-derived bicyclic compounds, are typical of forest trees including *Pinus* and *Eucalyptus*, where they are thought to protect the heartwood from fungal decay. They were subsequently found to be inducible by pathogens in sapwood of many conifers (Hart 1981). Wound-induction of the common stilbene pinosylvin was observed in *Pinus radiata*, although to lower levels than in pathogen-infected wood (Hillis and Inoue 1968). In the bark of *Picea abies*, wounding was shown to induced the biosynthetic enzyme stilbene synthase (Brignolas et al. 1995). While pinosylvin and other stilbenes have been investigated mostly for their antifungal attributes, they can have detrimental effects on insects (Hart 1981) and deter browsing by mammals (Bryant et al. 1991). Their antifungal activity may additionally protect the tree against the pathogenic fungi which are introduced into wounds by boring insects.

A very diverse and widespread group of phenylpropanoids are the flavonoids, which have significant ecological functions such as UV protection, flower pigmentation, phytoalexins, and microbial signaling. They appear to be much more commonly induced by abiotic rather than biotic stress. In the Fabaceae, however, isoflavonoids are pathogen-induced as phytoalexins. Chalcone synthase and chalcone isomerase, key biosynthetic enzymes specific to the flavonoid pathway are wound-inducible in some species, and there are a few examples of wound-induced flavonoids in the literature. Following mechanical wounding, *Petunia* stigmas accumulate kaempferol and related flavonols, (Vogt et al. 1994), and in Norway spruce bark wounding was shown to induce (+)-catechin. However, this compound is a precursor for condensed tannins which themselves may be herbivore-induced (see below). Flavonoids in general do not appear to have strong anti-herbivore activities, although there are prominent exceptions such the insecticidal rotenoids and estrogen-mimicking isoflavonoids (Harborne 1991).

Phenolic polymers: tannins and lignin. Tannins are large polyphenolics which range in molecular weight from 500-4000 Da and whose many hydroxyl groups interact with proteins, denaturing and precipitating them from solution. Tannins have a long association with chemical ecology and herbivore defense, and are often considered general feeding deterrents in plant-herbivore interactions (Swain 1979, Hagermann and Butler 1991). Although widespread, tannins are more prevalent in woody perennials than herbaceous plants (Swain 1979). They are generally grouped into one of two structural types: the condensed tannins, or proanthocyanidins, and the hydrolysable tannins, which are gallic acid or ellagic acid esters of various sugars. However, tannin structure varies greatly within each of these groups depending on species, and due caution must be employed when comparing the biological effects of tannins among different plants (Ayres et al. 1997). Tannins are often found at high constitutive levels in wood and bark, and typically at lower levels in leaves (Swain 1979). The finding that gypsy moth larvae induce elevated hydrolysable tannin levels in red oak leaves within several days (Schultz and Baldwin 1982) has been followed by reports of increased tannin levels in herbivore-damaged leaves of other plants, including other oak species, birch, willow, and *Acacia* (Wratten et al. 1984, Faeth 1991, Raupp and Sadof 1991, Furstenburg and van Hoven 1994). In other trees, for example trembling aspen, defoliation does not induce condensed tannin concentrations in remaining foliage above constitutive levels (Lindroth and Kinney 1998). In addition, tannin levels are often highly variable within a species and even in different leaves of the same individual, and thus the significance of smaller increases induced by herbivore attack has been questioned (Hartley and Lawton 1991). Ingestion of tannin-rich diets with more than 2% tannin content (per dry weight leaf tissue) has generally been considered to depress insect growth rate and food utilization efficiencies (Swain 1979, Hagerman and Butler 1991); it should be noted, however, that any anti-herbivore effects are likely to be dependent on specific tannin structures (Ayres et al. 1997). A deterrent effect on mammalian herbivores, proportional to tannin content of leaves, has also been demonstrated (Furstenberg and van Hoven 1994).

Lignin is a complex phenolic polymer found predominantly in the secondary cell wall of xylem and other vascular cells, where it functions to strengthen cell wall resistance to compression. Induced lignification has also been associated with wound repair (Kahl 1982, Rittinger et al. 1987) and as an active defense to reinforce cell walls against invading pathogens (Bostock and Stermer 1989, Nicholson and Hammerschmidt 1992). Wound-induced accumulation of lignin has been observed in leaves, but appears to be particularly prevalent in organs such as potato tubers (Borg-Olivier and Monties 1993), cucumber fruit (Hyodo et al. 1993) and the bark of several tree species (Rittinger et al. 1987, Biggs and Peterson 1990, Thomas et al. 1995). This is consistent with the observation that these organs have highly elaborated and complex wound repair mechanisms (Kahl 1982). The stimulation of lignin deposition by wounding may play an important protective role against herbivores, since leaf toughness and fiber content are important factors in determining the suitability of foliage for insect herbivores (see discussion of cell wall proteins).

Total phenolics. Many ecologists have used the Folin-Denis and other phenolic assay methods to quantify the total phenolic content of leaves (Waterman and Mole 1994). While such non-specific methods have obvious limitations, they are rapid and can provide an indication of chemical differences between samples. Herbivore-induced accumulation of total phenolics has been demonstrated in several woody plants including birch, oak and trembling aspen (Schultz and Baldwin 1982, Mattson and Palmer 1988, Bergelson and Lawton 1988, Hartley 1988). As discussed above for tannins, the induction often results in only a minor increase compared to the preexisting phenolic levels, which can themselves be quite variable. In addition, changes in specific compounds may not be detectable with non-specific phenolic assays, so that chemical analysis of specific phenolics is a preferable strategy. Excellent examples of this level of analytical work are Ayres et al. (1997) for tannins of 16 species, and Lindroth and Hwang (1996) for the phenolic glycosides of trembling aspen.

Terpenoids

Although the terpenoids are an extremely diverse family of phytochemicals, they share a common biosynthetic origin from isopentenoid precursors. Terpenoids are classified according to the number of ten-carbon units, giving rise to mono-, sesqui-, di, tri- and tetra-terpene chemicals. The functions for these chemicals are varied and include pigmentation and free radical scavenging, membrane components, growth regulators, and volatiles, as well as defensive compounds against pathogens and herbivores (Harborne 1993). Over 15,000 terpenoids have been described, and individual chemicals from all major groups of structures have been shown to have anti-herbivore activity (Gershenzon and Croteau 1991). A few of these defensive terpenoids have been shown to be herbivore-induced.

The release of oleoresin following bark beetle attacks and mechanical damage to bark is widespread among conifers and constitutes one of the best-characterized systems of induced phytochemical production (Raffa 1991). Depending on the species, the major oleoresin constituents are monoterpenes (including α– and β-pinene, α-phellandrene, sabinene, and limonene) and diterpene acids (Lewinsohn et al. 1991). The phenolic 4-allylanisole is also found in oleoresin of some conifers (Werner 1995). Whereas in some genera such as *Pinus*, the resin is preformed and flows into the wound zone, in *Abies* and *Picea* these terpenes are synthesized *de novo* in cells surrounding tissue damage (Lewinsohn et al. 1992). In *A. grandis* it was shown that the required mono-, sesqui-, and di-terpene cyclase enzymes and their mRNAs are induced by stem wounding (Steele et al. 1998). Following accumulation of the oleoresin in the wound, the mono- and sesquiterpene components are volatilized while the diterpene acids harden to seal off the wound and entrap the beetles; in addition, individual monoterpenes may be directly toxic to bark beetles (Raffa et al. 1985, Raffa 1991, Werner 1995). Increased monoterpene synthesis in response to herbivory was also observed in needles of some conifers, but there was no overall accumulation of terpenoids, which suggests an induction of volatile compounds (Litvak and Monson 1998). The release of damage-induced volatile terpenoids may be very common among plants. The discovery that parasitoids and predators of herbivores are specifically attracted to mixtures of these volatiles and use them to locate host insects has generated much interest (De Moraes et al. 1998, see Paré et al., Sabelis et al., this volume).

Several di- and tri-terpenoid phytochemicals with potent anti-insect activity are inducible by herbivory or wounding. The diterpenoid aldehydes gossypol and hemigossypolone in cotton have long been known to possess strong anti-herbivore activity, yet only recently was it shown that leaf damage can further induce hemigossypolone and terpenoid aldehyde accumulation in pigment glands (McAuslane et al. 1997). This induction correlated with increased deterrence of those leaves to beet armyworm. Highly effective triterpenoid defense phytochemicals also include the cucurbitacins, extremely bitter chemicals with both toxic and deterrent effects, which are typical of Cucurbitaceae. Although originally these phytochemicals were reported to be induced by cucumber beetle feeding (Tallamy 1985), in later reports this could not be corroborated (McCloud et al. 1995). Cucurbitacin induction in *Cucurbita maxima* by mechanical wounding, however, is dramatic, reaching a hundred-fold increase within 60 minutes (McCloud et al. 1995); the rapidity of this response suggests that cucurbitacins are translocated to the site of damage rather than synthesized *de novo.*

Other triterpenoid-derivatives induced by tissue damage include the toxic steroidal alkaloids solanidine and tomatidine, which accumulate in wounded potato tubers slices (Bostock and Stermer 1989). These molecules are synthesized *de novo* after wounding, and several of the required enzymes are also subject to induction (Bianchini et al. 1996). Interestingly, when challenged by pathogens in wound sites, the tuber activates a different terpenoid branch pathway leading to sesquiterpene phytoalexins, notably rishitin, while suppressing the pathway leading to the triterpenoid steroidal alkaloids. This was one of the first clear demonstrations of differential responses to pathogen versus wound stresses (Bostock and Stermer 1989). Phytoecdysteroids are another class of triterpenoid defensive plant chemicals, which can act as insect molting hormones and interfere with insect development. Although over 100 such compounds have been described from a variety of plant species (Harborne 1993), wound-induction appears to have been reported only in spinach, where 20-hydroxyecdysone accumulation in roots is stimulated by mechanical damage (Schmelz et al. 1998).

Alkaloids and Other Amino Acid-Derived Phytochemicals

The alkaloids contain some of the most potent bioactive phytochemicals known (Harborne 1993). Alkaloids as a group may be of divergent biosynthetic origins but are defined by the presence of a nitrogen-containing structure, which is important for many of the pharmacological effects. The toxicity of many alkaloids to animals, and especially their potential for interfering with nervous system functioning, suggests an important function for alkaloids in plant anti-herbivore defense (Hartmann 1991). An important example are the pyrrolizidine alkaloids, common in species of *Senecio*, which protect the plant against generalist herbivores; the sequestration of the alkaloids by some insects for their own defense underlines the defensive potential of these compounds (Hartmann et al. 1997).

Alkaloids are clearly important as constitutive defenses, yet as emphasized by Hartmann (1991), of the thousands of known alkaloids only a few examples of herbivore- and wound-induction have been reported. Several of these are from the Solanaceae, a plant family known for potent and diverse alkaloids. In *Nicotiana sylvestris* and *N. attenuata*, mechanical wounding, insect damage, or application of jasmonates can stimulate a five to ten-fold increase in nicotine alkaloid content of the leaves (reviewed in Baldwin 1999). Nicotine is synthesized in roots and transported to shoots, and the accumulation of the alkaloid in foliage correlates with decreased growth of *Manduca sexta* and *Trichoplusia ni* (Baldwin 1991). Furthermore, Baldwin (1998) showed that artificial induction of nicotine (and likely other defense responses) protected the induced plants from herbivory in a field situation, and that this came at a significant cost to the plant.

Other wound-inducible alkaloids from the Solanaceae include tropane alkaloids in *Atropa acuminata* which are induced in leaves by wounding or insect damage (Khan and Harborne 1991), and the steroidal alkaloids of potato discussed earlier. Many species of the Fabaceae contain potent alkaloids; *Lupinus polyphyllus* leaves accumulate quinolizidine alkaloids which are reported to be wound-induced (Wink 1983). Quinolizidine alkaloids in general have strong deterrent effects on mammalian and insect herbivores, and their toxic effects on livestock are well known (Hartmann 1991). In the Poaceae, the alkaloidal hydroxamic acids (DIMBOA and DIBOA) have been identified, and these also appear to be inducible by wounding or insect damage (Niemeyer 1988). Aphid damage on wheat leaves results in induction of the major hydroxamic acids several-fold (Niemeyer et al. 1989). As a general observation, alkaloid levels appear to be dependent on leaf and plant age, making wound-induction difficult to detect.

Plants of the Brassicaceae often accumulate glucosinolates, amino acid-derived sulfur-containing metabolites. Glucosinolate-derived breakdown products, including isothiocyanates, are rapidly released during tissue damage and maceration, and are considered toxic to non-specialist insects (Harborne 1993). Glucosinolates are typically expressed constitutively, although in some species, wound-, herbivore-, and jasmonate-induced increases have been observed (Koritsas et al. 1991, Birch et al. 1992, Bodnaryk 1992, 1994, Agrawal et al. 1999). It does not appear, however, that all glucosinolates are inducible; only the indole glucosinolate component of the glucosinolate profile is generally increased following damage, and not aliphatic and aromatic components.

Synthesis and Future Directions

This chapter has surveyed the spectrum of induced biochemical responses and herbivore defenses observed in higher plants. There is great diversity in defense mechanisms found in the plant kingdom; nevertheless, a few general trends can be discerned in the literature. Whether these are due to the evolutionary strategies of plants or the methodological biases of biologists is a question which remains to be answered.

1. Reports describing herbivore- and wound-induced defense proteins appear to be more prevalent than those focusing on induced phytochemicals. The protease inhibitors, in particular, stand out for their wide distribution and diversity, and their importance in defense has been clearly established. Many other herbivore-inducible proteins have also been described, yet their contribution to resistance is less certain. Demonstrating the significance of these induced proteins for plant defense against herbivores will be an important area for future research, one that will benefit from advances in transgenic plant technology.

Interestingly, a large number of phytochemicals with clear anti-herbivore activity are known, although few have been shown to be induced by herbivory. The induction of terpenoid oleoresin in wounded bark is typical of conifers, and could constitute the most widespread example of phytochemical induction. It is possible that future research will establish herbivore inducibility for other anti-herbivore phytochemicals. Progress in this area will require that the importance of developmental and environmental factors for defense responses is recognized and better understood.

2. Most protein-based defenses known to date have an anti-nutritive effect on herbivores, so that they destroy or prevent assimilation of nutrients by the insect and thereby slow its growth and development. This is in contrast to phytochemical defenses, which can act as toxins, feeding deterrents, and hormone mimics. However, since the roles of many inducible proteins still need to be defined, it is likely that novel mechanisms of anti-herbivore proteins still await discovery.

3. Overall there are no inherent differences between mechanisms of induced and constitutive anti-herbivore defense; many proteins such as protease inhibitors are constitutively expressed in tubers and seeds, but are synthesized only following herbivory or other damage in leaves. Likewise, induction of phytochemical defenses typically involve the induced accumulation of a pre-existing compound. This pattern supports the idea that more known constitutive anti-herbivore phytochemicals may turn out to be inducible under the right conditions or stages of plant development.

4. Although it appears likely all plants have some form of induced defense response to herbivory, no inducible herbivore defense mechanism described is universally distributed. Even though some defensive enzymes or pathways, such as polyphenol oxidase and the phenylpropanoid enzymes, are found in all higher plants, they are induced by herbivore damage in only some species. Clearly, different species have evolved distinct strategies for induced defense.

Given the diverse induced defense mechanisms found within the plant kingdom, the complexity of the defense response within a single species is an intriguing question. To date, there been few attempts to systematically identify all induced defenses within one plant. A plant for which this is being achieved is the tomato, where extensive work has shown that a broad battery of defense proteins is coordinately induced by wounding, wound signals, and herbivory (Bergey et al. 1996, Duffey and Stout 1996, Thaler, this volume). The induced defense includes four distinct protease inhibitor proteins, polyphenol oxidase and oxidative enzymes, prosystemin and other signaling-related genes, and several proteolytic enzymes of unknown defensive function. In addition to these proteins, the phenolic conjugates feruloyl-tyramine and coumaroyl-tyramine are also locally induced. Therefore, the induced defense of tomato shows both diversity and redundancy, perhaps an adaptation to a multitude of potential herbivores. A long-term challenge for future research in plant defense will be not only to discover novel defensive mechanisms in other species, but to determine how important diversity and redundancy is for inducible plant defense.

Acknowledgments

Research in the author's laboratory is supported by the Natural Sciences and Engineering Research Council (NSERC) of Canada.

Literature Cited

Abe, M., K. Abe, M. Kuroda, and S. Arai. 1992. Corn kernel cysteine proteinase inhibitor as a novel cystatin superfamily member of plant origin - Molecular cloning and expression studies. European Journal of Biochemistry 209:933-937.

Agrawal, A. A. 1998. Induced responses to herbivory and increased plant performance. Science 279:1201-1202.

Agrawal, A. A., S. Y. Strauss, and M. J. Stout. 1999. Costs of induced responses and tolerance to herbivory in male and female fitness components of wild radish. Evolution (In press).

Ayres, M.P., T.P. Clausen, S.F. MacLean, Jr., A.M. Redman, and P.B. Reichardt. 1997. Diversity of structure and antiherbivore activity in condensed tannins. Ecology 78:1696-1712.

Baldwin, I.T. 1991. Damage-induced alkaloids in wild tobacco. Pages 47-70 *in* D.W. Tallamy and M.J. Raupp, editors. Phytochemical Induction by Herbivores. John Wiley, New York.

Baldwin, I.T. 1998. Jasmonate-induced responses are costly but benefit plants under attack in native populations. Proceedings of the National Academy of Sciences USA 95:8113-8118.

Balwdin, I.T. 1999. Inducible nicotine production in native *Nicotiana* as an example of adaptive phenotypic plasticity. Journal of Chemical Ecology 25:3-30.

Bell, E., R.A. Creelman, and J. Mullet. 1995. A chloroplast lipoxygenase is required for wound-induced jasmonic acid accumulation in *Arabidopsis*. Proceedings of the National Academy of Sciences USA 92:8675-8679.

Bell, E. and J.E. Mullet. 1991. Lipoxygenase gene expression is modulated in plants by water deficit, wounding, and methyl jasmonate. Molecular and General Genetics 230:456-462.

Bell, E. and J.E. Mullet. 1993. Characterization of an *Arabidopsis* lipoxygenase gene responsive to methyl jasmonate and wounding. Plant Physiology 103:1133-1137.

Benchekroun, A., D. Michaud, B. Nguyen-Quoc, S. Overney, Y. Desjardins, and S. Yelle. 1995. Synthesis of active oryzacystatin I in transgenic potato plants. Plant Cell Reports 14:585-588.
Berenbaum, M.R. and A.R. Zangerl. 1996. Phytochemical diversity: adaptation or random variation? Pages 1-24 *in* J.T. Romeo, J.A. Saunders, and P. Barbosa, editors. Phytochemical Diversity and Redundancy in Ecological Interactions. Plenum Press, New York.
Bergelson, J.M. and J.H. Lawton. 1988. Does foliage damage influence predation on the insect herbivores of birch? Ecology 69:434-445.
Bergey, D.R., G.A. Howe, and C.A. Ryan. 1996. Polypeptide signaling for plant defensive genes exhibits analogies to defense signaling in animals. Proceedings of the National Academy of Sciences USA 93:12053-12058.
Bi, J.L., J.B. Murphy, and G.W. Felton. 1997. Antinutritive and oxidative components as mechanisms of induced resistance in cotton to *Helicoverpa zea*. Journal of Chemical Ecology 23:97-117.
Bianchini, G.M., B.A. Stermer, and N.L. Paiva. 1996. Induction of early mevalonate pathway enzymes and biosynthesis of end products in potato (*Solanum tuberosum*) tubers by wounding and elicitation. Phytochemistry 42:1563-1571.
Biggs, A.R. and C.A. Peterson. 1990. Effect of chemical applications to peach bark wounds on accumulation of lignin and suberin and susceptibility to *Leucostoma persoonii*. Phytopathology 80:861-865.
Birch, A.N.E., D.W. Griffiths, R.J. Hopkins, W.H.M. Smith, and R.G. McKinlay. 1992. Glucosinolate response of swede, kale, forage and oilseed rape to root damage by turnip root fly (*Delia floralis*) larvae. Journal of the Science of Food and Agriculture 60:1-9.
Bodnaryk, R.P. 1992. Effects of wounding on glucosinolates in the cotyledons of oilseed rape and mustard. Phytochemistry 31:2671-2677.
Bodnaryk, R. P. 1994. Potent effect of jasmonates on indole glucosinolates in oilseed rape and mustard. Phytochemistry 35:301-305.
Bol, J.F., H.J.M. Linthorst, and B.J.C. Cornelissen. 1990. Plant pathogenesis-related proteins induced by virus infection. Annual Review of Phytopathology 28:113-138.
Bolter, C.J. and M.A. Jongsma. 1995. Colorado potato beetles (*Leptinotarsa decemlineata*) adapt to proteinase inhibitors induced in potato leaves by methyl jasmonate. Journal of Insect Physiology 41:1071-1078.
Borg-Olivier, O. and B. Monties. 1993. Lignin, suberin, phenolic acids and tyramine in the suberized, wound-induced potato periderm. Phytochemistry 32:601-606.
Bostock, R.M. and B.A. Stermer. 1989. Perspectives on wound healing in resistance to pathogens. Annual Review of Phytopathology 27:343-371.
Botella, M.A., Y. Xu, T.N. Prabha, Y. Zhao, M.L. Narasimhan, K.A. Wilson, S.S. Nielsen, R.A. Bressan, and P.M. Hasegawa. 1996. Differential expression of soybean cysteine proteinase inhibitor genes during development and in response to wounding and methyl jasmonate. Plant Physiology 112:1201-1210.
Boulter, D. 1993. Insect pest control by copying nature using genetically engineered crops. Phytochemistry 34:1453-1466.
Bowles, D.J. 1990. Defense-related proteins in higher plants. Annual Review of Biochemistry 59:873-907.
Bradshaw, H.D.Jr., J.B. Hollick, T.J. Parsons, H.R.G. Clarke, and M.P. Gordon. 1989. Systemically wound-responsive genes in poplar trees encode proteins similar to sweet potato sporamins and legume Kunitz trypsin inhibitors. Plant Molecular Biology 14:51-59.
Breda, C., D. Buffard, R.B. van Huystee, and R. Esnault. 1993. Differential expression of two peanut peroxidase cDNA clones in peanut plants and cells in suspension culture in response to stress. Plant Cell Reports 12:268-272.
Brederode, F.T., H.J.M. Linthorst, and J.F. Bol. 1991. Differential induction of acquired resistance and PR gene expression in tobacco by virus infection, ethephon treatment, UV light and wounding. Plant Molecular Biology 17:1117-1125.
Brignolas, F., B. Lacroix, F. Lieutier, D. Sauvard, A. Drouet, A.C. Claudot, A. Yart, A.A. Berryman, and E. Christiansen. 1995. Induced responses in phenolic metabolism in two Norway spruce clones after wounding and inoculations with *Ophiostoma polonicum*, a bark beetle-associated fungus. Plant Physiology 109:821-827.

Broadway, R.M. 1996. Dietary proteinase inhibitors alter complement of midgut proteases. Archives of Insect Biochemistry and Physiology 32:39-53.

Broadway, R.M., C. Gongora, W.C. Kain, J.P. Sanderson, J.A. Monroy, K.C. Bennett, J.B. Warner, and M.P. Hoffmann. 1998. Novel chitinolytic enzymes with biological activity against herbivorous insects. Journal of Chemical Ecology 24:985-998.

Broekaert, W., H.I. Lee, A. Kush, N.-C. Chua, and N. Raikhel. 1990. Wound-induced accumulation of mRNA containing a hevein sequence in laticifers of rubber tree (*Hevea brasiliensis*). Proceedings of the National Academy of Sciences USA 87:7633-7637.

Brown, W.E., K. Takio, K. Titani, and C.A. Ryan. 1985. Wound-induced trypsin inhibitor in alfalfa leaves: identity as a member of the Bowman-Birk inhibitor family. Biochemistry 24:2105-2108.

Bryant, J.P., F.D. Provenza, J. Pastor, P.B. Reichardt, T.P. Clausen, and J.T. du Toit. 1991. Interactions between woody plants and browsing mammals mediated by secondary metabolites. Annual Review of Ecology and Systematics 22:431-446.

Butt, V.S. 1980. Direct oxidases and related enzymes. Pages 81-123 *in* E.E. Conn and P.K. Stumpf, editors. The Biochemistry of Plants, Vol.2. Academic Press, New York.

Cammue, B.P.A., W.J. Broekaert, and W. Peumans. 1990. Wheat germ agglutinin in wheat seedling roots: induction by elicitors and fungi. Plant Cell Reports 9:264-267.

Capellades, M., M.A. Torres, I. Bastisch, V. Stiefel, F. Vignols, W.R. Bruce, D. Peterson, P. Puigdomenech, and J. Rigau. 1996. The maize caffeic acid O-methyltransferase gene promoter is active in transgenic tobacco and maize plant tissues. Plant Molecular Biology 31:307-322.

Carpita, N.C. and D.M. Gibeaut. 1993. Structural models of primary cell walls in flowering plants: consistency of molecular structure with the physical properties of the walls during growth. Plant Journal 3:1-30.

Cassab, G. I. and J.E. Varner. 1988. Cell wall proteins. Annual Review of Plant Physiology and Plant Molecular Biology 39:321-353.

Clausen, T.P., P.B. Reichardt, J.B. Bryant, R.A. Werner, K. Post, and K. Frisby. 1989. Chemical model for short-term induction in quaking aspen (*Populus tremuloides*) foliage against herbivores. Journal of Chemical Ecology 15:2335-2346.

Coley, P.D. 1983. Herbivory and defensive characteristics of tree species in a lowland tropical forest. Ecological Monographs 53:209-233.

Collinge, D.B., K.M. Kragh, J.D. Mikkelsen, K.K. Nielsen, U. Rasmussen, and K. Vad. 1993. Plant chitinases. Plant Journal 3:31-40.

Constabel, C.P. and C.A. Ryan. 1998. A survey of wound- and methyl jasmonate-induced leaf polyphenol oxidase in crop plants. Phytochemistry 47:507-511.

Constabel, C.P., D.R. Bergey, and C.A. Ryan. 1995. Systemin activates synthesis of wound-inducible tomato leaf polyphenol oxidase via the octadecanoid defense signaling pathway. Proceedings of the National Academy of Sciences USA 92:407-411.

Constabel, C.P., D.R. Bergey, and C.A. Ryan. 1996. Polyphenol oxidase as a component of the inducible defense response in tomato against herbivores. Pages 231-252 *in* J.T Romeo, J.A. Saunders, and P. Barbosa, editors. Phytochemical Diversity and Redundancy in Ecological Interactions. Plenum Press, New York.

Cordero, M.J., D. Raventos, and B. San-Segundo. 1994. Expression of a maize proteinase inhibitor gene is induced in response to wounding and fungal infection: systemic wound-response of a monocot gene. Plant Journal 6:141-150.

Cramer, C., K. Edwards, M. Dron, X. Liang, S.L. Dildine, G.P. Bolwell, R.A. Dixon, C.J. Lamb, and W. Schuch. 1989. Phenylalanine ammonia-lyase gene organization and structure. Plant Molecular Biology 12:367-383.

De Moraes, C.M., W.J. Lewis, P.W. Pare, H.T. Alborn, and J.H. Tumlinson. 1998. Herbivore-infested plants selectively attract parasitoids. Nature 393:570-573.

Dey, P.M. and J.B. Harborne. 1997. Plant biochemistry. Academic Press, London.

Diaz, J. and F. Merino. 1998. Wound-induced shikimate dehydrogenase and peroxidase related to lignification in pepper (*Capsicum annuum*) leaves. Journal of Plant Physiology 152:51-57.

Ding, X., B. Gopalakrishnan, L.B. Johnson, F.F. White, X. Wang, T.D. Morgan, K.J. Kramer, and S. Muthukrishnan. 1998. Insect resistance of transgenic tobacco expressing an insect chitinase gene. Transgenic Research 7:77-84.

Dixon, R.A. and N. L. Paiva. 1995. Stress-induced phenylpropanoid metabolism. Plant Cell 7:1085-1097.
Douglas, C.J. 1996. Phenylpropanoid metabolism and lignin biosynthesis: from weeds to trees. Trends in Plant Science 1:171-178.
Duffey, S.S. and M.J. Stout. 1996. Antinutritive and toxic components of plant defense against insects. Archives of Insect Biochemistry and Physiology 32:3-37.
Duffey, S.S and G.W. Felton. 1991. Enzymatic antinutritive defenses of the tomato plant against insects. Pages 167-197 *in* P.A. Hedin, editor. Naturally occurring pest bioregulators. ACS Press, Washington, D.C.
Dyer, W.E., J.M. Henstrand, A.K. Handa, and K.M. Herrmann. 1989. Wounding induces the first enzyme of the shikimate pathway in Solanaceae. Proceedings of the National Academy of Sciences USA 86:7370-7373.
Edmonds, H.S., L.N. Gatehouse, V.A. Hilder, and J.A. Gatehouse. 1996. The inhibitory effects of the cysteine protease inhibitor, oryzacystatin, on digestive proteases and on larval survival and development of the southern corn rootworm (*Diabrotica undecimpunctata howardi*). Entomologia Experimentalis et Applicata 78:83-94.
Ellard-Ivey, M. and C. Douglas. 1996. Role of jasmonates in the elicitor- and wound-inducible expression of defense genes in parsley and transgenic tobacco. Plant Physiology 112:183-192.
Estruch, J.J., N.R. Carozzi, N.B. Desai, N.B. Duck, G.W. Warren, and M.G. Koziel. 1997. Transgenic plants: an emerging approach to pest control. Nature Biotechnology 15:137-141.
Faeth, S.H. 1991. Variable induced responses: direct and indirect effects on oak folivores. Pages 293-323 *in* D.W. Tallamy and M.J. Raupp, editors. Phytochemical Induction by Herbivores. John Wiley, New York.
Farmer, E.E. and C.A. Ryan. 1990. Interplant communication: airborne methyl jasmonate induces synthesis of proteinase inhibitors in plant leaves. Proceedings of the National Academy of Sciences USA 87:7713-7716.
Felton, G.W., J.L. Bi, C.B. Summers, A.J. Mueller, and S.S. Duffey. 1994a. Potential role of lipoxygenases in defense against insect herbivory. Journal of Chemical Ecology 20:651-666.
Felton, G.W., K.K. Donato, R.M. Broadway, and S.S. Duffey. 1992. Impact of oxidized plant phenolics on the nutritional quality of dietary protein to a noctuid herbivore, *Spodoptera exigua*. Journal of Insect Physiology 38:277-285.
Felton, G.W. and C.B. Summers. 1993. Potential role of ascorbate oxidase as a plant defense protein against insect herbivory. Journal of Chemical Ecology 19:1553-1568.
Felton, G.W., C.B. Summers, and A.J. Mueller. 1994b. Oxidative responses in soybean foliage to herbivory by bean leaf beetle and three-cornered alfalfa hopper. Journal of Chemical Ecology 20:639-650.
Felton, G.W., K.K. Donato, R.J. Del Vecchio, and S.S. Duffey. 1989. Activation of foliar oxidases by insect feeding reduces nutritive quality of dietary protein for foliage for noctuid herbivores. Journal of Chemical Ecology 15:2667-2693.
Frank, M.R., J.M. Deyneka, and M.A. Schuler. 1996. Cloning of wound-induced cytochrome P450 monooxygenases expressed in pea. Plant Physiology 110:1035-1046.
Fry, S.C. 1986. Cross-linking of matrix polymers in the growing cell walls of angiosperms. Annual Review of Plant Physiology and Plant Molecular Biology 37:165-186.
Fukasawa-Akada, T., S. Kung, and J.C. Watson. 1996. Phenylalanine ammonia-lyase gene structure, expression, and evolution in *Nicotiana*. Plant Molecular Biology 30:711-722.
Furstenburg, D, and W. van Hoven. 1994. Condensed tannin as anti-defoliate agent against browsing by giraffe (*Giraffa camelopardalis*) in the Kruger National Park. Comparative Biochemistry and Physiology 107A:425-431.
Galliard, T. and H.W.-S. Chan. 1980. Lipoxygenases. Pages 132-161 *in* E.E. Conn and P.K. Stumpf, editors. The Biochemistry of Plants, Vol. 4. Academic Press, New York.
Gatehouse, A.M.R., G.M. Davison, C.A. Newell, A. Merryweather, W.D.O. Hamilton, E.P.J. Burgess, R. Gilbert, and J.A. Gatehouse. 1995. Transgenic potato plants with enhanced resistance to the tomato moth, *Lacanobia oleracea*: growth room trials. Molecular Breeding 3:49-63.

Gershenzon, J. and R. Croteau. 1991. Terpenoids. Pages 165-219 *in* G.A. Rosenthal and M.R. Berenbaum, editors. Herbivores. Their Interaction with Secondary Plant Metabolites. Academic Press, San Diego.
Green, T.R. and C.A. Ryan. 1972. Wound-induced proteinase inhibitor in plant leaves: a possible defense mechanism against insects. Science 175:776-777.
Gundlach, H., M.J. Muller, T.M. Kutchan, and M.H. Zenk. 1992. Jasmonic acid is a signal transducer in elicitor-induced plant cell cultures. Proceedings of the National Academy of Sciences USA 89:2389-2393.
Hagerman, A.E. and L.G. Butler. 1991. Tannins and lignins. Pages 355-388 *in* G.A. Rosenthal and M.R. Berenbaum, editors. Herbivores. Their Interaction with Secondary Plant Metabolites. Academic Press, San Diego.
Hahlbrock, K. and D. Scheel. 1989. Physiology and molecular biology of phenylpropanoid metabolism. Annual Review of Plant Physiology and Plant Molecular Biology 40:347-369.
Hannapel, D.J. 1993. Nucleotide and deduced amino acid sequence of the 22-kilodalton cathepsin D inhibitor protein of potato (*Solanum tuberosum* L.). Plant Physiology 101:703-704.
Harborne, J.B 1991. Flavonoid pigments. Pages 389-429 *in* G.A. Rosenthal and M.R. Berenbaum, editors. Herbivores. Their Interaction with Secondary Plant Metabolites. Academic Press, San Diego.
Harborne, J.B. 1993. Introduction to Ecological Biochemistry. Academic Press, London.
Hart, J.H. 1981. Role of phytostilbenes in decay and disease resistance. Annual Review of Phytopathology 19:437-458.
Hartley, S. 1988. The inhibition of phenolic biosynthesis in damaged and undamaged birch foliage and its effect on insect herbivores. Oecologia 76:65-70.
Hartley, S.E. and J.H. Lawton. 1991. Biochemical aspects and significance of the rapidly induced accumulation of phenolics in birch foliage. Pages 105-132 *in* D.W. Tallamy and M.J. Raupp, editors. Phytochemical Induction by Herbivores. John Wiley, New York.
Hartmann, T. 1991. Alkaloids. Pages 79-121 *in* G.A. Rosenthal and M.R. Berenbaum, editors. Herbivores. Their Interaction with Secondary Plant Metabolites. Academic Press, San Diego.
Hartmann, T., L. Witte, A. Ehmke, C. Theuring, M. Rowell-Rahier, and J.M.Pasteels. 1997. Selective sequestration and metabolism of plant derived pyrrolizidine alkaloids by chrysomelid leaf beetles. Phytochemistry 45:489-497.
Heitz, T., D.R. Bergey, and C.A. Ryan. 1997. A gene encoding a chloroplast-targeted lipoxygenase in tomato leaves is transiently induced by wounding, systemin, and methyl jasmonate. Plant Physiology 114:1085-1093.
Hilder, V.A., A.M.R. Gatehouse, S.E. Sheerman, R. F. Barker, and D. Boulter. 1987. A novel mechanism of insect resistance engineered into tobacco. Nature 330:160-163.
Hilder, V.A., K.S. Powell, A.M.R. Gatehouse, J.A. Gatehouse, L.N. Gatehouse, Y. Shi, W.D.O. Hamilton, A. Merryweather, C.A. Newell, and J.C. Timans. 1995. Expression of snowdrop lectin in transgenic tobacco plants results in added protection against aphids. Transgenic Research 4:18-25.
Hildmann, T., M. Ebneth, H. Peña-Cortés, J.J. Sánchez-Serrano, L. Willmitzer, and S. Prat. 1992. General roles of abscisic and jasmonic acids in gene activation as a result of mechanical wounding. Plant Cell 4:1157-1170.
Hillis, W.E. and T. Inoue. 1968. The formation of polyphenols in trees. IV. The polyphenols formed in *Pinus radiata* after *Sirex* attack. Phytochemistry 7:13-22.
Hyodo, H., C. Hashimoto, S. Morozumi, M. Ukai, and C. Yamada. 1993. Induction of ethylene production and lignin formation in wounded mesocarp tissue of *Cucurbita maxima*. Acta Horticulturae 34:264-269.
Irie, K., H. Hosoyama, T. Takeuchi, K. Iwabuchi, H. Watanabe, M. Abe, K. Abe, and S. Arai. 1996. Transgenic rice established to express corn cystatin exhibits strong inhibitory activity against insect gut proteinases. Plant Molecular Biology 30:149-157.
Ito, H., F. Kimizuka, A. Ohbayashi, H. Matsui, M. Honma, A. Shinmyo, Y. Ohashi, A.B. Caplan, and R.L. Rodriguez. 1994. Molecular cloning and characterization of two complementary DNAs encoding putative peroxidases from rice (*Oryza sativa* L.) shoots. Plant Cell Reports 13:361-366.

Johnson, R., J. Narvaez, G. An, and C.A. Ryan. 1989. Expression of proteinase inhibitors I and II in transgenic tobacco plants: Effects on natural defense against *Manduca sexta* larvae. Proceedings of the National Academy of Sciences USA 86:9871-9875.

Jongsma, M.A., P.L. Bakker, J. Peters, D. Bosch, and W.J. Stiekema. 1995. Adaptation of *Spodoptera exigua* larvae to plant proteinase inhibitors by induction of gut proteinase activity insensitive to inhibition. Proceedings of the National Academy of Sciences USA 92:8041-8045.

Joos, H.J. and K. Hahlbrock. 1992. Phenylalanine amonia-lyase in potato (*Solanum tuberosum* L.). Genomic complexity, structural comparison of two selected genes and modes of expression. European Journal of Biochemistry 204:621-629.

Kahl, G. 1982. Molecular biology of wound healing: the conditioning phenomenon. Pages 211-267 *in* G. Kahl and J. Schell, editors. Molecular Biology of Plant Tumors. Academic Press, New York.

Khan, M.B and J.B. Harborne. 1991. A comparison of the effect of mechanical and insect damage on alkaloid levels in *Atropa acuminata*. Biochemical Systematics and Ecology 19:529-534.

Kolattukudy, P.E. 1981. Structure, biosynthesis, and biodegradation of cutin and suberin. Annual Review of Plant Physiology and Plant Molecular Biology 32:539-567.

Koritsas, V. M., J. A. Lewis, and G. R. Fenwick. 1991. Glucosinolate responses of oilseed rape, mustard and kale to mechanical wounding and infestation by cabbage stem flea beetle (*Psylliodes chrysocephala*). Annals of Applied Biology 118:209-222.

Kuć, J. 1995. Phytoalexins, stress metabolism, and disease resistance in plants. Annual Review of Phytopathology 33:275-297.

Lee, D. and C.J. Douglas. 1996. Two divergent members of a tobacco 4-coumarate:Coenzyme A ligase (4CL) gene family. cDNA structure, gene inheritance and expression, and properties of recombinant proteins. Plant Physiology 112:193-205.

Lewinsohn, E., M. Gijzen, T.J. Savage, and R. Croteau. 1991. Defense mechanisms of conifers. Plant Physiology 96:38-43.

Lewinsohn , E., M. Gijzen, and R.B. Croteau. 1992. Regulation of monoterpene biosynthesis in conifer defense. Pages 8-17 *in* W.D. Nes, E.J. Parish, and J.M. Trzakos, editors. Regulation of Isopentenoid Metabolism. ACS Symposium Series 497, Washington, DC.

Li, Z., A. Sommer, T. Dingermann, and C. Noe. 1996. Molecular cloning and sequence analysis of cDNA encoding a cysteine proteinase inhibitor from *Sorghum bicolor* seedlings. Molecular and General Genetics 251:499-502.

Lim, C.O., S.I. Lee, W.S. Chung, S.H. Park, I. Hwang, and M.J. Cho. 1996. Characterization of a cDNA encoding cysteine proteinase inhibitor from Chinese cabbage (*Brassica campestris* L. ssp. pekinensis) flower buds. Plant Molecular Biology 30:373-379.

Lindroth, R.L. and S.-Y. Hwang. 1996. Diversity, redundancy, and multiplicity in chemical defense systems of aspen. Pages 25-56 *in* J.T Romeo, J.A. Saunders, and P. Barbosa, editors. Phytochemical Diversity and Redundancy in Ecological Interactions. Plenum Press, New York.

Lindroth, R.L. and K.K. Kinney. 1998. Consequences of enriched atmospheric CO_2 and defoliation for foliar chemistry and gypsy moth performance. Journal of Chemical Ecology 24:1677-1695.

Linthorst, H.J.M., F.T. Brederode, C. Van der Does, and J.F. Bol. 1993. Tobacco proteinase inhibitors are locally, but not systemically, induced by stress. Plant Molecular Biology 21:985-992.

Litvak, M.E. and R.K. Monson. 1998. Patterns of induced and constitutive monoterpene production in conifer needles in relation to insect herbivory. Oecologia 114:531-540.

Lois, R. and K. Hahlbrock. 1992. Differential wound activation of members of the phenylalanine ammonia-lyase and 4-coumarate:CoA ligase gene families in various organs of parsley plants. Zeitschrift für Naturforschung 47c:90-94.

Loiza-Velarde, J.G., F.A. Tomas-Barbera, and M.E. Saltveit. 1997. Effect of intensity and duration of heat-shock treatments on wound-induced phenolic metabolism in iceberg lettuce. Journal of the American Society for Horticultural Science 122:873-877.

Luckner, M. 1990. Secondary Metabolism in Microorganisms, Plants, and Animals. Gustav Fischer Verlag, Jena.

Matton, D.P. and N. Brisson. 1989. Cloning, expression, and sequence conservation of pathogenesis-related gene transcripts of potato. Molecular Plant-Microbe Interactions 2:325-331.

Mattson, W.J. and S.R Palmer. 1988. Changes in levels of foliar minerals and phenolics in trembling aspen, *Populus tremuloides*, in response to artificial defoliation. Pages 157-169 *in* W.J. Mattson, J. Levieux, and C. Bernard-Dagan, editors. Mechanisms of Woody Plant Defenses Against Insects. Springer Verlag, New York.

Mayer, A.M. and E. Harel. 1979. Polyphenol oxidases in plants. Phytochemistry 18:193-215.

McAuslane, H.J., H.T. Alborn, and J.P. Tott. 1997. Systemic induction of terpenoid aldehydes in cotton pigment glands by feeding of larval *Spodoptera exigua*. Journal of Chemical Ecology 23:2861-2879.

McCloud, E.S., D.W. Tallamy, and F.T. Halaweish. 1995. Squash beetle trenching behaviour: avoidance of cucurbitacin induction or mucilaginous plant sap? Ecological Entomology 20:51-59.

Mizutani, M., D. Ohta, and R. Sato. 1997. Isolation of a cDNA and a genomic clone encoding cinnamate 4-hydroxylase from *Arabidopsis* and its expression manner *in planta*. Plant Physiology 113:755-763.

Murdock, L.L., J.E. Huesing, S.S. Nielsen, R.C. Pratt, and R.E. Shade. 1990. Biological effects of plant lectins on the cowpea weevil. Phytochemistry 29:85-89.

Negrel, J., B. Pollet, and C. Lapierre. 1996. Ether-linked ferulic acid amides in natural and wound periderms of potato tuber. Phytochemistry 43:1195-1199.

Nicholson, R.L. and R. Hammerschmidt. 1992. Phenolic compounds and their role in disease resistance. Annual Review of Phytopathology 30:369-389.

Niemeyer, H.M. 1988. Hydroxamic acids (4-hydroxy-1,4-benzoxazin-3-ones), defence chemicals in the Gramineae. Phytochemistry 27:3349-3358.

Niemeyer, H.M., E. Pesel, S.V. Copaja, H.R. Bravo, S. Franke, and W. Francke. 1989. Changes in hydroxamic acid levels of wheat plants induced by aphid feeding. Phytochemistry 28:447-449.

Olson, M.M. and C.R. Roseland. 1991. Induction of the coumarins scopoletin and ayapin in sunflower by insect-feeding stress and effects of coumarins on the feeding of sunflower beetle (Coleoptera: Chrysomelidae). Environmental Entomology 20:1166-1172.

Orr, G.L., J.A. Strickland, and T.A. Walsh. 1994. Inhibition of *Diabrotica* larval growth by a multicystatin from potato tubers. Journal of Insect Physiology 40:893-900.

Pautot, V., F.M. Holzer, B. Reisch, and L.L. Walling. 1993. Leucine aminopeptidase: an inducible component of the defense response in *Lycopersicon esculentum* (tomato). Proceedings of the National Academy of Sciences USA 90:9906-9910.

Pearce, G., S. Johnson, and C.A. Ryan. 1993. Purification and characterization from tobacco (*Nicotiana tabacum*) leaves of six small, wound-inducible, proteinase isoinhibitors of the potato inhibitor II family. Plant Physiology 102:639-644.

Pearce, G., P.A. Marchand, J. Griswold, N.G. Lewis, and C.A. Ryan. 1998. Accumulation of feruloyltyramine and p-coumaroyltyramine in tomato leaves in response to wounding. Phytochemistry 47:659-664.

Peumans, W.J. and E.J.M. van Damme. 1995. Lectins as plant defense proteins. Plant Physiology 109:347-352.

Ponstein, A.S., S.A. Bres-Vloemans, M.B. Sela-Buurlage, P.J.M. van den Elzen, L.S. Melchers, and B.J.C. Cornelissen. 1994. A novel pathogen- and wound-inducible tobacco (*Nicotiana tabacum*) protein with antifungal activity. Plant Physiology 104:109-118.

Powell, K.S., A.M.R. Gatehouse, V.A. Hilder, and J.A. Gatehouse. 1993. Antimetabolic effects of plant lectins and plant and fungal enzymes on the nymphal stages of two important rice pests, *Nilaparvata lugens* and *Nephotettix cinciteps*. Entomologia Experimentalis et Applicata 66:119-126.

Raffa, K.F. 1991. Induced defensive reactions in conifer-bark beetle systems. Pages 245-276 *in* D.W. Tallamy and M.J. Raupp, editors. Phytochemical induction by herbivores. John Wiley, New York.

Raffa, K.F., A.A. Berryman, J. Simasko, W. Teal, and B.L. Wong. 1985. Effects of grand fir monoterpenes on the fir engraver, *Scolytus ventralis* (Coleoptera: Scolytidae), and its symbiotic fungus. Environmental Entomology 14:552- 556.

Raikhel, N.V., H.-I. Lee, and W.F. Broekaert. 1993. Structure and function of chitin-binding proteins. Annual Review of Plant Physiology and Plant Molecular Biology 44:591-615.
Raupp, M.J. and C.S. Sadof. 1991. Responses of leaf beetles to injury-related changes in their salicaceous hosts. Pages 183-204 *in* D.W. Tallamy and M.J. Raupp, editors. Phytochemical Induction by Herbivores. John Wiley, New York.
Richardson, M. 1991. Seed storage proteins: the enzyme inhibitors. Methods in Plant Biochemistry 5:259-305.
Rittinger, P.A., A.R. Biggs, and D.R. Peirson. 1987. Histochemistry of lignin and suberin deposition in boundary layers formed after wounding in various plant species and organs. Canadian Journal of Botany 65:1886-1892.
Robison, D.J. and K.F. Raffa. 1994. Characterization of hybrid poplar clones for resistance to the forest tent caterpillar. Forest Science 40:686-714.
Rohrmeier, T. and L. Lehle. 1993. WIP1, wound-inducible gene from maize with homology to Bowman-Birk proteinase inhibitors. Plant Molecular Biology 22:783-792.
Royo, J., G. Vancanneyt, A.G. Perez, C. Sanz, K. Störmann, S. Rosahl, and J.J. Sánchez-Serrano. 1996. Characterization of three potato lipoxygenases with distinct enzymatic activities and different organ-specific and wound-regulated expression patterns. Journal of Biological Chemistry 271:21012-21019.
Rumeau, D., E.A. Maher, A. Kelman, and A.M. Showalter. 1990. Extensin and phenylalanine ammonia-lyase gene expression altered in potato tubers in response to wounding, hypoxia, and *Erwinia carotovora* infection. Plant Physiology 93:1134-1139.
Ryan, C.A. 1990. Protease inhibitors in plants: genes for improving defenses against insects and pathogens. Annual Review of Phytopathology 28:425-449.
Saravitz, D.M. and J.N. Siedow. 1996. The differential expression of wound-inducible lipoxygenase genes in soybean leaves. Plant Physiology 110:287-299.
Schaller, A. and C.A. Ryan. 1995. Systemin-a polypeptide defense signal in plants. BioEssays 18:27-33.
Schmelz, E.A., R.J. Grebenok, D.W. Galbraith, and W.S. Bowers. 1998. Damage-induced accumulation of phytoecdysteroids in spinach: a rapid root response involving the octadecanoic acid pathway. Journal of Chemical Ecology 24:339-360.
Schultz, J.C. and I.T. Baldwin. 1982. Oak leaf quality declines in response to defoliation by gypsy moth larvae. Science 217:149-151.
Sembdner, G. and B. Parthier. 1993. The biochemistry and the physiological and molecular actions of jasmonates. Annual Review of Plant Physiology and Plant Molecular Biology 44:569-589.
Sherman, T.D. K.C. Vaughn, and S.O. Duke. 1991. A limited survey of the phylogenetic distribution of polyphenol oxidase. Phytochemistry 30:2499-2506.
Showalter, A.M. 1993. Structure and function of plant cell wall proteins. Plant Cell 5:9-23.
Siedow, J. 1991. Plant lipoxygenase: structure and function. Annual Review of Plant Physiology and Plant Molecular Biology 42:145-188.
Smith, C.G., M.W. Rodgers, A. Zimmerlin, D. Ferdinando, and G.P. Bolwell. 1994. Tissue and subcellular immunolocalisation of enzymes of lignin synthesis in differentiating and wounded hypocotyl tissue of French bean (*Phaseolus vulgaris* L.). Planta 192:155-164.
Sommer-Knudsen, J., A. Bacic, and A.E. Clarke. 1998. Hydroxyproline-rich plant glycoproteins. Phytochemistry 47:483-497.
Stanford, A., M. Bevan, and D. Northcote. 1989. Differential expression within a family of novel wound-induced genes in potato. Molecular and General Genetics 215:200-208.
Steele, C.L., S. Katoh, J. Bohlmann, and R. Croteau. 1998. Regulation of oleoresinosis in grand fir (*Abies grandis*). Plant Physiology 116:1497-1504.
Stout, M.J., J. Workman, and S.S. Duffey. 1994. Differential induction of tomato foliar proteins by arthropod herbivores. Journal of Chemical Ecology 20:2575-2594.
Stout, M.J., K. Workman, R.M. Bostock, and S.S. Duffey. 1998. Specificity of induced resistance in the tomato, *Lycopersicon esculentum*. Oecologia 113:74-81.
Stout, M.J., K. Workman, and S.S. Duffey. 1996. Identity, spatial distribution, and variability of induced chemical responses in tomato plants. Entomologia Experimentalis et Applicata 79:255-271.

Swain, T. 1979. Tannins and lignins. Pages 657-682 *in* G.A. Rosenthal and D.H. Janzen, editors. Herbivores. Their Interaction with Secondary Plant Metabolites. Academic Press, San Diego.
Taipalensuu, J., A. Falk, B. Ek, and L. Rask. 1997. Myrosinase-binding proteins are derived from a large wound-inducible and repetitive transcript. European Journal of Biochemistry 243:605-611.
Tallamy, D.W. 1985. Squash beetle feeding behaviour: an adaptation against induced cucurbit defenses. Ecology 66:1574-1579.
Thaler, J. S., M. J. Stout, R. Karban, and S. S. Duffey. 1996. Exogenous jasmonates simulate insect wounding in tomato plants (*Lycopersicon esculentum*) in the laboratory and field. Journal of Chemical Ecology 22:1767-1781.
Thomas, V., D. Premakumari, C.P. Reghu, A.O.N. Panikkar, and C.K. Saraswathy-Amma. 1995. Anatomical and histochemical aspects of bark regeneration in *Hevea brasiliensis*. Annals of Botany 75:421-426.
Vogt,T., P. Pollak, N. Tarlyn, L.P. Taylor. 1994. Pollination- or wound-induced kaempferol accumulation in Petunia stigmas enhances seed production. Plant Cell 6:11-23.
Walsh, T. A.and J.A. Strickland. 1993. Proteolysis of the 85-kilodalton crystalline cysteine proteinase inhibitor from potato releases functional cystatin domains. Plant Physiology 103:1227-1234.
Walters, D.S. and J.C. Steffens. 1990. Branched chain amino acid metabolism in the biosynthesis of *Lycopersicon pennelli* glucose esters. Plant Physiology 93:1544-1551.
Warner, S.A.J., R. Scott, and J. Draper. 1992. Characterisation of a wound-induced transcript from the monocot *Asparagus* that shares similarity with a class of intracellular pathogenesis-related (PR) proteins. Plant Molecular Biology 19:555-561.
Waterman, P.G. and S. Mole. 1994. Analysis of phenolic plant metabolites. Blackwell Scientific, Oxford.
Weiler, E.W. 1997. Octadecanoid mediated signal transduction in higher plants. Naturwissenschaften 84:340-349.
Werner, R.A. 1995. Toxicity and repellency of 4-allylanisole and monoterpenes from white spruce and tamarack to the spruce beetle and eastern larch beetle (Coleoptera: Scolytidae). Environmental Entomology 24:372-379.
Wink, M. 1983. Wounding-induced increase of quinolizidine alkaloid accumulation in lupin leaves [*Lupinus polyphyllus*]. Zeitschrift für Naturforschung 38c:905-909.
Wratten, S.D., P.J. Edwards, and I. Dunn. 1984. Wound-induced changes in the palatability of *Betula pubescens* and *Betula pendula*. Oecologia 61:372-375.
Wycoff, K.L., P.A. Powell, R.A. Gonzales, D.R. Corbin, C. Lamb, and R.A. Dixon. 1995. Stress activation of a bean hydroxyproline-rich glycoprotein promoter is superimposed on a pattern of tissue-specific developmental expression. Plant Physiology 109:41-52.
Xiang, C.B. and D.J. Oliver. 1998. Glutathione metabolic genes coordinately respond to heavy metals and jasmonic acid in *Arabidopsis*. Plant Cell 10:1539-1550.
Ye, Z.H. and J.E. Varner. 1991. Tissue-specific expression of cell wall proteins in developing soybean tissues. Plant Cell 3:23-37.
Yeh, K., J. Chen, M. Lin, Y. Chen, and C. Lin. 1997. Functional activity of sporamin from sweet potato (*Ipomoea batatas* Lam.): a tuber storage protein with trypsin inhibitory activity. Plant Molecular Biology 33:565-570.
Zangerl, A.R. 1990. Furanocoumarin induction in wild parsnip: evidence for an induced defense against herbivores. Ecology 71:1926-1932.
Zhou, J., D. Rumeau, and A.M. Showalter. 1992. Isolation and characterization of two wound-regulated tomato extensin genes. Plant Molecular Biology 20:5-17.

Induced Plant Volatiles: Biochemistry and Effects on Parasitoids

Paul W. Paré, W. Joe Lewis, and James H. Tumlinson

Abstract

Wounding of plant tissue by insect feeding triggers the release of volatile compounds, which have been shown to attract natural enemies of insect herbivores. These volatile semiochemicals are derived from several biosynthetic pathways. Although some volatile compounds are released from storage immediately when cells or glands are damaged, the induced compounds are only synthesized and released during periods of light-driven photosynthesis. Plants release the induced compounds from undamaged as well as damaged leaves. These plant volatiles include terpenoids, green-leaf lipoxygenase products, and the nitrogen containing compound indole. In some plant species that we have examined, the blend of volatiles synthesized and released differs between artificial and herbivore-mediated damage, as well as between separate herbivore species. Differential attraction of parasitoids can also be correlated with distinctive volatile plant emissions from different herbivore species feeding on leaf tissue.

Plant Volatile Chemicals as Attractants for Beneficial Insects

The induction of plant defensive compounds in response to herbivory directly impacts the amount of insect damage a plant receives by triggering the accumulation of toxic or antifeedant compounds (Constabel, this volume). Insect damage also triggers the release of plant volatiles that can serve as an indirect plant defense response; these volatile compounds signal parasitoids of herbivores during initial stages of host location and habitat selection (Vinson 1998). Unlike insect pollinators that can capitalize on visual cues to seek out well marked flower targets, parasitoids rely on odor cues in searching for small herbivores that are usually well camouflaged, residing on the undersides of leaves, in leaf folds or in the interior of flower buds. Both McCall et al. (1993)

and Steinberg et al. (1993) have shown by wind tunnel flights and gas chromatographic (GC) analysis that herbivore odors alone provide parasitoids with weak signals. In contrast, the chemicals released from herbivore-damaged plants appear to contain critical chemical information that draws parasitoids, in the laboratory, to air streams spiked with these plant odors, and in the field to damaged plants placed among a group of undamaged neighbors. Once a female parasitoid has located her host, she injects her eggs into the herbivore by stinging; the fate for a stung caterpillar is that its reproductive cycle is terminated while a new generation of wasps is set in motion.

Volatile Emissions in Plants

An undamaged plant maintains a baseline level of volatile metabolites that are released from the surface of the leaf and/or from accumulated storage sites in the leaf. These constitutive chemical reserves often accumulate to high levels in specialized glands or trichomes (Paré and Tumlinson 1997a). They include monoterpenes, sesquiterpenes, and aromatics. This pattern of constitutive compounds has been analyzed in the field for perennial plants including beech (Tollsten and Müller 1996) and ash (Markovic et al. 1996) trees as well as under glasshouse conditions for many herbaceous annuals including Brussels sprouts (Mattiacci et al. 1994) and cucumber (Takabayashi et al. 1994).

With damage, the profile of volatiles emitted from the foliage markedly changes. In cotton (*Gossypium hirsutum*), breakage of leaf glands causes stored terpenes to be released in much higher levels and the emissions of lipoxygenase pathway green leaf volatiles also climbs. The release of these metabolites correlates closely with leaf damage from insect feeding (Loughrin et al. 1994). This is in contrast to a subset of terpenes, indole, and hexenyl acetate that are also released in much higher levels with insect feeding but have a diurnal cycle that is decoupled from short term insect damage (e.g., Loughrin et al. 1994). One proposed mechanism for regulating the emission of these plant volatiles is that these metabolites are stored as non-labile glycoside coupled compounds. With herbivore damage, the sugars are cleaved and the volatile components are released (Boland et al. 1992).

Chemical labeling studies have established that, at least in some species, plant volatiles are not released by cleavage of sugars from stored volatiles. In cotton, damage of plant tissue by beet armyworm (*Spodoptera exigua*) triggers the *de novo* synthesis of volatile compounds (Paré and Tumlinson 1997b). These cotton volatiles, linalool and (*E*) α-ocimene (monoterpenes), (*E,E*) β-farnesene, and (*E*) α-farnesene (sesquiterpenes), nonatriene and tridecatetraene (homoterpenes), indole, and (*Z*) 3-hexenyl acetate, have emissions profiles that follow the light:dark cycle with low emissions at night and high emissions during the periods of maximal photosynthesis. Chemical labeling studies have also established that these induced compounds are synthesized specifically in response to insect damage

(Paré and Tumlinson 1997b). Plants rapidly incorporated a high level of labeled ^{13}C when plants damaged by feeding caterpillars were held in volatile collection chambers under an atmosphere containing ^{13}C carbon dioxide. In cotton, approximately equal numbers of volatile terpenes incorporated very low or nondetectable amounts as high levels of labeled carbon when plants were exposed to $^{13}CO_2$. Compounds with low enrichment levels included: β-pinene, α-pinene, (*E*)-α-caryophyllene, and β-humulene. These volatiles were either released from storage upon insect wounding, or were synthesized from other precursor pools and thus did not incorporate the labeled carbon. The monoterpenes limonene and myrcene contained intermediate levels of ^{13}C label after 30 hours of $^{13}CO_2$ exposure. There is a consistent delay of several hours between onset of insect feeding and the emissions of these induced volatiles. This observation substantiates the hypothesis that a series of biochemical reactions such as protein assembly and/or enzyme induction is required for the synthesis and release of the induced volatile compounds.

Similar compounds are emitted in response to insect herbivore damage in several agricultural species that have been studied including cucumber, apple, lima bean, corn and cotton (reviewed by Paré and Tumlinson 1999). In bean, infestation by spider mites (*Tetranychus urticae*) results in terpenoids and homoterpenes being synthesized *de novo* in response to herbivore damage (Takabayashi and Dicke 1996). Across plant species, it is not known whether the blend of volatile compounds is induced through a common signaling pathway or if their emissions is triggered by different signaling mechanisms. In any case, these induced compounds, when driven by the photosynthesis cycle, are released independent of minor feeding interruptions and can serve as reliable chemical attractants to parasitoids searching for their hosts (Loughrin et al. 1994).

Systemic Biosynthesis and Release of Volatiles

In addition to the release of volatiles at the site of herbivore feeding, analysis of volatile emissions from unharmed leaves of insect damaged plants has established that there can be a systemic response. In both corn (Turlings and Tumlinson 1992) and cotton (Röse et al. 1996), leaves distal to the site of herbivore feeding showed an increase in the release of volatiles. The chemical blend of volatiles from undamaged cotton leaves differs from the volatiles collected from the entire plant (Röse et al. 1996). Products of the lipoxygenase pathway, including hexenals and hexenols, which are released from freshly cut or damaged tissue, are not detected in the systemically released volatiles, except for (*Z*)-3-hexenyl acetate. One hypothesis is that these six-carbon compounds can only be released from undamaged leaf tissue when converted to the acetate form (Paré and Tumlinson 1998). In cotton, some of the monoterpenes and sesquiterpenes, as well as indole and isomeric hexenyl butyrates and 2-methyl butyrates, are only released from damaged leaves (Röse et al. 1996). The terpenoids that are synthesized *de novo* in cotton leaves in response to herbivore damage (Paré and Tumlinson 1997a,b) are also released systemically from undamaged leaves of herbivore injured plants. Thus, only the terpenoids that incorporated ^{13}C were released systemically.

These results raised the question of whether compounds released systemically are synthesized *de novo* in the leaves from which they are released, or are synthesized in the damaged leaves and transported to the undamaged leaves. Volatiles were collected from leaves in the upper portion of plants, after the bottom leaves of the plants, which were outside the volatile collection chamber, had been damaged by beet armyworm feeding. During volatile collection, the upper leaves were exposed to air containing $^{13}CO_2$. Analysis of the systemically released volatiles by gas chromatography-mass spectroscopy showed that a high level of ^{13}C was rapidly incorporated into these compounds, indicating that they are synthesized *de novo* at the site of release (Paré and Tumlinson 1998). This suggests that a signal is transmitted from the site of damage to distal, undamaged leaves, to trigger synthesis and release of volatile compounds. The transfer of such a signal to undamaged portions of the plant serves as a mechanism for amplifying the message that the plant is under herbivore attack.

Biosynthesis of Plant Volatiles in Response to Insect Herbivory

Though there are numerous reports on the release of volatiles following insect feeding, as well as the effect of these volatiles on parasitoids, there is little information on the way in which plants regulate the blend of compounds emitted. At least four biosynthetic pathways are responsible for the blend of volatiles that are released. An outline of the biosynthetic route for plant volatile emissions is shown in Fig. 1. The isoprenoid precursor isopentenyl pyrophosphate serves as a substrate for monoterpenes and sesquiterpenes, the fatty acid/lipoxygenase pathway generates green leaf volatiles and jasmone, and the shikimic acid/tryptophan pathway results in the nitrogen containing product indole (Mann 1987).

The monoterpene and sesquiterpene pathways share in common the isopentenyl pyrophosphate (IPP) intermediate, although catalysis is physically separated from monoterpene synthesis located in plastids and sesquiterpene assembly in the cytosol (Lichtenthaler et al. 1997). Half of the IPP is enzymatically converted to the isomer dimethylallyl pyrophosphate and the two isomers condense via prenyltransferases to form geranyl pyrophosphate (GPP). GPP can be channelled into monoterpene biosynthesis catalyzed by cyclase enzymes (Alonso and Croteau 1993) resulting in metabolites including the acyclic structures linalool and (*E*)-α-ocimene, the monocyclic structure limonene, and the bicyclic structure β-pinene, all of which are released with insect feeding in corn.

GPP can also be condensed with another IPP molecule to form the C_{15} molecule, farnesyl pyrophosphate, the precursor of sesquiterpenes such as the

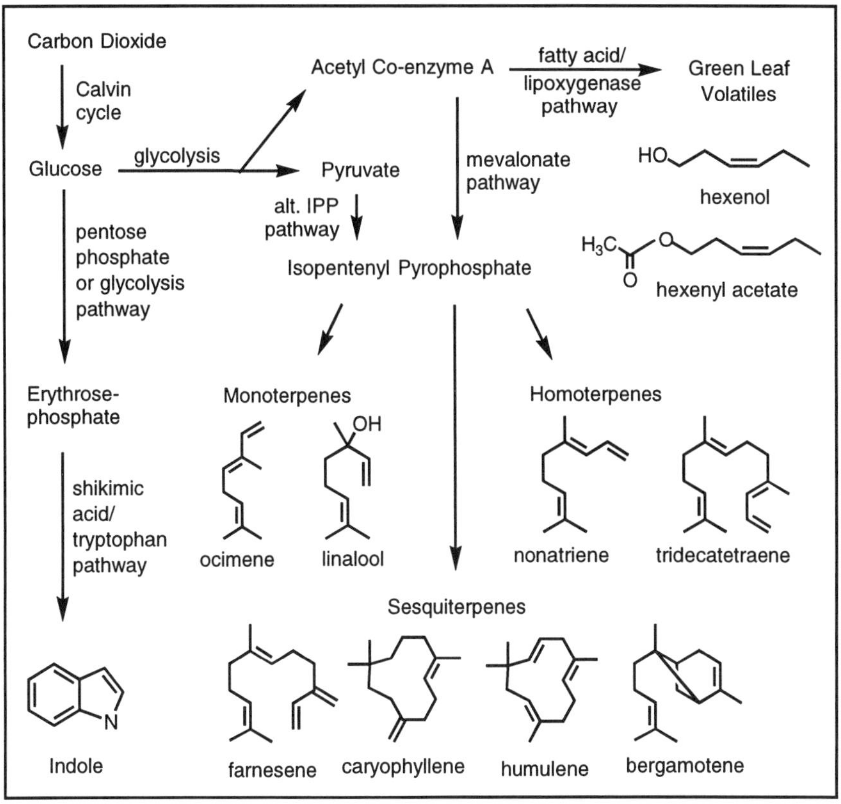

Figure 1: Metabolic pathways leading to volatile emissions from herbivore damaged plants and structural examples of volatile components in corn seedlings.

acyclic compounds nerolidol, β-farnesene and α-farnesene, the monocyclic compounds humulene and 8-bisabolene, and the bicyclic compound caryophyllene, all of which are also released from corn after insect feeding (Turlings et al. 1991).

For sesquiterpenes, a key intermediate in the isoprenoid pathway is mevalonic acid, which is formed by the condensation of three acetyl-CoA molecules to (3S)-*3*-hydroxy-3-methylglutaryl CoA and a two step reduction to mevalonic acid (Goodwin and Mercer 1990). As shown in Fig. 2, IPP the five carbon building block of sesquiterpenes, is then formed via decarboxylation of mevalonic acid pyrophosphate (Gershenzon and Croteau 1989). For monoterpenes, IPP is synthesized according to a non-mevalonate glyceraldehyde 3-phosphate (GAP)/pyruvate pathway that has been recently discovered (Fig. 3). Rohmer et al. (1993) proposed that the GAP/pyruvate conversion to IPP proceeds via 1-deoxy-D-xylulose-5-phosphate and a C-methyl-branched D-tetrose-4-phosphate as key intermediates.

In cotton, the induced terpenes are early biosynthetic products of the isoprenoid pathway. (*E,E*)-β- and (*E*)-α-farnesene as well as (*E*)-α-ocimene are likely formed via an ionization-isomerization-elimination reaction, analogous to the proposed mechanism of monoterpene syntheses (Savage et al. 1994), with farnesyl- and geranyl-pyrophosphate as respective precursors (see Fig. 4). Linalool, another induced metabolite, is structurally similar to ocimene except that the former compound undergoes an addition reaction with the loss of the pyrophosphate moiety to form the tertiary alcohol instead of proton elimination to form an additional double bond. In the case of the induced C_{11} homoterpene, (*E*)-4,8-dimethyl-1,3,7-nonatriene and C_{16} homoterpene (3*E*,7*E*)-4,8,12-trimethyl-1,3,7,11-tridecatetraene, a one step oxidative cleavage of the sesquiterpenoid nerolidol and the diterpenoid geranyllinalool, respectively results in their formation (Gäbler et al. 1991, Boland et al. 1992).

Why acyclic terpenes are synthesized at the time of herbivore damage while more biosynthetically complex derivatives are constitutively stored is not clear. One explanation is that unequal enzyme activation increases the flux of terpenoid biosynthesis at the level of IPP formation, however insufficient cyclase activity limits formation of the cyclic terpenes. At least in cotton, the terpenes synthesized *de novo* are the same compounds that Loughrin et al. (1994) observed to have a diurnal cycling pattern in their release.

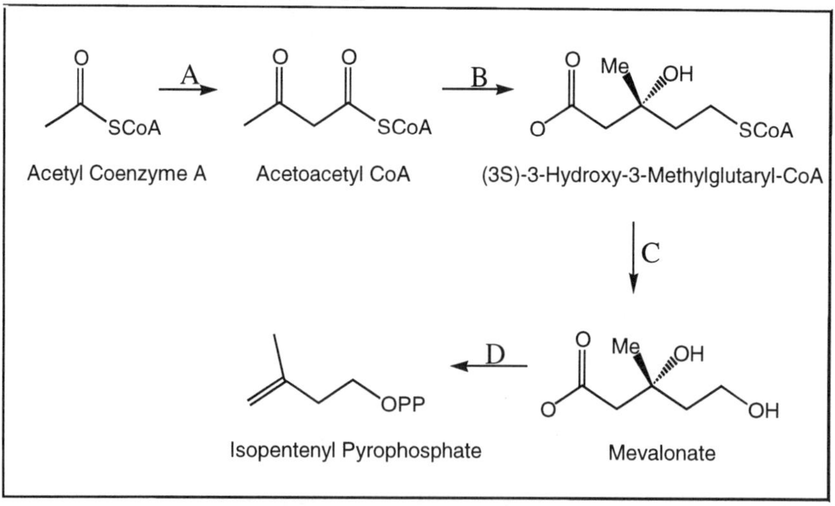

Figure 2: Formation of sesquiterpene precursor isopentenyl pyrophosphate from acetyl CoA. Enzymes in the pathway include: acetyl-CoA transferase (A), hydroxymethyl-glutaryl-CoA synthase (B), hydroxymethylglutaryl-CoA reductase (C), mevalonate kinase, phosphomevalonate kinase and pyrophosphomevalonate decarboxylase (D).

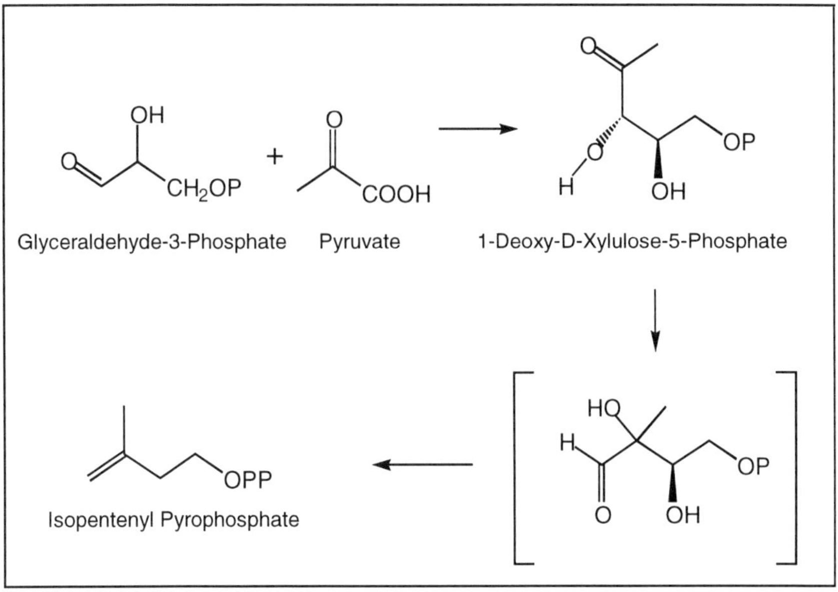

Figure 3: Non-mevalonate GAP/pyruvate pathway conveivable proceeding via 1-deoxy-D-xylulose-5-phosphate and a C-methyl-branched D-tetrose-4-phosphate as key intermediates.

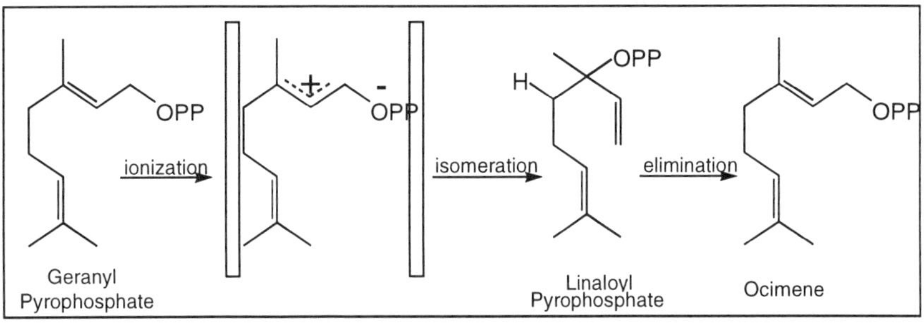

Figure 4: A biosynthetic mechanism for the formation of ocimene from geranyl pyrophosphate (OPP=pyrophosphate).

The pathway responsible for the synthesis of the leafy green (C_6) volatiles is the fatty acid/lipoxygenase pathway, in which lipids are broken down into short chain volatile compounds. The C_6 volatiles and jasmone are formed from the polyunsaturated octadecenoid fatty acids (*Z,Z*)-9,12-octadecadienoic acid (linoleic acid) or (2, 2, 2) 9,12,15-octadecatrienoic acid (linolenic acid). The substrates for lipoxygenase, linoleic acid and linolenic acid are major components in plant membranes but are not readily acted upon by the enzyme lipoxygenase until in the free acid form (Croft et al. 1993). In the formation of

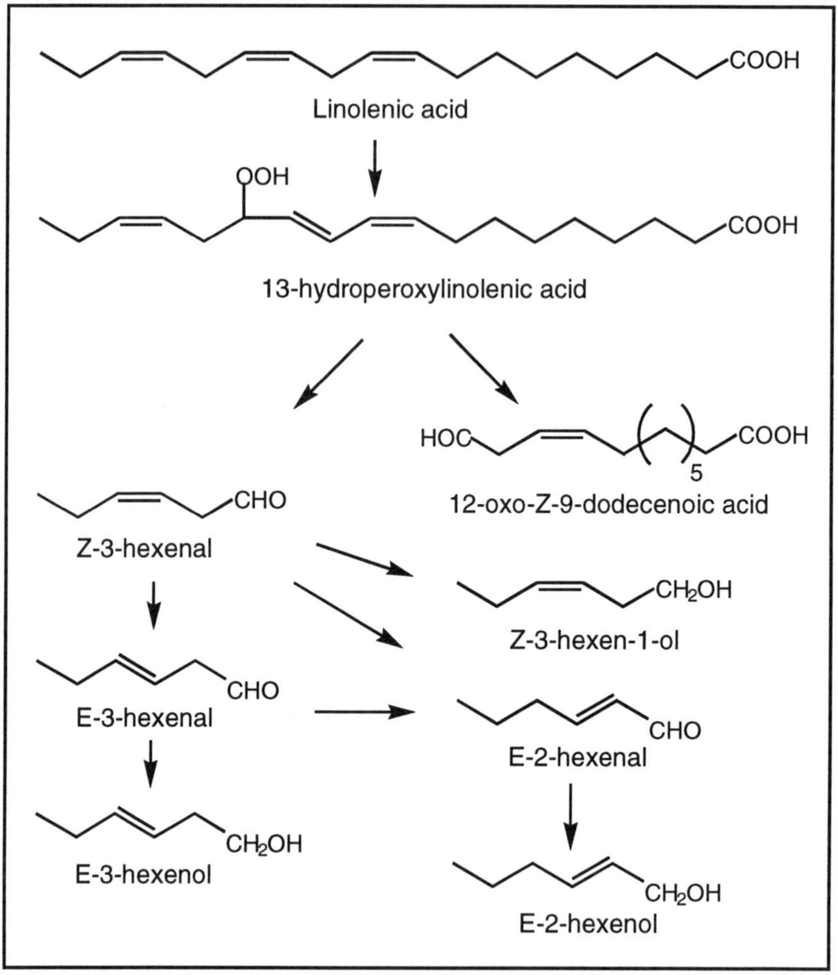

Figure 5: Green leaf odors consisting of a blend of unsaturated six carbon alcohols, and aldehydes produced by autolytic oxidative breakdown of the plant membrane lipid linolenic acid (Croft et al. 1993).

the C_6 volatiles, linolenic acid is oxidized to either 9- or 13-hydroperoxylinolenic acid or a mixture of both. Hydroperoxide lyase then catalyzes the formation of the C_6 Z-3-hexenal and the C_{12} 12-oxo-Z-9-dodecenoic acid (see Fig. 5). Rearrangement, reduction and/or esterification of the C_6 product result in volatiles such as (*Z*)-3-hexenal, (*E*)-2-hexenal, (*Z*)-3-hexenol, and (*Z*)-3-hexenyl acetate, all of which have been identified in the head space collections of insect damaged cotton leaves (Röse et al. 1996).

The shikimic acid pathway provides the substrate for synthesis of tryptophan as well as other aromatic plant volatiles that have been identified in association with herbivore feeding such as methyl salicylate and phenylacetonitrile (Takabayashi et al. 1994, Loughrin et al. 1995). Indole, a nitrogen containing volatile detected in the tryptophan pathway from plants attacked by insects, is the penultimate intermediate in the biosynthesis of tryptophan.

Efficacy of Plant Volatiles as Semiochemicals

For parasitoids, induced plant volatile emissions is a major factor in influencing host selection behavior. Chemically mediated tritrophic interactions have been well documented in several agroecosystems. For example, Dicke and coworkers have shown that leaves from beans (*Phaseolus lunatus*) (Dicke et al. 1990) cucumbers (Takabayashi et al. 1994) and apple trees (Takabayashi et al. 1991) attacked by phytophagous mites release volatiles that attract predatory mites. Turlings et al. (1990, 1991, 1993) demonstrated that corn seedlings damaged by beet armyworm larvae released a blend of terpenoids and indole that was attractive to *Cotesia marginiventris*, a generalist parasitoid.

To examine what role systemically released chemicals alone play in the attraction of wasps, Cortesero et al. (1997) removed herbivore-damaged leaves immediately before flight tests. Wind tunnel experiments showed that volatiles released from undamaged portions of cotton plants emitted sufficient quantities and/or an appropriate blend of volatiles to direct parasitoids to their hosts. Similar results were found in the field with cotton and tobacco with 95% of native female wasps (*Cardiochiles nigriceps*) landing on tobacco budworm-damaged versus undamaged plants (with the damaged leaves intact or cut off) in both crops (DeMoraes et al. 1998). This specialist parasitic wasp, *C. nigriceps*, can distinguish plants infested by her host *Heliothis virescens* from those infested by *Helicoverpa zea,* a closely related non-host herbivore species, based on chemical cues released by the plant. It appears that at least with tobacco, cotton, and corn, each plant produces a herbivore-specific blend of volatile components in response to a particular herbivore species, and that these differences are observable by GC chemical analyzes as well as detectable by parasitic wasps (De Moraes et al. 1998).

Although the volatile compounds released by insect herbivore damage are similar among the several plant species studied thus far, the specific blends are quite distinct, varying in both the number of compounds and the actual structures produced. Thus, the task of finding a host is more complicated for the parasitoid when the host feeds on several different plant species. The wasps can overcome this obstacle, because they have the ability to learn chemical cues associated with the presence of a host (Lewis and Tumlinson 1988). A successful host experience increases the wasp's responsiveness to host-associated chemicals. For example, oviposition experience for the aphid parasitoid *Aphidius ervi* on the plant-host complex significantly increases the oriented flight and landing reponses of females relative to those that are not allowed to sting, but are exposed to undamaged plants or host damaged plants (Du et al. 1997). This underscores the importance of the oviposition experience in combination with host-damaged plant cues. Interestingly, female wasps can also learn volatile odors associated with food sources and use them to locate necessary food (Lewis and Takasu 1990).

Herbivore Signals that Trigger Volatile Release

Substances in the oral secretion of herbivores are key to the emissions of plant signals used as cues by foraging parasitoids (see also, Felton and Eichenseer, this volume). Recent work suggests that volatile emissions, as well as other plant defense responses, are potentiated by a component or components associated with the feeding herbivore that allows the plant to differentiate between general wounding and damage due to chewing insects. In cotton, induced volatiles that are synthesized in response to wounding are released in greater quantities due to caterpillar feeding than they are due to mechanical damage alone (Paré and Tumlinson 1997a). In tobacco, higher concentrations of the defense signaling molecule jasmonic acid result from herbivore damage by hornworm caterpillars than from mechanical damage designed to mimic herbivory (McCloud and Baldwin 1998). At the transcriptional level, potato

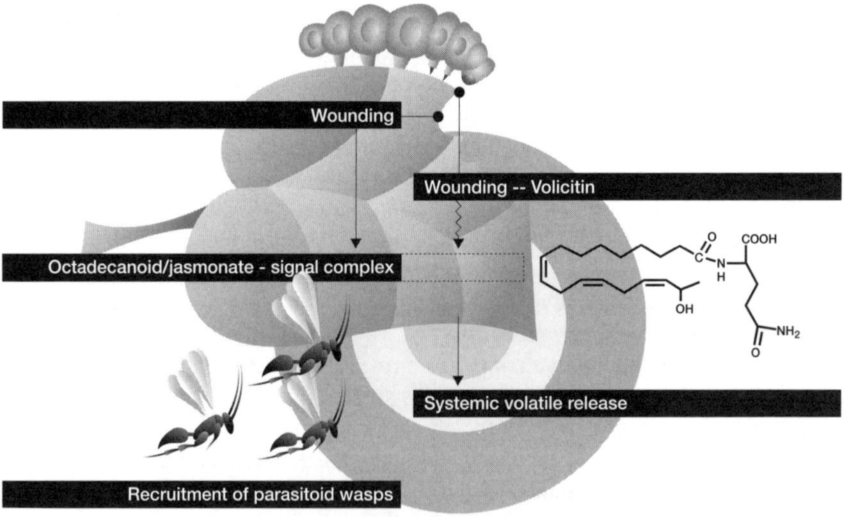

Figure 6: A schematic representation indicating an increase of volatile compounds released by plants in response to insect feeding triggered by an interaction of elicitors such as volicitin in the oral secretions of insect herbivores with damaged plant tissue. Volatile semiochemicals are then used by natural enemies of herbivores such as parasitiod wasps to locate their hosts.

mRNAs involved in plant defense accumulate more rapidly with insect-derived elicitor(s) in contact with the damaged leaves than with mechanical damage alone (Korth and Dixon 1997).

Thus far, two oral secretion products from chewing insects have been identified that augment the release of plant volatiles. β-Glucosidase, an enyzme present in the regurgitant of *Pieris brassicae* caterpillars, triggers the same emissions of volatiles in cabbage plants as are induced by feeding caterpillars (Mattiacci et al. 1995). Since enzyme activity in the regurgitant is retained when caterpillars are fed on α-glucosidase free diet, enzyme activity does not appear to be plant derived. Presumably the enzyme acts to cleave sugars coupled to organic compounds that then become more volatile and are released. Volicitin is a low molecular weight fatty acid derivative, *N*-(17-hydroxylinolenoyl)-L-glutamine (for structure see Fig. 6). It has been identified from the oral secretion of beet armyworm caterpillars and induces corn seedlings to release volatile chemical signals (Alborn et al. 1997).

Analysis of volicitin from beet armyworms fed ^{13}C labeled corn seedlings demonstrated that the caterpillar synthesizes this elicitor by adding a hydroxyl group and glutamine to linolenic acid obtained directly from the plant on which the caterpillar feeds (Paré et al. 1998). Thus, although the precursor of volicitin is obtained from plants, the bioactive product has only been found in the caterpillar. This strongly suggests that these molecules play an important, yet still unknown, role either in the metabolism or some other process critical to the life of herbivorous insects. Although it is known that the plant provides linolenic acid, which is essential for most lepidopteran larvae (Stanley-Samuelson 1994), it would seem detrimental to the insect to chemically convert this fatty acid into an elicitor that triggers plant defense. The full implications of this are not yet understood.

Synthesis and Future Directions

There is still much to learn about the chemical interactions among plants and insect herbivores that lead to the synthesis and release of volatiles by plants. We know that damage of a plant by different herbivore species can induce the release of volatile blends with different proportions of constituents, suggesting that distinct responses are induced by elicitors of different structures from different herbivore species. How such elicitors biochemically trigger synthesis and release of plant volatiles is unknown. Do they interact with the octadecanoid signaling pathway (Fig. 6, see Thaler, this volme), and if so, how? Do they regulate the release of linolenic acid, the production of jasmonic acid or the activation of the oxidative burst, all associated with the wounding of plant tissue? Also, we have no knowledge of the mechanism leading to the systemic release of volatiles. Does the original, herbivore-produced elicitor serve as a mobile messenger triggering whole plant volatile synthesis, or are secondary messengers employed to transmit the signal to sites distal from the damaged site?

Furthermore, why do herbivores produce compounds that activate plant chemical defenses? What function, if any, do these compounds serve in herbivore metabolism or defense? What role does enzyme level regulation play in volatile storage and emissions of these herbivore induced volatiles? Many of the protocols used to study other inducible plant responses can readily be adapted to investigate plant signaling involving the release of volatiles triggered by herbivore feeding.

Acknowledgements

The authors would like to kindly thank Anurag Agrawal, Nancy Epsky, Heather McAuslane, and Herb Oberlander for their constructive comments on earlier drafts of this manuscript.

Literature Cited

Alborn H. T., T. C. J. Turlings, T. H. Jones, G. Stenhagen, J. H. Loughrin, and J. H. Tumlinson. 1997. Isolation and identification of a plant volatile elicitor from beet armyworm oral secretion. Science 276:945-949.

Alonso, W.R., and R. Croteau. 1993. Prenyltransferases and cyclases. Methods in Plant Biochemistry 9:239-260.

Beckage N.E. 1997. The parasitic wasp's secret weapon. Scientific American 277:82-87.

Boland, W., Z. Feng, J. Donath, and A. Gäbler. 1992. Are acyclic C_{11} and C_{16} homoterpenes plant volatiles indicating herbivory? Naturwissenswchaften 79:368-371.

Cortesero A.M., C.D. De Moraes, J.O. Stapel, J.H. Tumlinson, and W.J. Lewis. 1997. Comparisons and contrasts in host-foraging strategies of two larval parasitoids with different degrees of host specificity. Journal of Chemical Ecology 23:1589-1606.

Croft, K.P., R. Juttner, and A.J. Slusarenko. 1993. Volatile products of the lipoxygenase pathway evolved from *Phaseolus vulgaris* (L.) leaves inoculated with *Pseudomonas syringae* pv. *phaseolicola*. Plant Physiology 101:13-24.

De Moraes, C. M., W. J. Lewis, P. W. Paré, H. T. Alborn, and J. H. Tumlinson. 1998. Herbivore-infested plants selectively attract parasitoids. Nature 393:570-573.

Dicke, M., T.A. van Deek, M.A. Posthumus, N. Ben Dom, H. Van Dokhoven, and A.E. De Groot. 1990. Isolation and identification of volatile kairomone that affects acarine predator-prey interactions. Journal of Chemical Ecology.16:381-396.

Du Y, G.M. Poppy, W. Powell, L.J. Wadhams. 1997. Chemically mediated associative learning in the host foraging behavior of the apid parasitoid *Aphidius ervi* (Hymenoptera: Braconidae). Journal of Insect Behavior 10:509-522.

Gäbler, A., W. Boland, U. Preiss, and H. Simon. 1991. Stereochemical studies on homoterpene biosynthesis in higher plants; mechanistic, phylogenetic and ecological aspects. Helvetica Chemica Acta 74:1773-1789.

Gershenzon, J., and R. Croteau. 1989. Regulation of monoterpene biosynthesis in higher plants. Pages 99-160 *in* G.H.N. Towers and H.A. Stafford, editors. Biochemistry of the Mevalonic Acid Pathway to Terpenoids. Plenum Publishing Co., New York.

Goodwin, T.W., and E.I. Mercer. 1990. Introduction to Plant Biochemistry. 2nd ed. Pergamon Press, Oxford, England.

Korth KL, and R.A. Dixon. 1997. Evidence for chewing insect-specific molecular events distinct from a general wound response in leaves. Plant Physiology 115:1299-1305.

Lewis WJ, and K. Takasu. 1990. Use of learned odours by a parasitic wasp in accordance with host and food needs. Nature 348:635-636.
Lewis WJ, and J.H. Tumlinson. 1988. Host detection by chemically mediated associative learning in a parasitic wasp. Nature 331:257-259.
Lichtenthaler H.K, M. Rohmer, and J. Schwender. 1997. Two independent biochemical pathways for isopentenyl diphosphate and isoprenoid biosynthesis in higher plants. Physiologia Plantarum 101:643-652.
Loughrin, J. H., A. Manukian, R. R. Heath, and J. H. Tumlinson. 1995. Volatiles emitted by different cotton varieties damaged by feeding beet armyworm larvae. Journal of Chemical Ecology 21:1217-1227.
Loughrin, J. H., A. Manukian, R. R. Heath, T. C. J. Turlings, and J. H. Tumlinson. 1994. Diurnal cycle of emission of induced volatile terpenoids by herbivore-injured cotton plants. Proceedings of the National Academy of Sciences USA 91:11836-11840.
Mann, J. 1987. Secondary Metabolism. Clarendon Press. Oxford, England.
Markovic I., D.M. Norris, J.K. Phillips, and F.X .Webster. 1996. Volatiles involved in the nonhost rejection of *Fraxinus pennsylvanica* by *Lymantria dispar* larvae. Journal of Agricultural and Food Chemistry. 44:929-935.
Mattiacci L, M. Dicke, and M.A. Posthumus. 1994. Induction of parasitoid attracting synomone in brussels sprouts plants by feeding of *Pieris brassicae* larvae: role of mechanical damage and herbivore elicitor. Journal of Chemical Ecology 20:2229-2247.
Mattiacci, L., M. Dicke, and M. A. Posthumus. 1995. β-Glucosidase: an elicitor of herbivore-induced plant odor that attracts host-searching parasitic wasps. Proceedings of the National Academy of Sciences USA 92:2036-2040.
McCall PJ, T..C. Turlings, W.J. Lewis, and J.H. Tumlinson. 1993. Role of plant volatiles in host location by the specialist parasitoid *Microplitis croceipes* cresson (Braconidae: Hymenoptera). Journal of Insect Behavior 6:625-639.
McCloud ES, I.T. Baldwin. 1998. Herbivory and caterpillar regurgitants amplify the wound-induced increases in jasmonic acid but not nicotine in *Nicotiana sylvestris*. Planta 203: 430-435.
Paré, P. W. and J. H. Tumlinson. 1997a. Induced synthesis of plant volatiles. Nature 385:30-31.
Paré, P. W. and J. H. Tumlinson. 1997b. *De novo* biosynthesis of volatiles induced by insect herbivory in cotton plants. Plant Physiology 114:1161-1167.
Paré, P. W. and J. H. Tumlinson. 1998. Cotton volatiles synthesized and released distal to the site of insect damage. Phytochemistry 47:521-526.
Paré, P. W. and J. H. Tumlinson. 1999. Plant volatiles as a defense against insect herbivores. Plant Physiology (in press).
Paré, P. W., H. T. Alborn, and J. H. Tumlinson. 1998. Concerted biosynthesis of an insect elicitor of plant volatiles. Proceedings of the National Academy of Sciences USA 95:13971-13975.
Rohmer M., M. Knani, P. Simonin, B. Sutter, and H. Sahm. 1993. Isoprenoid biosynthesis in bacteria: a novel pathway for early steps leading to isopentenyl diphosphate. Biochemistry Journal 295:517-524.
Röse, U.S.R, A. Manukian, R.R. Heath, and J.H. Tumlinson. 1996. Volatile semiochemicals released from undamaged cotton leaves: a systemic response of living plants to caterpillar damage. Plant Physiology 111:487-495.
Savage, T.J., M.W. Hatch, and R. Croteau. 1994. Monoterpene syntheses of *Pinus contorta* and related conifers. Journal of Biological Chemistry 269:4012-4020.
Stanley-Samuelson, D. W. 1994. Prostaglandins and related eicosanoids in insects. Advances in Insect Physiology 94:115-212.
Steinberg S., M. Dicke, and L.E. Vet. 1993. Relative importance of infochemicals from first and second trophic level in long-range host location by the larval parasitoid *Cotesia glomerata*. Journal of Chemical Ecology 19:47-59.
Takabayashi, J., M. Dicke, and M. A. Posthumus. 1991. Variation in composition of predator attracting allelochemicals emitted by herbivore-infested plants: relative influence of plant

and herbivore. Chemoecology 2:1-6.
Takabayashi, J., M. Dicke, S. Takahashi, M. A. Posthumus, and T. A. Van Beek. 1994. Leaf age affects composition of herbivore-induced synomones and attraction of predatory mites. Journal of Chemical Ecology 20:373-386.
Takabayashi J., and M. Dicke. 1996. Plant-carnivore mutualism through herbivore-induced carnivore attractants. Trends in Plant Science 1:109-113.
Tollsten L, P.M. Müller. 1996. Volatile organic compounds emitted from beech leaves. Phytochemistry 43:759-762.
Turlings, T. C. J., J. H. Tumlinson, and W. J. Lewis. 1990. Exploitation of herbivore-induced plant odors by host-seeking parasitic wasps. Science 250:1251-1253.
Turlings, T.C.J., J.H. Tumlinson, R.R. Heath, A.T. Proveaux, and R.E. Doolittle. 1991. Isolation and identification of allelochemicals that attract the larval parasitoid, *Cotesia marginiventris* (Gesson), to the microhabitat of one of its hosts. Journal of Chemical Ecology 17:2235-2251.
Turlings, T. C. J., and J. H. Tumlinson. 1992. Systemic release of chemical signals by hervbivore-injured corn. Proceedings of the National Academy of Sciences USA 89:8399-8402.
Turlings, T. C. J., P. J. McCall, H. T. Alborn, and J. H. Tumlinson. 1993. An elicitor in caterpillar oral secretions that induces corn seedlings to emit chemical signals attractive to parasitic wasps. Journal of Chemical Ecology 19:411-425.
Vinson, S.B. 1998. The general host selection behavior of parasitoid hymenoptera and a comparison of initial strategies utilized by larvaphagous and oophagous species. Biological Control 11:79-96.

PART II

Ecology and Evolution

Specificity of Induced Responses to Arthropods and Pathogens

Michael J. Stout and Richard M. Bostock

Abstract

Two aspects of specificity of induced responses, specificity of response and specificity of effect, are reviewed. The former is defined as the ability of a plant to distinguish between different types of biotic challenges and to produce different sets of chemicals or gene products in response to each type of challenge. It is argued that plants possess a greater capacity for specificity of response than is generally acknowledged, and numerous examples of differential responses to different challenges are presented. Requirements for the production of specific (differential) responses by plants are also discussed. Specificity of effect refers to the differential effects of a chemical response on different arthropods and pathogens. Most induced responses to arthropods and pathogens appear to be general (but not comprehensive) in their effect on subsequent organisms, although the issue of specificity of effect is rarely addressed systematically or comprehensively. Several recent studies have demonstrated the potential for plant-mediated interactions between pathogens and arthropods. A greater understanding of 1) the production of elicitors during plant-phytophage interactions, 2) signal transduction following pathogen and arthropod attack, and 3) the susceptibility of diverse pathogens and arthropods to diverse inducible chemicals is needed before generalizations about specificity of induced responses can be made.

Introduction

Plants, because they are sessile, must cope with variable environments by making adaptive biochemical and physiological changes (Bradshaw and Hardwick 1989). It is not surprising, then, that plants respond biochemically to attack by both arthropod herbivores and microbial pathogens, and that these biochemical responses are often accompanied by changes in resistance to subsequent arthropods and pathogens. Induced responses to pathogens (e.g., accumulation of phytoalexins and systemic acquired resistance) have been studied for over 65 years (Uknes et al. 1995, Kuć 1995, Ryals et al. 1996). The study of induced responses to arthropods is a younger discipline, with intensive study of the phenomenon initiated only after Green and Ryan's (1972) demonstration that tomato plants respond to wounding by producing large amounts of proteinase inhibitors, but the field has flourished in a relatively short time (Karban and Myers 1989, Tallamy and Raupp 1991, Karban and Baldwin 1997). Unfortunately, the study of plant responses to pathogens and the study of plant responses to arthropods have progressed largely independently, and each discipline has its own methodologies, experimental approaches, and definitions (Hatcher 1995, Stout et al. 1999). The integration of plant responses to pathogen invasion and arthropod feeding is therefore a virtually unexplored topic.

A typical plant is faced with a great diversity of threats. Leaf-chewing and leaf-mining arthropods, vertebrate grazers, phloem and xylem feeders, lesion-inducing fungi, bacteria, and viruses, gall-forming nematodes and insects, and the like (collectively referred to below as *phytophages*) may all attack a plant within its lifetime. A plant must also counter various abiotic stresses such as ultraviolet light, water stress, and nutrient stress. Because these stresses can severely reduce plant fitness (Hatcher 1995), and because multiple stresses can impinge simultaneously, plant responses to pathogens and arthropods must be consolidated into a larger system capable of dealing simultaneously in an effective, adaptive manner with multiple biotic and abiotic stresses.

Two extreme strategies can be envisioned by which a plant might integrate and coordinate its responses to phytophages. On the one hand, a plant may produce unique responses to each of its phytophages, with each response active against only a few phytophages. Such specificity in induced responses might allow a plant to tailor its defensive phenotype to its environment; i.e., to produce chemicals that are particularly effective against the inducing agents present in its environment. On the other hand, a plant may be capable of only one or a few types of responses, each initiated by a wide variety of inducing agents, and each with broad antibiotic activity. Such overlap or non-specificity in induced responses may be favored if many phytophages attack a plant simultaneously and the plant lacks the resources to mount specific defenses against each.

The specificity of induced responses is thus an important aspect of the integration of plant responses to pathogens and arthropods and is the subject of this review. For heuristic purposes, we distinguish two aspects of specificity (Karban and Baldwin 1997). *Specificity of response* refers to the ability of a plant to generate distinct chemical responses to different attackers or abiotic stresses; i.e., the ability of a plant to distinguish between different types of challenges and to respond in a manner particular to each. *Specificity of effect* refers to the differential effect of a chemical response on different herbivores and pathogens. These two aspects are not necessarily associated. In what follows, we review data relevant to both aspects of specificity and show how recent biochemical and genetic approaches are providing definition to and critical assessment of steps in signaling pathways and responses associated with defense.

Specificity of Plant Response

A single plant species may contain dozens of pathogen- or arthropod-responsive enzymes, low molecular-weight organic compounds ("secondary compounds"), non-catalytic proteins, and structural compounds (Constabel, this volume, Hammerschmidt and Nicholson, this volume). A plant response to pathogen invasion or herbivory involves the coordinate induction of sets of this total array of defense-related chemicals and gene products (Ward et al. 1991, Stout et al. 1996). *Specificity of plant response may be defined as the induction of different sets of chemicals or gene products by different biotic challenges.* Differential induction may be manifested as differences in patterns of accumulation of secondary chemicals, of mRNA transcripts, or of isoforms of a protein.

Requirements for Specificity in Plant Response

The production of responses to infection or herbivory by a plant is the result of a process that can be divided into four phases. First, the interaction of the phytophage with the plant must result in the production of one or more signal molecules (*elicitors*); second, these elicitors must be recognized by the plant; third, the recognition of the elicitor must lead to the activation of a signal transduction pathway(s) for local and, if applicable, systemic signaling; fourth, chemical compounds must be synthesized and transported to sites of expression. Certain requirements must be met at each step of this process for differential responses to be produced.

Various aspects of this process have been the subject of recent reviews, albeit usually from a plant pathological but not entomological perspective. The critical determinant of specificity in response is probably the generation of elicitors during plant-phytophage interactions. This step has received some attention from plant pathologists, but very little from entomologists and ecologists. Accordingly, it is emphasized below.

Generation and recognition of elicitors during plant/phytophage interactions. Elicitors are signals released during plant-phytophage interactions that trigger plant responses such as cell death, the production of phytoalexins and other secondary chemicals, and fortification of cell walls (Ebel and Cosio 1994, Hahn 1996, Staskawicz 1995). These elicitors may originate in the phytophage itself (i.e., exogenous elicitors), or they may be of plant origin and be released as a consequence of the plant-phytophage interaction (i.e., endogenous elicitors) (Ebel and Cosio 1994).

Irrespective of the origin of elicitors, the production of distinct elicitors or combinations of elicitors by different phytophages is necessary for the phytophages to be recognized as different by a plant and for differential responses to be produced. In the simplest case, two phytophages may be distinguished by a plant if their interactions with the plant result in the release of two distinct elicitors. However, a phytophage-plant interaction almost certainly results in the release of multiple signals, and phytophages are probably distinguished on the basis of the combination of signals that they produce. An arthropod or pathogen may secrete numerous digestive/degradative enzymes during feeding or invasion, respectively, and the action of these enzymes may lead to the generation of more than one elicitor. Other general elicitors, as well as more specific elicitors—for example, products of *avr* (avirulence) genes—may be released in combination with digestive/degradative enzymes. Furthermore, both pathogen invasion and arthropod feeding involve physical trauma to the plant in addition to the release of chemical factors. Physical trauma itself may be multifaceted, with several types of physical forces brought to bear on leaf tissue by a single biotic agent. Thus, feeding and pathogen invasion are both likely to release an array of endogenous and exogenous elicitors; the elicitors released by different phytophages may partially overlap and partially differ.

Very little is known about the generation of elicitors during feeding by insects. Plant toxins *per se* have not been isolated from arthropods, as they have from some plant pathogens (e.g., *Alternaria* spp.), but some compounds capable of eliciting responses in plants have been isolated from arthropods (see Felton and Eichenseer, this volume, for a more complete review). Elicitors isolated from arthropod saliva include elicitors of volatile production from the caterpillar species *Pieris brassicae* and *Spodoptera exigua* (Mattiacci et al. 1995, Alborn et al. 1997), an apparent inhibitor of systemic nicotine production in tobacco from *Manduca sexta* (McCloud and Baldwin 1997), and an assortment of pectinesterases, polygalacturonases, and cellulases (functionally similar to those produced by many plant pathogens) from many phloem- and parenchyma-feeding arthropods (Ma et al. 1990, Miles 1990).

Elicitors are also produced as a result of the physical injury to plant tissues caused by insect feeding. The potential clearly exists for the differential release of elicitors during feeding by different types of arthropods, because feeding of arthropods with different modes of feeding causes different kinds of physical injury to plants. For example, the damage caused by the extension of an aphid proboscis intercellularly to phloem tissue is very different from the shearing and crushing of cells that occurs when a chewing insect feeds on a leaf (Fidantsef et al. 1999). Even arthropods with ostensibly similar modes of feeding (e.g., two types of chewing insects) can differ markedly in the ways that they remove leaf material or cell contents from a plant (Parella et al. 1985, Bernays and Janzen 1988). Furthermore, arthropods that feed on different tissues of a plant (e.g., parenchyma feeders versus phloem feeders) may cause the release of different endogenous elicitors by virtue of differences in the biochemical and structural components of different plant tissues.

The most extensive research on the release of endogenous elicitors by wounding has been conducted using tomato as a model. Feeding by chewing insects on or mechanical wounding of tomato leaves results in the systemic, *de novo* synthesis of proteinase inhibitors and over a dozen other defense-related proteins (Schaller and Ryan 1995). Oligogalacturonides (pectin fragments) were found to elicit production of proteinase inhibitors when supplied in relatively high concentrations to cut petioles of tomato leaves (Bishop et al. 1984, Ryan 1992, Bergey et al. 1996). These oligogalacturonides are perhaps derived in an as-yet unknown fashion from wounded plant cell walls, although the enzymes necessary for their release are not found in tomato cell walls (but see Bergey et al. 1999). The same oligogalacturonides were found to be elicitors of phytoalexin synthesis and of several other defense-related responses in other plants (Walker-Simmons et al. 1983, Ryan 1992) and are thus general elicitors of plant responses. Oligogalacturonides were later shown to be immobile in tomato plants and cannot be responsible for systemic signaling of wound responses (Baydoun and Fry 1985, Bergey et al. 1996). Because of the low activity and mobility of oligogalacturonides, and because their natural derivation from wounded cell walls is unclear, other elicitors of proteinase inhibitor synthesis were sought and found. One of these is systemin, an octadecapeptide active at femto- to pico-molar levels that is rapidly generated in injured tomato leaves and that can move through the vascular system to uninjured leaves (Schaller and Ryan 1995, Bergey et al. 1996). Systemin induces jasmonic acid (JA) and its volatile derivative, methyl jasmonate, both ubiquitous in plants, which are potent inducers of proteinase inhibitor production in tomato (Bergey et al. 1996). The current model favors the generation of a hydraulic signal immediately following injury to a leaf that triggers the release within 60-90 minutes of systemin (derived from a larger precursor protein, prosystemin) into the vascular system. Receptors on the plasma membranes of distal cells interact with the translocated systemin, resulting in the intracellular release of linolenic acid from membrane phospholipids, via phospholipases. The linolenic acid is then enzymatically converted into JA which, in turn, causes the transcriptional activation of genes for proteinase inhibitors, polyphenol oxidase, and other defense-related proteins (Schaller and Ryan 1995, Bergey et al. 1996).

There is limited direct evidence for the release of elicitors during pathogen infection, although processing from bound or polymeric precursors is inferred by detailed structure-activity studies of highly purified and potent forms (Hahn 1996). Ricker and Bostock (1992) provided direct evidence for release and uptake of eicosanoid-derived elicitors during infection of potato by *Phytophthora infestans* using microautoradiography. Arachidonate or its metabolites from ^{14}C-labelled sporangia could be detected in leaf cells at a distance from the inoculation site within 9-12 hours after inoculation. Since many elicitors that have been characterized can be derived from cell walls of the host or pathogen, it is likely that the mix of elicitor-active signals present at the host-pathogen interface is as complicated as the potential array of wall-degrading/modifying enzymes also present. This, coupled with the evidence for suppressors and synergists of elicitor action (Darvill and Albersheim 1984, Preisig and Kuć 1985, Bostock and Stermer 1989, Hahn 1996), has created a confusing picture that is subject to generalization at only a rather superficial level. If each mixture elicits a unique signature response, then it is not surprising that the potential for signal interactions – both positive and negative - is great. The capacity for different signals to alter the cellular sensitivity to a biologically active compound – an issue well recognized in phytohormone research (Trevawas 1991) – has not received adequate attention in elicitor research.

The generation of distinct signals during plant-phytophage interactions is not a sufficient condition for the production of differential responses. For differential responses to be produced, distinct signals must also be recognized as distinct by the plant (Hahn 1996, Ebel and Cosio 1994, Baker et al. 1997). In many cases, recognition is accomplished via the interaction of elicitors with receptors located on plasma membranes or in the cytoplasm of cells. Several studies have provided direct evidence for interaction of certain elicitors with proteins that bind them with high specificity and affinity (Cosio et al. 1992, Ji et al. 1998). However, binding *per se* does not establish that the ligand-protein interaction determines a receptor-mediated response to the elicitor, and molecular genetic studies are needed to validate that the binding proteins are functionally relevant. In other cases, binding of receptors and ligands may not be necessary for recognition to occur. For example, the interactions of many phytophages with their host plants result in the generation of active oxygen species, and these species may elicit plant responses (e.g., cell death and hypersensitive responses) without the participation of a receptor (Alvarez et al. 1998, Jennings et al. 1998, Watanabe and Sakai 1998). Regardless of how recognition is accomplished, elicitors from different phytophages must be recognized as distinct in order for consistently different responses to be produced.

Elicitor-receptor interactions vary in their specificity. For example, a wide range of plant species appear to possess receptors for oligosaccharide fragments of chitin, whereas receptors for hepta-β-glucoside appear to be restricted to plants of the Fabaceae (Hahn 1996). Gene-for-gene interactions represent the extreme in elicitor-receptor specificity. In such interactions, the direct or indirect product of an *avr* gene which is present only in certain races of a pathogen (race-specific elicitor) is presumed to interact with the product of a corresponding resistance (*R*) gene that is present only in certain cultivars of a plant species, resulting in the induction of a hypersensitive response and restriction of the pathogen (Ebel and Cosio 1994, Staskawicz et al. 1995, Tang et al. 1996, Buell, this volume). There is evidence that protein products of some *R* genes interact directly with the products of *avr* genes in a race-specific manner, such as that reported for the *avrPto* and *Pto* genes that confer specificity in the bacterial speck disease of tomato (Scofield et al. 1996).

Signal transduction and production of defense-related compounds. Recognition by a plant of an elicitor results in the activation of one or more signal transduction pathways and ultimately in the synthesis of defense-related compounds, usually via transcriptional activation of defense-related genes. Signal transduction pathways involved in responses to pathogens are currently the subject of intensive research (Dixon and Harrison 1990, Enyedi et al. 1992, Ebel and Cosio 1994, Dixon et al. 1994, Hammond-Kosack and Jones 1996, Baker et al. 1997). Very rapid, local events common to many signal transduction pathways include influx of calcium ions, alteration of membrane potentials, production of active oxygen species, and phosphorylation of proteins (Ebel and Cosio 1994, Dixon et al. 1994, Ryals et al. 1996, Hammond-Kosack and Jones 1996, Baker et al. 1997, Romeis et al. 1999). Later events (but prior to synthesis of defensive compounds) include synthesis of salicylic acid (SA) and other signal compounds (see below) (Hammond-Kosack and Jones 1996). If the response is systemic, a systemic signal (e.g., systemin) must be produced and translocated (Enyedi et al. 1992, Bergey et al. 1996, Ryals et al. 1996). Importantly, if differential responses to different elicitors are to be produced, signal transduction pathways must be at least partially distinct. This does not, however, exclude the possibility of interactions between components of signal transduction pathways ("crosstalk") (Romeis et al. 1999). Both inhibitory (Doares et al. 1995) and synergistic (Graham and Graham 1996, Schweizer et al. 1997) interactions between pathways have been reported.

Specificity may be lost during signal transduction because signal transduction pathways initiated by separate elicitor-receptor interactions may converge downstream to produce the same chemical response. Such appears to be the case for many *R-Avr* interactions in *Arabidopsis* (Baker et al. 1997). Specificity of a kind may also be gained during signal transduction if interactions between signaling pathways lead to the production of qualitatively different responses than are produced by each pathway in isolation (Graham and Graham 1996).

With regard to systemic responses to pathogens and arthropods, the majority of recent work has focused on two signal transduction pathways, one mediated in part by SA, the other mediated in part by the octadecanoid pathway, of which JA is a component. The SA pathway is generally associated with pathogen infection; the octadecanoid pathway, with wounding or feeding by chewing insects (but see below for recent studies with *Arabidopsis*). Endogenous levels of SA and JA increase following pathogen infection and wounding, respectively (Ryals et al. 1996, Bergey et al. 1996, Hammerschmidt and Smith-Becker, this volume, Staswick and Lehman, this volume). Increases in SA are associated with the expression of a number of pathogenesis-related (PR) proteins and of broad spectrum resistance to pathogens (SAR), whereas increases in JA are associated with increases in the levels of several compounds, including proteinase inhibitors and polyphenol oxidase, and with increases in levels of resistance to chewing insects (Schaller and Ryan 1995, Mauch-Mani and Métraux 1998). Plants compromised in their ability to produce SA show reduced induction of PR proteins and increased susceptibility to pathogens; likewise, plants compromised in their ability to produce JA expressed reduced levels of proteinase inhibitors and do not show increases in resistance to insects following wounding (Schaller and Ryan 1995, Mauch-Mani and Métraux 1998, McConn et al. 1997). The presence of this fundamental dichotomy has been confirmed in several plant species.

Examples of Differential Responses

The few integrated models of plant responses to pathogens and arthropods proposed thus far have emphasized the wounding (octadecanoid) / pathogen infection (SA) dichotomy discussed in the previous section (Enyedi et al. 1992, Farmer and Ryan 1992, Ebel and Cosio 1994). Indeed, there is plentiful, though not unequivocal, evidence for differences in responses to wounding and pathogen infection. This emphasis in the literature on the wounding/infection dichotomy has led to the perception that plants possess a strictly bifurcated phytophage recognition system (Enyedi et al. 1992) that distinguishes only between "wounding", meant to encompass all types of arthropod feeding, and pathogen invasion. However, several lines of evidence suggest that phytophage recognition in plants is more sophisticated and more specific than this simple dichotomy suggests. In the following sections, evidence for differential responses to pathogen infection and wounding, as well as evidence that undermines a simplistic, dichotomous view of plant responses, will be presented.

Pathogen infection and mechanical wounding. Examples of differences in induction by pathogen infection and mechanical wounding pervade the literature. Characterizations of differential induction by wounding and pathogen infection have been conducted using both monocots and dicots, with the phenomenon most thoroughly characterized in leguminous and solanaceous plants.

Nonetheless, plant responses to pathogen infection and mechanical wounding also share much in common. *De novo* synthesis of phenolic compounds, deposition of structural compounds such as lignin and hydroxyproline-rich glycoproteins, and increases in the activities of oxidative enzymes such as lipoxygenase and peroxidase are common consequences of both pathogen infection and wounding (Bostock and Stermer 1989, Hatcher 1995). This apparent contradiction is partially reconciled by the fact that many of these similarities are only superficial: when many ostensibly similar responses to wounding and pathogen infection are examined more closely (e.g., when patterns of accumulation of isozymes of a wound- and pathogen-induced enzyme activity are examined), subtle differences in responses to wounding and pathogens can often be discerned. However, some responses to infection and wounding are identical even at the level of isozyme expression or gene expression; for example, genes encoding basic PR proteins are often induced by both wounding and pathogen infection (Brederode et al. 1991, Chang et al. 1995, Després et al. 1995). Thus, the wounding/pathogen dichotomy, while evident in many contexts, does not apply to all plant responses.

Stress-induced isoprenoid metabolism in Solanaceous plants illustrates how responses to wounding and pathogens can be subtly regulated at the level of differential gene expression, but with profound consequences at the levels of metabolic pathway products and cellular response. Wounding induces sterol and steroid glycoalkaloid accumulation in potato and tobacco with a corresponding activation of key enzyme activities in the acetate-mevalonate pathway – HMG CoA reductase (HMGR) and squalene synthase (Bostock and Stermer 1989, Chappell 1995). Pathogen infection or elicitor treatment can trigger a hypersensitive response and results in redirecting the pathway towards sesquiterpenoid phytoalexins by activation of a different HMGR activity and a novel sesquiterpene cyclase, with a concomitant inhibition of squalene synthase (Tjamos and Kuć 1984, Stermer and Bostock 1987, Zook and Kuć 1991). Choi et al. (1992) showed that the effects on HMGR activities were evident at the level of differentially regulated isoforms of HMGR gene family members, and that these isoforms responded differentially to methyl jasmonate and salicylates (Choi et al. 1994). These studies indicate that signals generated during wounding and pathogen attack can have inverse effects on the expression of activities that appear functionally identical. These studies also provide evidence for the presence of a complex organization of plant isoprenoid metabolism involving possibly discrete channels ("metabolons", Chappell 1995) or other forms of compartmentation.

The effects of wounding and pathogen infection on the metabolism of phenylpropanoid phytoalexins in legumes have been extensively studied (Bowles 1990). In bean hypocotyls and cotyledons, addition of elicitor preparations from *Colletotrichum lindemuthianum* and wounding both induced the accumulation of the isoflavonoids kievitone and phaseollin, but induction following elicitor treatment was more marked. Moreover, wounding caused an increase in activities of phenylalanine ammonia lyase, but elicitor treatment did not (Whitehead et al. 1982). Later studies showed that wounding and infection by *C. lindemuthianum* of *Phaseolus vulgaris* hypocotyls induced the accumulation of different patterns of both chalcone synthase and phenylalanine ammonia lyase transcripts (Ryder et al. 1987, Liang et al. 1989). In soybean roots, induction following treatment with a glucan elicitor from *Phytophthora sojae* was over 60-fold, while induction following wounding was approximately 10-fold (Krause et al. 1995). In soybean leaves, use of a minimal-wound cotyledon assay allowed Graham and Graham (1996) to distinguish the effects of wounding and treatment with an elicitor from *P. sojae*. Wounding alone induced the deposition of phenolic polymers, and elicitor treatment alone induced the accumulation of conjugates of the isoflavone daidzein, a precursor of glyceollin. The addition of both a chemical wound exudate and elicitor to cotyledons resulted in enhanced deposition of phenolic polymers and induction of glyceollin.

Differential effects of wounding and pathogen infection are not restricted to phytoalexin responses. Genes encoding acidic, extracellular PR proteins in tobacco were strongly induced by tobacco mosaic virus (TMV) infection but were not induced by wounding, whereas genes encoding basic, intracellular PR proteins were induced by both wounding and TMV infection (Brederode et al. 1991). Induction of acidic but not basic PR proteins was systemic. Incidentally, the fact that acidic but not basic PR proteins are differentially induced by wounding and TMV infection demonstrates that different signal transduction pathways from the same plant may differ in their specificity. Similarly, in potato, mRNAs for three osmotin-like proteins were differentially expressed following wounding and infection by *Phytophthora infestans.* Proteins corresponding to the osmotin mRNAs accumulated only in fungus-infected plants (Zhu et al. 1995). Wounding of tobacco leaves induced the expression of several cationic isozymes of the oxidative enzyme peroxidase, while infection by TMV induced two anionic isozymes of peroxidase (Lagrimini and Rothstein 1987). Three hydroxyproline-rich glycoprotein transcripts exhibited differential patterns of accumulation in bean hypocotyls in response to excision and treatment with a fungal elicitor (Corbin et al. 1987).

Beyond the wounding/pathogen dichotomy. The wounding/pathogen infection dichotomy is clearly important and pervasive. But is this the only distinction of which plants are capable? A definitive answer to this question is hindered by the scarcity of studies that have directly compared the responses of a plant to different types of biotic inducing factors and that moreover have simultaneously examined multiple responses at a level adequate for detecting differences in response. Nonetheless, several lines of recent evidence strongly suggest that the wounding/pathogen infection dichotomy is not sufficient to characterize plant responses to pathogens and arthropods.

First, mechanical wounding and arthropod feeding are often not synonymous elicitation events; thus, a dichotomy in responses to pathogens and arthropods cannot be inferred from the existence of such a dichotomy in responses to pathogens and mechanical wounding. A number of quantitative comparisons of induction by wounding and arthropod feeding have been made, with various results (Baldwin 1990). Some studies have found larger increases following mechanical wounding than following arthropod feeding (Hildebrand et al. 1989, Baldwin 1991, McCloud et al. 1995), some have found smaller increases following wounding (Kendall and Bjostad 1990, Lin et al. 1990, Stout et al. 1994), and some have found no difference (Bodnaryk 1992). Qualitative differences in responses to wounding and insect feeding have also been found.

Lima bean, for example, emits different blends of volatile compounds following mite feeding and mechanical damage (Takabayashi and Dicke 1996).

Strong evidence for differences in responses to mechanical wounding and feeding by chewing insects comes from recent studies of elicitation by oral secretions from several caterpillar species. Volatile blends released by corn seedlings following *S. exigua* feeding differ substantially from blends released following mechanical wounding (Turlings et al. 1993) An elicitor from *S. exigua* regurgitant was found to be responsible for this difference (Alborn et al. 1997). Korth and Dixon (1997) found that transcripts for HMGR and a proteinase inhibitor accumulated more quickly following damage by *M. sexta* than following mechanical damage; addition of *M. sexta* regurgitant to mechanical wounds resulted in patterns of transcript accumulation similar to those observed following *M. sexta* feeding. McCloud and Baldwin (1997) found that allowing *M. sexta* caterpillars to damage tobacco plants, or adding regurgitant to mechanical wounds, resulted in a greater induction of endogenous jasmonic acid than did mechanical wounding alone.

Qualitative and quantitative differences in responses to different types of arthropods have also been found. Leaves of apple, tobacco, cotton, and maize emit different blends of volatiles following injury by different arthropods and even following injury by different instars of the same insect (Takabayashi and Dicke 1996, De Moraes et al. 1998). Tomato plants respond differentially to feeding by different herbivores (Stout et al. 1994, 1996, 1998). Feeding by mites and aphids resulted in increased foliar activities of the enzymes lipoxygenase and peroxidase; feeding by leafminers induced only slight increases in peroxidase activity; feeding by caterpillars induced lipoxygenase activity locally and polyphenol oxidase and proteinase inhibitor activities locally and systemically. Similarly, in soybean, feeding by the three-cornered alfalfa hopper caused increases in the activities of several oxidative enzymes, including lipoxygenase, peroxidase, ascorbate oxidase, and polyphenol oxidase, whereas feeding by the bean leaf beetle caused increases in lipoxygenase activity only (Felton et al. 1994). Finally, Hartley and Lawton (1991) reported differential effects of leafminers, chewing insects, and mechanical damage on phenolics and phenolic metabolism in birch.

Plants likewise respond to different types of pathogens differently. Differences in response to compatible and incompatible pathogens, of course, are well-documented (Dixon et al. 1994, Heath 1997). Often, the responses induced by compatible and incompatible pathogens are qualitatively similar, but responses to compatible pathogens are delayed and sometimes attenuated in magnitude relative to responses to incompatible pathogens. However, different pathogens may also induce qualitatively different responses in their hosts. A comparison of gene expression in young rice plants infected with the compatible pathogen *Magnaporthe grisea* and with the incompatible pathogen *Pseudomonas syringae* pv. *syringae* revealed significant differences in response to the two pathogens (Schweizer et al. 1997a). In *Arabidopsis*, several genes encoding proteins of unknown function are specifically activated following different *R-Avr* gene interactions (Hammond-Kosack and Jones 1996). Fidantsef et al. (1999) and Stout et al. (1999) examined the responses of tomato leaves to *Phytophthora infestans*, the causal agent of late blight, and *P. syringae* pv. *tomato* (*Pst*), the causal agent of bacterial speck. They found that infection by both pathogens induces the expression of genes encoding a PR protein and lipoxygenase, but that infection by *Pst* induced an additional and marked increase in polyphenol oxidase activity and a gene for a proteinase inhibitor (responses normally associated with wounding).

Third, there have been a number of reports of induction by arthropod feeding of responses normally associated with pathogen infection. A few arthropods have been shown to elicit hypersensitive responses in their hosts similar to the hypersensitive responses elicited in incompatible plant-pathogen interactions (Fernandes 1990). Feeding by the gall mite *Aceria cladophthirus* caused a hypersensitive response on resistant varieties of *Solanum dulcamara*, and the hypersensitive response was accompanied by an increase in the expression of several PR proteins and an increase in chitinase and glucanase activities (Bronner et al. 1991). Chitinase activities increased in citrus roots damaged by larvae of the citrus root weevil, *Diaprepes abbreviatus* (Mayer et al. 1995). Interestingly, most of the examples of similarities between pathogens and arthropods involve arthropods with a piercing/sucking mechanism of feeding (mites, aphids, etc.). This suggests that injury caused by these types of feeding may be similar to injury caused by some types of pathogen invasion.

Finally, when responses to pathogens and arthropods have been directly compared, both similarities and differences have been found. One of two proteins induced by feeding of the silverleaf whitefly, *Bemisia argentifolii*, on *Cucurbita pepo* was also induced upon infection by Zucchini yellow mosaic virus (Jimenez et al. 1995). In tomato, aphid feeding and infection by both *Pst* and *P. infestans* induced expression of PR proteins and peroxidase activity (Fidantsef et al. 1999). In addition, infection by *Pst* induced expression of a gene for a proteinase inhibitor, a response also observed following caterpillar feeding. All four organisms induced lipoxygenase gene expression to varying degrees, and both wounding and *Pst* infection induced polyphenol oxidase activity (Fidantsef et al. 1999, Stout et al. 1999).

Plant responses to diverse biotic threats are seemingly idiosyncratic, alike in some respects but different in others, with the nature and extent of overlap in responses dependent on the type of pathogens and herbivores compared. This seeming idiosyncrasy is not surprising given the lack of detailed comparisons of responses to different phytophages, and given that induced responses are the results of a complex interplay of multiple signaling events and signal transduction pathways. In order to formulate generalizations about the conditions that result in differential induction, characterization of the responses of plants to diverse phytophages must be combined with basic research on signal transduction. In particular, three features of signal transduction pertain directly to the issue of specificity:

(i) A plant possesses signal transduction pathways for response to pathogens and herbivores in addition to the octadecanoid and SA pathways. Strong evidence for this hypothesis comes from recent work in tomato and *Arabidopsis*. In tomato, mechanical wounding was shown to induce a 10-fold increase in levels of feruloyltyramine and *p*-coumaroyltyramine in addition to the well-documented increases in levels of proteinase inhibitors (Pearce et al. 1998). Increases in these phenolic compounds were not induced by JA and occurred in plants deficient in the octadecanoid signaling pathway. These data strongly suggest the existence of a wound response pathway which is distinct from the octadecanoid pathway (and distinct from the SA pathway, which is not activated by mechanical wounding in tomato). In *Arabidopsis*, multiple signal transduction pathways mediate both wound responses and pathogen responses. Many wound-induced genes are controlled by a JA-dependent pathway, but others are controlled by a JA-independent pathway (McConn et al. 1997). Race-specific responses in *Arabidopsis* to several races of *Peronospora parasitica* are mediated by at least two signal transduction pathways (Aarts et al. 1998). Other data are suggestive of transduction pathways in *Arabidopsis* apart from those involving JA and SA (Pieterse et al. 1996, Clarke et al. 1998, Thomma et al. 1998).

(ii) A biotic elicitation event usually leads to the production of more than one elicitor and to the activation of more than one signal transduction pathway. An example of the activation of multiple transduction pathways by a single phytophage is the response of tomato foliage to *Pst*, which appears to involve the activation of both the SA pathway, signified by the production of a PR protein, and the octadecanoid pathway, signified by the production of proteinase inhibitors (Fidantsef et al. 1999). Another example is the activation of both SA-dependent and JA-dependent classes of antifungal proteins by *Alternaria brassicicola* infection in *Arabidopsis* (Pennincyx et al. 1996).

(iii) Interactions of signals and transduction pathways may result in responses to combinations of elicitors that differ quantitatively and qualitatively from responses to single elicitors. Combinations of ethylene and JA caused the synergistic induction of PR-protein transcripts in tobacco seedlings (Xu et al. 1994) and proteinase inhibitors in tomato (O'Donnell et al. 1996). SA inhibits wound responses in rice (Schweizer et al. 1997), tomato (Doares et al. 1995), and other plants. In soybean leaves, as mentioned above, glyceollin accumulates only after the addition of both a wound exudate and an elicitor preparation (Graham and Graham 1996).

Specificity of Effect

Attack by a phytophage may induce *specific resistance* (effective against only the inducing species and perhaps a few related organisms) or it may induce *general resistance* (effective against a broad spectrum of organisms). Understandably, most investigations of specificity of induced resistance have examined the effect of arthropod feeding on subsequent arthropods or the effect of pathogen infection on subsequent pathogens, although increasing attention has been given to the possibility of plant-mediated interactions between pathogens and arthropods (*induced cross-resistance*). However, no detailed, systematic investigations of the breadth of resistance induced by a single phytophage against multiple pathogens and arthropods have been performed. As with investigations of response specificity, this lack of data hinders the formation of generalizations about the specificity of effect of induced responses.

A priori Considerations

The induction of a specific chemical response by pathogens or arthropods is often viewed as *prima facie* evidence for functional significance of specificity (Xu et al. 1994), i.e., specificity in response is generally thought to lead to specificity in effect, and specific responses are thought to be somehow more effective than general responses. For example, the differential regulation of compartmentalized isoforms of HMGR by pathogen infection and wounding is thought to allow more specific and appropriate responses to wounding and pathogen infection (Bostock and Stermer 1989, Choi et al. 1992).

Opposing this hypothetical argument for specificity in effect is the fact that most inducible phytochemicals have broad antibiotic effects that probably limit the potential for specific effects (Hatcher 1995, Karban and Baldwin 1997). All the major classes of secondary chemicals that are inducible by arthropods have been shown to possess antimicrobial properties, and the converse is also true (Nicholson and Hammerschmidt 1992, Osbourn 1996, Karban and Baldwin 1997, Constabel, Hammerschmidt and Nicholson, this volume). Damage-inducible quinolizidine alkaloids from lupines, for example, exhibit antibiotic or deterrent activities against fungi, bacteria, insects, mammals, and molluscs (Wink 1984). Nicotine inhibits the growth of both pathogens and insects (Krischik et al. 1991). Isoflavonoid phytoalexins from soybean and other legumes are deterrent to insects in addition to possessing antimicrobial properties (Lwande et al. 1985, Kogan and Fischer 1991). Inducible terpenes and phenolics from conifers are repellent to several beetle species and inhibit the growth of two species of fungi (Klepzig et al. 1996). Chitinases (identified in some species as PR proteins) reduce the growth or survival of several insect species when incorporated into artificial diet, perhaps by damaging peritrophic envelopes (Mayer et al. 1995, Broadway et al. 1998).

Another factor limiting the potential for specific effects is the manifold nature of plant responses. A plant response to a phytophage is generally comprised of changes in levels or activities of numerous primary and secondary metabolites, structural compounds, enzymes, and defense-related proteins (Bowles 1990, Ward et al. 1991, Stout et al. 1996). In tomato, for example, mechanical wounding of leaves results in the systemic induction of the mRNA transcripts of over 15 proteins (Bergey et al. 1996). It seems likely that at least one, if not more, of these component changes will be capable of affecting almost any given phytophage. Moreover, there may additive, synergistic, and antagonistic effects of multiple induced chemicals on phytophages.

Specificity of Induced Resistance to Pathogens

SA-mediated protection observed in SAR provides a general, broad-based, but not all-inclusive, resistance to many different types of pathogens, including those caused by viruses, bacteria, fungi and nematodes, an aspect that has been adequately reviewed elsewhere (Kuć 1987, Hammerschmidt 1993, Delaney et al. 1994, Mauch-Mani and Métraux 1998). This broad-spectrum resistance apparently results from the coordinate expression of a large battery of PR proteins (Cao et al. 1998). Relatively few plant species have been studied intensively, but the literature does suggest that SAR occurs broadly among flowering plants. Although acquired resistance to many diseases is a general phenomenon, there are certainly exceptions, and certain aggressive pathogens, such as *Sclerotinia sclerotiorum*, seem to be unaffected. Also, diseases important in post-harvest settings, such as brown rot of stone fruits, are not effectively controlled by this strategy (Adaskaveg, personal communication). Late blight disease of potato and tomato caused by *P. infestans* is not significantly reduced by SA and synthetic SA mimics (Fidantsef et al. 1999).

Recent research with *Arabidopsis* supports the presence of additional signaling pathways in induced resistance to pathogens, and some specificity of effect has been demonstrated. Different pathogens induce different sets of putatively antimicrobial proteins in *Arabidopsis*. Infection with *Alternaria brassicicola* activates a JA-dependent pathway, resulting in expression of the genes PDF1.2, PR-3 (a basic chitinase), and PR-4, whereas infection with *Peronospora parasitica* activates a SA-dependent pathway resulting in expression of PR-1, PR-2 , and other genes (Thomma et al. 1998). The former pathway is necessary for expression of resistance to *A. brassicocola* and *Botrytis cinerea* but not to *P. parasitica*; expression of the latter pathway is necessary for expression of resistance to *P. parasitica* but not to *A. brassicocola* and *B. cinerea.* (Thomma et al. 1998). Other studies with *Arabidopsis* mutants indicate that nonpathogenic rhizobacteria can induce systemic resistance by an SA-dependent pathway but also by pathways requiring the action of JA and ethylene (Mauch-Mani and Métraux 1998, van Loon et al. 1998).

Specificity of Induced Resistance to Insects

In several systems, resistance induced by arthropod feeding apparently affects a wide range of subsequent arthropods. Injury to soybean plants induces a systemic increase in resistance to both a lepidopteran (soybean looper) and a coleopteran (Mexican bean beetle) (Kogan and Fischer 1991). Cotton seedlings exposed to spider mites (*Tetranychus urticae* or *T. turkestani*) as cotyledons are more resistant to a number of herbivores than are cotton plants not exposed to mites as cotyledons. The induced resistance extends not only to the inducing mite species and to other, related mite species, but also to an insect with an entirely different feeding mechanism, *S. exigua* (Karban 1991). In tomato, damage inflicted by larvae of the corn earworm (*Helicoverpa zea*) to single leaflets of young plants was shown to induce systemic resistance not only against caterpillars (*H. zea* and *S. exigua*) but also against a species of aphid (*Macrosiphum euphorbiae*), a mite (*T. urticae*), and a leafminer (*Liriomyza trifolii*) (Stout and Duffey 1996, Stout et al. 1998). In contrast, larvae of *S. exigua* grew at a slightly faster rate on leaflets previously damaged by aphid feeding (Stout et al. 1998). In both the mite/cotton system and caterpillar/tomato systems, arthropod feeding also induces resistance against certain pathogens (see below).

Some specificity in effect has been demonstrated in other systems. In birch, four species of caterpillars preferred undamaged leaves to leaves damaged by leafminers. However, only one of three caterpillar species preferred undamaged birch leaves to leaves damaged by a chewing insect, and two of four caterpillars actually preferred mechanically-damaged leaves to undamaged leaves (Hartley and Lawton 1991). In a study of induced resistance in mountain birch, the growth of only two of six species of sawfly was significantly reduced by rapidly-induced responses to feeding by *Epirrita* larvae (Hanhimäki 1989).

Another possible example of specificity in induced resistance was reported by Raffa and Berryman (1987). Attempted colonization of conifers by several species of bark beetles and their associated fungal symbionts induces the formation of a lesion around the invasion site in resistant trees. These lesions contain monoterpenes and other compounds. A close correlation was found between the percent increase following invasion/inoculation and the toxicity of induced monoterpenes to inducing bark beetles and fungal symbionts, with the more toxic monoterpenes induced to a greater extent than less toxic monoterpenes.

Finally, some arthropods induce chemical responses in their host plants to which they are nearly impervious but which increase the resistance of the plants to other herbivores (Karban and Baldwin 1997). A recent example was provided by Bolser and Hay (1998). They found that feeding by the specialist beetle *Galerucella nymphaeae* on the water lily *Nuphar luteum macrophyllum* decreased attractiveness of water lily tissue to a generalist herbivore (a crayfish) but had no effect on attractiveness of tissues to subsequent *G. nymphaeae*.

Damage-inducible cucurbitacins from cucurbits have no negative effects on squash beetles (*Epilachna borealis*) or striped cucumber beetles (*Acalymma vittata*) (McCloud et al. 1995), but do have negative effects on other herbivores (e.g., spider mites; A. Agrawal, personal communication). The arthropods in these examples are apparently able to tolerate or detoxify the chemicals they induce; these chemicals are toxic or deterrent to a wide range of other organisms.

Interactions between pathogens and arthropods

Local effects of wounding and pathogen infection. Damaged tissues and diseased tissues undergo many biochemical and physiological changes related to wound repair and infection (Bostock and Stermer 1989). Such local changes can have many consequences for arthropods and pathogens attempting to feed on or invade damaged or infected tissues. A detailed treatment of the local effects of infection and wounding on plant resistance is beyond the scope of this review, and the reader is referred to reviews by Hammond and Hardy (1988), Bostock and Stermer (1989), Hammerschmidt (1993), Hatcher (1995), and Zangerl (this volume).

Plant-mediated interactions between pathogens and arthropods. The breadth of SAR to pathogens and induced resistance to insects has prompted several researchers to ask whether systemic resistance induced by arthropods extends to pathogens, and conversely, whether systemic resistance induced by pathogens extends to arthropods. Some experiments that have tested for induced cross resistance between pathogens and arthropods have found such interactions, but others have not. The following review is restricted to direct tests of plant-mediated interactions between pathogens and arthropods in which the plant part or parts subjected to the inducing treatment (inoculation with a pathogen or bout of herbivory) was spatially separated from the plant parts used subsequently to assess resistance to pathogens or arthropods.

An early report of induced cross-resistance involving pathogens and arthropods was made by McIntyre et al. (1981). In their experiments, inoculation of tobacco with an isolate of TMV which caused a hypersensitive response was shown to induce systemic resistance against TMV, against two species of fungi and a bacterium, and against the green peach aphid, *Myzus persicae*. The resistance against the aphid species was manifested as a decrease in fecundity. Interestingly, initial infection with cucumber mosaic virus induced resistance against TMV but not against the other pathogens. Using an identical experimental protocol, Hare (1983) reported that TMV infection also caused a slight systemic reduction in the suitability of tobacco leaves for the tobacco hornworm, *M. sexta*. In contrast to these results, Ajlan and Potter (1992) found that tobacco plants inoculated on their lower leaves with TMV were no more resistant to the tobacco aphid (*M. nicotianae*) or *M. sexta* than uninoculated plants. Also, prior feeding by *M. sexta* larvae did not induce resistance against

TMV. These negative results were not attributable to a general non-responsiveness on the part of the plant: in the same experiments, prior TMV inoculation was found to increase systemic resistance of plants to TMV, and prior feeding by *M. sexta* induced resistance against aphids.

In cotton, damage to cotyledons caused by mite feeding (earlier shown to induce resistance against several arthropod species) resulted in increased resistance of cotton seedlings to *Verticillium dahliae*, the causal agent of bacterial wilt in cotton (Karban et al. 1987). This increased resistance was expressed as a decreased likelihood that previously-damaged plants would develop symptoms of wilt. In addition, populations of spider mites on plants infected with *V. dahliae* were smaller than populations of mites on healthy plants. However, because infection by the wilt fungus was systemic and concurrent with mite feeding, direct interspecific interactions between the mite and the fungus cannot be discounted.

Several investigations of plant-mediated interactions between pathogens and arthropods have been conducted using cucumber as a model. Restricted infection of lower leaves of cucumber plants with the anthracnose fungus, *Colletotrichum lagenarium*, had no effect on suitability of foliage for spider mites, for larvae of the fall armyworm, or for melon aphids, despite a demonstrable SAR response (Ajlan and Potter 1991). Prior herbivory by mites or fall armyworm larvae did not induce resistance against anthracnose. More recently, Moran (1998) found that cucumber plants inoculated with the fungal pathogen *Cladosporium cucumerinum* were systemically protected against subsequent infection by *C. orbiculare*, but no systemic effects on resistance to melon aphids or spotted cucumber beetles were found. Local effects (both positive and negative) of fungal infection on insects were found in the same study. Apriyanto and Potter (1990) found that inoculation of cucumber plants with tobacco necrosis virus (TNV) protected plants against subsequent infection by the anthracnose fungus but did not increase resistance of plants to spider mites, fall armyworm larvae, or whiteflies. However, a positive effect of prior inoculation with TNV on the feeding preference of the striped cucumber beetle was found.

Field treatment of cucumber plants with plant growth-promoting rhizobacteria (PGPR) protected plants against infestation by cucumber beetles (*Acalymma vittatum* and *Diabrotica undecimpunctata howardi*) and hence against infection by a cucurbit wilt pathogen (*Erwinia tracheiphila*) vectored by the beetles. Treatment with PGPR also protected cucumber plants against multiple diseases (Zehnder et al., this volume). Recent research in *Arabidopsis* indicates that induced resistance caused by PGPR treatment is regulated by transduction pathways distinct from the pathways that regulate SAR (Pieterse et al. 1998).

In tomato, a variety of plant-mediated interactions between pathogens and arthropods have been demonstrated (Stout et al. 1998, 1999). As previously noted, feeding by *H. zea* larvae localized to a single leaflet of young tomato plants causes a systemic increase in plant resistance to a wide variety of arthropod herbivores. The same type and extent of damage also induces a slight resistance to *Pst*. Localized infection by *Pst* induces systemic wound responses as well as SAR-related responses (see above) and causes a systemic decrease in suitability of foliage for larvae of the noctuid *H. zea* and for *Pst*. Thus, systemic cross resistance involving *H. zea* and *Pst* in tomato is reciprocal, although the two phytophages do not induce an identical battery of responses. In contrast, infection of single tomato leaflets by *P. infestans* had no effect on systemic resistance to *H. zea*. Feeding by aphids on tomato leaves induces the local expression of PR proteins; protection against *Pst* associated with this expression of PR proteins is inconsistent (Stout, Fidantsef, and Bostock, unpublished data).

Synthesis

Plants are capable of producing specific responses to different phytophages. The extent of this capacity to produce specific responses is not completely known, nor are the factors that determine the composition and nature of a response. With the limited data available, the nature of chemical responses to a given phytophage appear to be idiosyncratic. A more detailed understanding of a) the generation of elicitors during plant-phytophage interactions and b) signal transduction following pathogen and arthropod attack is needed before generalizations concerning specificity in induced responses can be made.

Most induced chemical responses appear to be somewhat general in their effects but are probably never effective against all potential phytophages. Thus, the functional significance of specificity of response is unclear. It may be that a chemical response to a phytophage is more likely to affect the inducing phytophage and closely related organisms than more distantly-related organisms, although this has yet to be clearly demonstrated. If these types of parallels do exist, they are not strict, and plants clearly do not respect taxonomic boundaries. Disparate phytophages can use similar biochemical and physical mechanisms to attack a plant; thus, they sometimes induce similar responses. Diverse phytophages are often sensitive to the same types of secondary chemicals, and they can possess similar detoxicative mechanisms; thus, they can sometimes be affected by similar chemical responses. A phytophage can be affected by a chemical response that it does not induce, and a phytophage may induce a chemical response in its host that it tolerates but that has negative impacts on a wide variety of other phytophages. Plant-mediated interactions between pathogens and arthropods can therefore be expected to occur in many systems, but the conditions under which these interactions are likely to occur have not been elucidated.

Future Directions

Host-plant resistance will be an increasingly important component of pest management programs in the future, and development of crop cultivars with broad-spectrum resistance against pathogens and arthropods is an important goal for agricultural scientists in the next few decades. Among other studies, the recent finding by Rossi et al. (1998) that resistance to a nematode and an aphid has a common genetic basis in tomato demonstrates the potential for breeding broad-spectrum resistance into plants. Research directed toward elucidating common genetic or biochemical bases of resistance to diverse pathogens and arthropods is critical to the attainment of this goal, but the subject has received inadequate attention. Inducible responses, because their expression can be partly controlled by the investigator, are ideal candidates for investigating questions about the integration of plant defenses against different phytophages, and for comparing and contrasting the biochemical and genetic bases of resistance to various pathogens and arthropods.

As a first step, more studies comparing the spectra of chemical responses and resistance induced by different phytophages must be performed. Such studies should ideally include a wide variety of pathogens and herbivores as both inducing agents and as testers of resistance. These experiments should be done systematically, with factors such as plant age and growth conditions held constant. The simultaneous measurement of induced resistance and induced chemical responses should provide insight into the similarities and differences between biochemical mechanisms of resistance to different types of pathogens and herbivores. All of this, of course, will require a greater degree of cooperation between plant pathologists and entomologists than is common. Moreover, a more mechanistic approach must be taken that recognizes that plants are essentially blind to the taxonomy of their attackers and that looks for the mechanistic bases for similarities and differences in induced responses to pathogens and arthropods.

Cataloguing of responses to various phytophages must be combined with research on signal transduction following both pathogen and arthropod attack. With respect to the dissection of signal transduction pathways, research on responses to arthropods lags far behind research on responses to pathogens. Use of both arthropods and pathogens in studies with plants that are genetically modified in signal generation, perception and response coupling should yield insights into the number of pathogen- and arthropod-responsive pathways operative, as well as how these response pathways are coordinated and functionally integrated during challenge from diverse biotic threats (McConn et al. 1997, Aarts et al. 1998).

Acknowledgements

This manuscript is dedicated with fond remembrance to the late Sean S. Duffey, who helped shape many of the ideas presented here. The manuscript benefited from discussions with Jennifer Thaler, Ana Fidantsef, and Rick Karban and from critical readings by Anurag Agrawal, Jim Ottea, and an anonymous reviewer. Our work on induced responses to pathogens and arthropods has been supported by grants from the USDA-NRICGP.

Literature Cited

Aarts, N., M. Metz, E. Holub, B.J. Staskawicz, M.J. Daniels, and J.E. Parker. 1998. Different requirements for EDS1 and NDR1 by disease resistance genes define at least two R gene-mediated signaling pathways in *Arabidopsis*. Proceedings of the National Academy of Sciences USA 95:10306-10311.

Ajlan, A.M. and D.A. Potter. 1991. Does immunization of cucmber against anthracnose by *Colletotrichum lagenarium* affect host suitability for arthropods? Entomologia Experimentalis et Applicata 58:83-91.

Ajlan, A.M. and D.A. Potter. 1992. Lack of effect of tobacco mosaic virus-induced systemic acquired resistance on arthropod herbivores in tobacco. Phytopathology 82:647-651

Alborn, H.T., T.C.J. Turlings, T.H. Jones, G. Stenhagen, J.H. Loughrin, and J.H. Tumlinson. 1997. An elicitor of plant volatiles from beet armyworm oral secretion. Science 276:945-949.

Alvarez, M.E., R.I. Pennell, P. Meijer, A. Ishikawa, R.A. Dixon, and C. Lamb. Reactive oxygen intermediates mediate a systemic signal network in the establishment of plant immunity. Cell 92:773-784.

Apriyanto, D. and D.A. Potter. 1990. Pathogen-activated induced resistance of cucumber: response of arthropod herbivores to systemically protected leaves. Oecologia 85:25-31.

Baker, B., P. Zambryski, B. Staskawicz, and S.P. Dinesh-Kumar. 1997. Signaling in plant-microbe interactions. Science 276:726-733.

Baldwin. I.T. 1990. Herbivory simulations in ecological research. Trends in Ecology Evolution 5:91-93.

Baldwin, I.T. 1991. Damage-induced alkaloids in wild tobacco. Pages 47-69 *in* D. Tallamy and M. J. Raupp, editors. Phytochemical Induction by Herbivores. John Wiley and Sons, Inc., New York.

Baydon, E.A.-H. and S.C. Fry. 1985. The immobility of pectic substances in injured tomato leaves and its bearing on the identity of the wound hormone. Planta 165:269-276.

Bergey, D.R., G.A. Howe, and C.A. Ryan. 1996. Polypeptide signaling for plant defensive genes exhibits analogies to defensive signaling in animals. Proceedings of the National Academy of Sciences USA 93:12053-12058.

Bergey, D.R., M. Orozco-Cardenas, D.S. De Moura, and C.A. Ryan. 1999. A wound- and systemin-inducible polygalacturonase in tomato leaves. Proceedings of the National Academy of Sciences USA 96:1756-1760.

Bernays, E.A. and D.H. Janzen. 1988. Saturniid and sphingid caterpillars: two ways to eat leaves. Ecology 69:1153-1160.

Bishop, P.D., G. Pearce, J.E. Bryant, and C.A. Ryan. 1984. Isolation and characterization of the proteinase inhibitor-inducing factor from tomato leaves. Identity and activity of poly- and oligogalacturonide fragments. Journal of Biological Chemistry 259:13172-13177.

Bodnaryk, R.P. 1992. Effects of wounding on glucosinolates in the cotyledons of oilseed rape and mustard. Phytochemistry 31:2671-2677.

Bolser, R.C. and M.E. Hay. 1998. A field test of inducible resistance to specialist and generalist herbivores using the water lily *Nuphar luteum*. Oecologia 116:143-153.

Bostock, R.M. and B.A. Stermer. 1989. Perspectives on wound healing in resistance to pathogens. Annual Review of Phytopathology 27:343-371.

Bowles, D.J. 1990. Defense-related proteins in higher plants. Annual Review of Biochemistry 59:873-907.

Bradshaw, A.D. and K. Hardwick. 1989. Evolution and stress—genotypic and phenotypic components. Biological Journal Linnean Society 37:137-155.

Brederode, F.T., H.J.M. Linthorst, and J.F. Bol. 1991. Differential induction of acquired resistance and PR gene expression in tobacco by virus infection, ethephon treatment, UV light and wounding. Plant Molecular Biology 17:1117-1125.

Broadway, R. M., C. Gongora, W.C. Kain, J.P. Sanderson, J.A. Monroy, K.C. Bennett, B. Warner, and M.P. Hoffmann. 1998. Novel chitinolytic enzymes with biological activity against herbivorous insects. Journal of Chemical Ecology 24:985-998.

Bronner, R., E. Westphal, and F. Dreger. 1991. Pathogenesis-related proteins in *Solanum dulcamara* L. resistant to the gall mite *Aceria cladophthirus* (Nalepa) (syn *Eriophyes cladophthirus* Nal.). Physiological and Molecular Plant Pathology 38:93-104.

Cao, H., X. Li, and X. Dong. 1998. Generation of broad-spectrum disease resistance by overexpression of an essential regulatory gene in systemic acquired resistance. Proceedings of the National Academy of Sciences USA 95:6531-6536.

Chang, M-M., D. Horovitz, D. Culley, L.A. Hadwiger. 1995. Molecular cloning and characterization of a pea chitinase gene expressed in response to wounding, fungal infection and the elicitor chitosan. Plant Molecular Biology 28:105-111.

Chappell, J. 1995. Biochemistry and molecular biology of the isoprenoid biosynthetic pathway in plants. Annual Review of Plant Physiology and Plant Molecular Biology 46:521-547.

Choi, D., B.L. Ward, and R.M. Bostock. 1992. Differential induction and suppression of potato 3-hydroxy-3-methylglutaryl coenzyme A reductase genes in response to *Phytophthora infestans* and to its elicitor arachidonic acid. Plant Cell 4:1333- 1344.

Choi, D., R.M. Bostock, S. Avdiushko, and D. Hildebrand. 1994. Lipid-derived signals that discriminate wound- and pathogen-responsive isoprenoid pathways in plants: methyl jasmonate and the fungal elicitor arachidonic acid induce different HMG-CoA reductase genes and antimicrobial isoprenoids in *Solanum tuberosum* L. Proceedings of the National Academy of Sciences USA 91:2329-2333.

Clarke, J.D., Y. Liu, D.F. Klessig, and X. Dong. 1998. Uncoupling PR gene expression from NPR1 and bacterial resistance: characterization of the dominant *Arabidopsis cpr6-1* mutant. Plant Cell 10:557-569.

Corbin, D.R., N. Sauer, and C.J. Lamb. 1987. Differential regulation of a hydroxyproline-rich glycoprotein gene family in wounded and infected plants. Molecular Cellular Biology 7:4337-4344.

Cosio, E.G., T. Frey, and J. Ebel. 1992. Identification of a high-affinity binding protein for a hepta-β-glucoside phytoalexin elicitor in soybean. European Journal of Biochemistry 204:1115-1123.

Darvill A.G. and P. Albersheim. 1984. Phytoalexins and their elicitors – a defense against microbial infection in plants. Annual Review of Plant Physiology 35:243-275.

Delaney T.P., S. Uknes, B. Vernooij, L. Friedrich, K. Weymann, et al. 1994. A central role of salicylic acid in plant disease resistance. Science 266:1247-50.

De Moraes, C.M., W.J. Lewis, P.W. Paré, H.T. Alborn, and J.H. Tumlinson. 1998. Herbivore-infested plants selectively attract parasitoids. Nature 393:570-572.

Despres, C., R. Subramaniam, D.P. Matton, and N. Brisson. 1995. The activation of the potato PR-10a gene requires the phosphorylation of the nuclear factor PBF-1. Plant Cell 7:589-598.

Dixon, R.A. and M. J. Harrison. 1990. Activation, structure, and organization of genes involved in microbial defense in plants. Advances in Genetics 28:165-234.

Dixon, R.A., M.J. Harrison, and C.J. Lamb. 1994. Early events in the activation of plant defense responses. Annual Review of Phytopathology 32:479-501.

Doares S.H., J. Narvaez-Vasquez, A. Conconi, and C.A. Ryan 1995. Salicylic acid inhibits sythesis of proteinase inhibitors induced by systemin and jasmonic acid. Plant Physiology 108:1741-1746.
Ebel, J. and E.G. Cosio. 1994. Elicitors of plant defense responses. International Review of Cytology 148:1-36.
Enyedi, A.J., N. Yalpani, P. Silverman, and I. Raskin. 1992. Signal molecules in systemic plant resistance to pathogens and pests. Cell 70:879-886.
Farmer, E.E. and C.A. Ryan. 1992. Octadecanoid precursors of jasmonic acid activate the synthesis of wound-inducible proteinase inhibitors. Plant Cell 4:129-134.
Felton, G.W., C.B. Summers, and A.J. Mueller. 1994. Oxidative responses in soybean foliage to herbivory by bean leaf beetle and three-cornered alfalfa hopper. Journal of Chemical Ecology 20:639-649
Fernandes, G.W. 1990. Hypersensitivity: a neglected plant resistance mechanism against insect herbivores. Environmental Entomology 19:1173-1182.
Fidantsef, A.L., M.J. Stout, J.S. Thaler, S.S. Duffey, and R.M. Bostock. 1999. Signal interactions in pathogen and insect attack: expression of lipoxygenase, proteinase inhibitor II, and pathogenesis-related protein P4 in the tomato, *Lycopersicon esculentum.* Physiological and Molecular Plant Pathology (in press).
Graham, T.L. and M.Y. Graham. 1996. Signaling in soybean phenylpropanoid responses - dissection of primary, secondary, and conditioning effects of light, wounding, and elicitor treatments. Plant Physiology 110:1123-1133.
Green, T.R. and C.A. Ryan, C.A. 1972. Wound induced proteinase inhibitor in plant leaves: a possible defense mechanism against insects. Science 175:776-777.
Hahn, M.G. 1996. Microbial elicitors and their receptors in plants. Annual Review of Phytopathology 34:387-412.
Hammond, A.M. and T.N. Hardy. 1988. Quality of diseased plants as hosts for insects. Pages 381-432 *in* E.A. Heinrichs, editor. Plant Stress-insect Interactions. John Wiley and Sons, New York.
Hammond-Kosack, K.E. and J.D.G. Jones. 1996. Resistance gene-dependent plant defense responses. Plant Cell 8:1773-1791.
Hammerschmidt, R. 1993. The nature and generation of systemic signals induced by pathogens, arthropod herbivores, and wounds. Advances in Plant Pathology 10:307-337.
Hanhimäki, S. 1989. Induced resistance in mountain birch: defence against leaf-chewing insect guild and herbivore competition. Oecologia 81:242-248.
Hare, J.D. 1983. Manipulation of host suitability for herbivore pest management. Pages 655-680 *in* R. F. Denno and M. S. McClure, editors. Variable Plants and Herbivores in Natural and Managed Systems. Academic Press, New York.
Hartley, S.E. and J.H. Lawton. 1991. Biochemical aspects and significance of the rapidly induced accumulation of phenolics in birch foliage. Pages 205-222 *in* D.W. Tallamy and M.J. Raupp, editors. Phytochemical Induction by Herbivores. John Wiley and Sons, New York.
Hatcher, P.E. 1995. Three-way interactions between plant pathogenic fungi, herbivorous insects and their host plants. Biological Reviews 70:639-694.
Heath, M.C. 1997. Evolution of plant resistance and susceptibility to fungal parasites. Pages 257-276 *in* Carroll and Tudzynski, editors. The Mycota V Part B - Plant Relationships. Springer-Verlag Berlin.
Hildebrand, D.F., J.G. Rodriguez, C.S. Legg, G.C. Brown, and G. Bookjans. 1989. The effects of wounding and mite infestation on soybean leaf lipoxygenase levels. Zeitschrift fuer Naturforschung 44c:655-659.
Jennings, D.B., M. Ehrenshaft, D.M. Pharr, and J.D. Williamson. 1998. Roles for mannitol and mannitol dehydrogenase in active oxygen-mediated plant defense. Proceedings of the National Academy of Sciences USA 95:15129-15133.
Ji, C., C. Boyd, D. Slaymaker, Y. Okinaka, Y. Takeuchi, S.L. Midland, J.J. Sims, E. Herman, and N. Keen. 1998. Characterization of a 34-kDa soybean binding protein for the syringolide elicitors. Proceedings of the National Academy of Sciences USA 95:3306-3311.

Jimenez, D.R., R.K. Yokomi, R.T. Mayer, and J.P. Shapiro. 1995. Cytology and physiology of silverleaf whitefly-induced squash silverleaf. Physiological and Molecular Plant Pathology 46:227-242.

Karban, R., R. Adamchak, and W.C. Schnathorst. 1987. Induced resistance and interspecific competition between spider mites and a vascular wilt fungus. Science 235:678-680.

Karban R. and I.T. Baldwin. 1997. Induced Responses to Herbivory. Chicago University Press, Chicago, IL.

Karban R. and J.H. Myers. 1989. Induced plant responses to herbivory. Annual Review of Ecology and Systematics 20:331-348.

Karban, R. 1991. Inducible resistance in agricultural systems. Pages 402-420 *in* D.W. Tallamy and M.J. Raupp, editors. Phytochemical Induction by Herbivores. John Wiley and Sons, Inc., New York.

Kendall, D.M. and L.B. Bjostad. 1990. Phytohormone ecology -- Herbivory by *Thrips tabaci* induces greater ethylene production in intact onions than mechanical damage alone. Journal of Chemical Ecology 16:981-991.

Klepzig, K.D., E.B. Smalley, and K.F. Raffa. 1996. Combined chemical defenses against an insect-fungal complex. Journal of Chemical Ecology 22:1367-1385.

Kogan, M. and D.C. Fischer. 1991. Inducible defenses in soybean against herbivorous insects. Pages 347-378 *in* D.W. Tallamy and M.J. Raupp, editors. Phytochemical Induction by Herbivores. John Wiley and Sons, Inc., New York.

Korth, K. L. and R.A. Dixon. 1997. Evidence for chewing insect-specific molecular events distinct from a general wound response in leaves. Plant Physiology 115:1299-1305.

Krause, C., G. Spiteller, A. Mithofer, and J. Ebel. 1995. Quantification of glyceollins in non-elicited seedlings of *Glycine max* by gas chromatography-mass spectrometry. Phytochemistry 40:739-743.

Krischik V.A., R.W. Goth, and P. Barbosa. 1991. Generalized plant defense: effects on multiple species. Oecologia 85:562-571.

Kuć, J. 1987. Plant Immunization and its applicability for disease control. Pages 255-274 *in* I. Chet, editor. Innovative Approaches to Plant Disease Control. John Wiley and Sons, New York.

Kuć, J. 1995. Phytoalexins, stress metabolism, and disease resistance in plants. Annual Review of Phytopathology 33:275-297.

Lagrimini, L.M. and S. Rothstein. 1987. Tissue specificity of tobacco peroxidase isozymes and their induction by wounding and tobacco mosaic virus infection. Plant Physiology 84:438-442.

Liang, X., M. Dron, C.L. Cramer, R.A. Dixon, and C.J. Lamb. 1989. Differential regulation of phenylalanine ammonia lyase genes during plant development and by environmental cues. Journal of Biological Chemistry 264:14486-14492.

Lin, H., M. Kogan, and D. Fischer. 1990. Induced resistance in soybean to the Mexican Bean Beetle (Coleoptera: Coccinellidae): comparison of inducing factors. Environmental Entomology 19:1852-1857.

Lwande, W., A. Hassanali, P.W. Njoroge, M.D. Bentley, F.D. Monache, and J.I. Jondiko. 1985. A new 6a-hydroxypterocarpan with insect antifeedant and antifungal properties from the roots of *Tephrosia hildebrandtii* Vatke. Insect Science and its Application 6:537-541.

Ma, R., J.C. Reese, W.C. Black IV, and P. Bramel-Cox. 1990. Detection of pectinesterase and polygalacturonase from salivary secretions of living greenbugs, *Schizaphis graminum* (Homoptera: Aphididae). Journal of Insect Physiology 36:507-512.

Mattiacci, L., M. Dicke, and M.A. Posthumus. 1995. β-glucosidase: an elicitor of herbivore-induced plant odor that attracts host-searching parasitic wasps. Proceedings of the National Academy of Sciences USA 92:2036-2040.

Mauch-Mani, B. and J.-P. Métraux. 1998. Salicylic acid and systemic acquired resistance to pathogen attack. Annals of Botany 82:535-540.

Mayer, R.T., J.P. Shapiro, E. Berdis, C.J. Hearn, T.G. McCollum, R.E. McDonald, and J. Doostdar. 1995. Citrus rootstock responses to herbivory by larvae of the sugarcane rootstock borer weevil (*Diaprepes abbreviatus*). Physiologia Plantarum 94:164-173.

McCloud, E.S. and I.T. Baldwin. 1997. Herbivory and caterpillar regurgitants amplify the wound-induced increases in jasmonic acid but not nicotine in *Nicotiana sylvestris*. Planta 203:430-435.

McCloud, E.S., D.W. Tallamy, and F.T. Halaweish. 1995. Squash beetle trenching behaviour: avoidance of cucurbitacin induction or mucilaginous plant sap? Ecological Entomology 20:51-59.

McConn, M., R.A. Creelman, E. Bell, J.E. Mullet, and J. Browse. 1997. Jasmonate is essential for insect defense in *Arabidopsis*. Proceedings of the National Academy of Sciences USA 94:5473- 5477.

McIntyre, J.L., J.A. Dodds, and J.D. Hare. 1981. Effects of localized infections of *Nicotiana tabacum* by tobacco mosaic virus on systemic resistance against diverse pathogens and an insect. Phytopathology 71:297-301.

Miles, P.W. 1990. Aphid salivary secretions and their involvement in plant toxicoses. Pages 131-148 *in* R.K. Campbell and R.D. Eikenbary, editors. Aphid-plant Genotype Interactions. Elsevier Press, Amsterdam.

Moran, P.J. 1998. Plant-mediated interactions between insects and a fungal plant pathogen and the role of plant chemical responses to infection. Oecologia 115:523-530.

Nicholson, R.L. and R. Hammerschmidt. 1992. Phenolic compounds and their role in disease resistance. Annual Review of Phytopathology 30:369-389.

O'Donnell, P.J., C. Calvert, R. Atzorn, C. Wasternack, H.M.O. Leyser, and D.J. Bowles. 1996. Ethylene as a signal mediating the wound response of tomato plants. Science 274:1914-1917.

Osbourn, A.E. 1996. Preformed antimicrobial compounds and plant defense against fungal attack. Plant Cell 8:1821-1831.

Parrella, M.P., V.P. Jones, R.R. Youngman, and L.M. Lebeck. 1985. Effect of leaf mining and leaf stippling of *Liriomyza* spp. on photosynthetic rates of chrysanthemum. Annals of the Entomological Society of America 78:90-93.

Pearce, G., P.A. Marchand, J. Griswold, N.G. Lewis, and C.A. Ryan. 1998. Accumulation of feruloyltyramine and *p*-coumaroyltyramine in tomato leaves in response to wounding. Phytochemistry 47:659-664.

Penninckx, I.A.M.A., K. Eggermont, F.R.G. Terras, B.P.H.J. Thomma, G.W. De Samblanx, A. Buchala, J-P. Métraux, J.M. Manners, and W.F. Broekaert. 1996. Pathogen-induced systemic activation of a plant defensin gene in *Arabidopsis* follows a salicylic acid-independent pathway. Plant Cell 8:2309-2323.

Pieterse, C.M., S.C.M. van Wees, E. Hoffland, J.A. van Pelt, and L.C. van Loon. 1996. Systemic resistance in *Arabidopsis* induced by biocontrol bacteria is independent of salicylic acid accumulation and pathogenesis-related gene expression. Plant Cell 8:1225-1237.

Pieterse, C.M., S.C.M. van Wees, J.A. van Pelt, M. Knoester, R. Laan, H. Gerrits, P.J. Weisbeek, and L.C. van Loon. 1998. A novel signalling pathway controlling induced systemic resistance in *Arabidopsis*. Plant Cell 10:1571-1580

Preisig, C. and J.A. Kuć. 1985. Arachidonic acid-related elicitors of the hypersensitive response in potato and enhancement of their activities by glucans from *Phytophthora infestans* (Mont.) de Bary. Archives of Biochemistry and Biophysics 236:379-389.

Raffa, K.F. and A.A. Berryman. 1987. Interacting selective pressures in conifer-bark beetle systems: a basis for reciprocal adaptations? American Naturalist 129:234-262.

Ricker, K. E. and R.M. Bostock. 1992. Evidence for release of the elicitor arachidonic acid and its metabolites from sporangia of *Phytophthora infestans* during infection of potato. Physiological and Molecular Plant Pathology 41:61-72.

Romeis, T., P. Piedras, S. Zhang, D.F. Klessig, H. Hirt, and J.D.G. Jones. 1999. Rapid Avr9- and Cf-9-dependent activation of MAP kinases in tobacco cell cultures and leaves: convergence of resistance gene, elicitor, wound, and salicylate responses. Plant Cell 11:273-287.

Rossi, M., F.L. Goggin, B. Milligan, I. Kaloshian, D.E. Ullman, and V.M. Williamson. 1998. The nematode resistance gene *Mi* of tomato confers resistance against the potato aphid. Proceedings of the National Academy of Sciences USA. 95:9750-9754.

Ryals, J.A., U.H. Neuenschwander, M.G. Willits, A. Molina, H-Y. Steiner, and M.D. Hunt. 1996. Systemic acquired resistance. Plant Cell 8:1809-1819.

Ryan, C.A. 1992. The search for the proteinase inhibitor-inducing factor, PIIF. Plant Molecular Biology 19:123-133.

Ryder, T.B., S.A. Hedrick, J.N. Bell, X. Liang, S.D. Clouse, and C.J. Lamb. 1987. Organization and differential activation of a gene family encoding the plant defense enzyme chalcone synthase in *Phaseolus vulgaris*. Molecular and General Genetics 210:219-233.

Schaller, A. and C.A. Ryan. 1995. Sytemin—a polypeptide defense signal in plants. BioEssays 18:27-33.

Schweizer, P., A. Buchala, P. Silverman, M. Seskar, I. Raskin, and J.-P. Métraux. 1997. Jasmonate-inducible genes are activated in rice by pathogen attack without a concomittant increase in endogenous jasmonate levels. Plant Physiology 114:79-88.

Scofield, S.R., C.M. Tobias, J.P. Rathjen, J.H. Chang, D.T. Lavelle, R.W. Michelmore, and B.J. Staskawicz. 1996. Molecular basis of gene-for-gene specificity in bacterial speck disease of tomato. Science 274:2063-2065.

Staskawicz, B.J., F.M. Ausubel, B.J. Baker, J.G. Ellis, and J.D.G. Jones. 1995. Molecular genetics of plant disease resistance. Science 268:661-669.

Stermer, B.A. and R.M. Bostock. 1987. Involvement of 3-hydroxy-3-methylglutaryl coenzyme A reductase in the regulation of sesquiterpenoid phytoalexin synthesis in potato. Plant Physiology 84:404-408.

Stout, M.J. and S.S. Duffey. 1996. Characterization of induced resistance in tomato plants. Entomologia Experimentalis et Applicata 79:273-283.

Stout, M.J., A.L. Fidantsef, S.S. Duffey, and R.M. Bostock. 1999. Plant- mediated interactions between pathogens and herbivores of the tomato, *Lycopersicon esculentum*. Physiological and Molecular Plant Pathology (in press).

Stout M.J., J. Workman , and S.S. Duffey. 1994. Differential induction of tomato foliar proteins by arthropod herbivores. Journal of Chemical Ecology 20:2575-2594

Stout M.J., K.V. Workman, and S.S. Duffey. 1996. Identity, spatial distribution and variability of induced chemical responses in tomato plants. Entomologia Experimentalis et Applicata 79:255-271.

Stout M.J., K.V. Workman, R.M. Bostock, and S.S. Duffey. 1998. Specificity of induced resistance in the tomato, *Lycopersicon esculentum*. Oecologia 113:74-81.

Takabayashi, J. and M. Dicke. 1996. Plant-carnivore mutualism through herbivore-induced carnivore attractants. Trends in Plant Science 1:109-113.

Tallamy, D.W. and M.J. Raupp, editors. 1991. Phytochemical Induction by Herbivores. John Wiley and Sons, Inc., New York.

Tang, X., R.D. Frederick, J. Zhou, D.A. Halterman, Y. Jia, G.B. Martin. 1996. Initiation of plant disease resistance by physical interaction of AvrPto and Pto kinase. Science 274:2060-2062.

Thomma, B., K. Eggermont, I. Penninckx, B. Mauch-Mani, R. Vogelsang, B. Cammue, and W. Broekaert. 1998. Separate jasmonate-dependent and salicylate-dependent defense-response pathways in *Arabidopsis* are essential for resistance to distinct microbial pathogens. Proceedings of the National Academy of Sciences USA 95:15107-15111.

Tjamos, E.C. and J.A. Kuć. 1982. Inhibition of steroid glycoalkaloid accumulation by arachadonic and eicosapentaenoic acids in potato. Science 217:542-544.

Trewavas, A. 1991. How do plant growth substances work? II. Plant Cell and Environment 14:1-12.

Turlings, T.C.J., P.J. McCall, H.T. Alborn, and J.T. Tumlinson. 1993. An elicitor in caterpillar oral secretions that induces corn seedlings to emit chemical signals attractive to parasitic wasps. Journal of Chemical Ecology 19:411-424.

Uknes, S., B. Vernooij, S. Williams, D. Chandler, K. Lawton, T. Delaney, L., Friedrich, K. Weymann, D. Negrotto, T. Gaffney, M. Gut-Rella, H. Kessman, D. Alexander, Ward, and J. Ryals. 1995. Systemic acquired resistance. HortScience 30:962-963.

van Loon L.C., P.A.H.M. Bakker, and C.M.J. Pieterse. 1998. Systemic resistance induced by rhizosphere bacteria. Annual Review of Phytopathology 36:453-83.

Whitehead, I.M., P.M. Dey, and R.A. Dixon. 1982. Differential patterns of phytoalexin accumulation in wounded and elicitor-treated tissues of *Phaseolus vulgaris*. Planta 154:156-164.

Walker-Simmons, M., L. Hadwiger, and C.A. Ryan. 1983. Chitosans and pectic polysaccharides both induce the accumulation of the antifungal phytoalexin pisatin in pea pods and antinutrient proteinase inhibitors in tomato leaves. Biochemical and Biophysical Research Communications 110:194-199.

Ward, E.R., S.J. Uknes, S.C. Williams, S.S. Dincher, D.L. Wiederhold, D.C. Alexander, Ahl-Goy, J.-P. Métraux, and J.A. Ryals. 1991. Coordinate gene activity in response to agents that induce systemic acquired resistance. Plant Cell 3:1085-1094.

Watanabe, T. and S. Sakai. 1998. Effects of active oxygen species and methyl jasmonate on expression of the gene for a wound-inducible 1-aminocyclopropane-1-carboxylate synthase in winter squash (*Cucurbita maxima*). Planta 206:570-576.

Wink, M. 1984. Chemical defense of Leguminosae. Are quinolizidine alkaloids part of the antimicrobial defense systems of lupins. Zeitschrift fuer Naturforschung 39c:548-552.

Xu, Y., P.L. Chang, D. Liu, M.L. Narasimhan, K.G. Raghothama, P.M. Hasegawa, and R.A. Bressan. 1994. Plant defense genes are synergistically induced by ethylene and methyl jasmonate. Plant Cell 6:1077-1085.

Zhu, B., T.H.H. Chen, and P.H. Li. 1995. Expression of three osmotin-like protein genes in response to osmotic stress and fungal infection in potato. Plant Molecular Biology 28:17-26.

Zook, M.N. and J.A. Kuć. 1991. Induction of sesquiterpene cyclase and suppression of squalene synthetase activity in elicitor-treated or fungal-infected potato tuber tissue. Physiological and Molecular Plant Pathology 39:377-390.

The Influence of Induced Plant Resistance on Herbivore Population Dynamics

Nora Underwood

Abstract

Verbal and mathematical theory suggest that induced resistance in plants may affect the long-term population dynamics of herbivores. However, there is little direct empirical evidence that induced resistance affects herbivore dynamics in the field. This chapter reviews both theoretical and empirical evidence that induced resistance may affect herbivore population dynamics. Methods for gathering new evidence, including the use of chemical elicitors, natural damage and genotypic variation to manipulate induced resistance, are also discussed. Finally, a simulation model of the interaction of induced resistance and herbivore population dynamics that includes plant population dynamics is presented. This model suggests that the relative number of herbivore to plant generations in a system should strongly affect the influence of induced resistance on herbivore population dynamics.

Introduction

Induced plant resistance to damage has clearly been shown to affect both the behavior and performance of individual herbivores. Induced resistance may also contribute to aspects of herbivore population dynamics such as outbreaks, cycles, and population regulation that have long fascinated ecologists (Cappuccino 1995). From the perspective of understanding herbivore population dynamics, induced resistance can be defined as any change in a plant resulting from damage and having a negative effect on herbivores (Karban and Baldwin 1997). Induced resistance is likely to be important for long-term herbivore dynamics because in at least some systems, the strength of induced resistance is a density dependent function of the amount of damage the plant receives (see below). This density dependence makes induced resistance a potential source of the negative feedback necessary to regulate herbivore populations (Rhoades 1985, Turchin 1990). Induced resistance that is delayed relative to the time of damage might also increase the likelihood of fluctuations in herbivore populations, because delayed density dependence has been shown to drive cycles in population dynamics models (May 1973, Berryman et al. 1987). Although it has frequently been suggested that induced resistance could be responsible for regulating or driving cycles in herbivore populations (Benz 1974, Haukioja 1980, Rhoades 1985, Myers 1988), very few empirical or theoretical studies have directly addressed the effects of induced resistance on long-term herbivore population dynamics.

Empirical Evidence that Induced Resistance Affects Herbivore Population Dynamics

Several kinds of indirect evidence can help us understand how induced resistance may affect the population dynamics of herbivores. For example, we can determine whether induced resistance affects herbivore characters such as population growth rate or mortality, which influence population dynamics. We can also ask whether the effects of induced resistance on herbivores are of the type that might regulate herbivore populations (density dependent effects) or produce cycles in those populations (delayed density dependence). However, even if indirect evidence indicates that induced resistance has the potential to influence herbivore populations, direct evidence is still required to demonstrate that induced resistance alters long-term patterns of herbivore population dynamics. In this section, direct and indirect empirical evidence for effects of induced resistance on herbivore population dynamics are briefly reviewed (see Karban and Baldwin 1997 for a recent review), and the pros and cons of methods available for gathering further evidence are discussed.

Studies in both the lab and the field have demonstrated that induced resistance affects herbivore performance characters (such as feeding preference, growth rate, fecundity, and survival) that should influence herbivore population size (Karban and Myers 1989, Karban and Baldwin 1997). This evidence suggests that induced resistance has the potential to influence herbivore population dynamics. Studies of herbivore populations over a few generations have also demonstrated negative effects of induced resistance on aspects of short-term dynamics such as population size or population growth rate (Karban and Baldwin 1997).

There is also evidence that induced resistance is density dependent in many systems (e.g., Karban and English-Loeb 1988, Baldwin and Schmelz 1994), and thus has the potential to regulate herbivore populations. For induced resistance *alone* to regulate herbivore populations, it should be able to reduce herbivore population growth rates to zero (Underwood 1999). How often induced resistance is this strong is not yet clear, though in most studies effects on herbivore performance are not very large (Karban and Baldwin 1997). However, induced resistance could work in conjunction with other density dependent influences, such as predation, to regulate herbivore populations (Underwood 1999). There have as yet been no experimental demonstrations that induced resistance regulates, or contributes significantly to regulating, any population. While the influence of density dependence has been demonstrated in herbivorous insect populations (e.g., Woiwood and Hanski 1992), this density dependence has not been linked to induced resistance.

Theory suggests that for induced resistance to drive cycles in herbivore populations, the effect of induced resistance on herbivores must be delayed relative to herbivore generation time so that damage by herbivores in one generation influences the performance of later generations (e.g., Edelstein-Keshet and Rausher 1989, Underwood 1999). Induced resistance can be delayed either by a lag between damage and the induction of resistance or by slow decay of induced resistance in the absence of further herbivory. Existing data on the timing of induced resistance support the idea that induced resistance could contribute to cycles in some herbivore populations (Karban and Baldwin 1997). Studies of induced resistance in annual plants (e.g., Edwards et al. 1985, Baldwin 1988a, Malcolm and Zalucki 1996, Underwood 1998) indicate that lag and decay times are generally fairly short relative to herbivore generation times, although even short lags might contribute to cycles in herbivores with very short generation times (e.g., aphids or mites). Studies in several cycling forest systems indicate that the decay of induced resistance can be substantially longer than herbivore generation times (decays taking several years) (Baltensweiler 1985, Haukioja and Neuvonen 1987). This indirect evidence suggests that induced resistance may contribute to cycles in herbivore populations.

Although indirect evidence thus suggests that induced resistance might affect long-term herbivore dynamics in at least some systems, there are very few direct observations of effects on population dynamics in either the lab or the field. This lack of data is understandable given the difficulty of studying population dynamics directly (especially in the forest systems that may be strongly influenced by induced resistance). The demonstrated effects on short-term population size do indicate that induced resistance may be useful in controlling pests on crops within a season (Karban 1991), and that induced resistance has the potential to influence long-term dynamics in agricultural and natural communities. However, these data cannot tell us whether induced resistance actually affects the aspects of long-term population dynamics that have excited the most interest in this field (regulation and cyclic fluctuation).

Techniques for Gathering Further Evidence

The ideal approach to evaluating the relative importance of induced resistance and other factors in influencing the population dynamics of herbivorous insects would be to conduct controlled experiments at the scale of whole populations, in which induced resistance is manipulated and effects on long-term population dynamics are observed. Unfortunately, experiments of this sort would be prohibitively difficult in most systems. There are however, more feasible approaches that can provide useful information about the effect of induced resistance on long-term dynamics, including correlational studies, direct observation of rapid-cycling herbivores, and density manipulation experiments.

Most studies that have explicitly addressed induced resistance and long-term herbivore dynamics in the field have been carried out in forest systems using techniques such as correlating herbivore dynamics with changes in plant-quality (e.g., Benz 1974), and transplanting herbivores between outbreaking and non-outbreaking populations (e.g., Myers 1981). These techniques allow a high level of realism by examining herbivore dynamics at very large scales, and comparing current conditions with existing long-term census data. However, because induced resistance cannot be manipulated, the precision with which the effects of induced resistance can be isolated from other factors is limited in these studies.

Studying herbivores with very short generation times (such as aphids, mites, whiteflies and thrips) might be one practical way to combine direct observations of long-term dynamics with manipulation of induced resistance. Several studies of induced resistance in systems with rapidly cycling herbivores do contain data over what must be many herbivore generations (e.g., Karban 1986, Shanks and Doss 1989). However, these studies sample population size only a few times - making it difficult to use these data to examine long-term population dynamics. Many census data points are required to adequately describe patterns of population dynamics, and particularly to assess effects on regulation (Turchin 1997). More frequent sampling might allow for characterization of within-season dynamics in these populations. However, since populations of these insects are also strongly affected by seasonal patterns, census data over many years may still be necessary to provide a true picture of their dynamics (Harrison and Cappuccino 1995). Once long-term census data are gathered, regardless of the generation time of the herbivore, several kinds of information can be extracted. A number of methods can be used to detect the regulating action of density dependence in census data, though this field has been very contentious (Turchin 1997). The presence and periodicity of cycles can be assessed from the autocorrelation function for the data (Berryman and Turchin 1997). Finally, a population dynamic model can be fit to the data and used to analyze the stability of the population dynamics (Edelstein-Keshet 1988).

In cases where direct observation of population dynamics is not possible, density manipulation experiments allow estimation of long-term dynamics from short-term (as little as one generation) population data. In this type of design, populations with a range of initial densities are followed for one or more generations, and data on initial density and density in the next generation are used to construct a recruitment curve. This curve can be fit with a discrete population dynamics model (such as the Ricker (Edelstein-Keshet 1988) or Hassell (Hassell 1975) models), and the model can be analyzed to determine the equilibrium population size, the stability of the equilibrium and tendency of the population to cycle. For organisms with overlapping or continuous generations, a similar approach could be used by fitting continuous population dynamic models to census data, with the power of the method being improved by combining data from populations initiated at several initial densities (Pascual and Kareiva 1996).

Combination of density-manipulation with manipulation of induced resistance would allow examination of the effect of induced resistance (or even different levels or types of induced resistance) on the estimated long-term dynamics of herbivore populations. Depending on the choice of system, this design could incorporate effects at the scale of the whole population, or at much smaller scales. Density manipulation approaches are only beginning to be used in ecological experiments (Belovsky and Joern 1995, Underwood 1997), though related techniques are standard in the management of fisheries (Roughgarden 1998). Harrison and Cappuccino (1995) discuss guidelines for this type of experiment, and suggest that this technique should be more widely used to search for population regulation.

Estimating long-term dynamics using density manipulation and model fitting also has several drawbacks, one of which is that the experiments can be very labor intensive. A more serious drawback is that using models fit to field data in one generation to estimate dynamics over long periods of time assumes that the parameters of the model will not change over time. This assumption may be more valid for some systems than for others. For instance, the characteristics of crop plants are likely to remain fairly constant from year to year, as opposed to natural perennial plants whose characteristics may change with age between years. Estimation of long-term dynamics is thus not suitable for precise prediction of the size of particular populations over time, but should allow examination of qualitative differences among populations subject to different conditions.

Manipulating Induced Resistance

Whether population dynamics are directly observed or estimated, the most rigorous experimental examination of the effects of induced resistance requires the experimenter to manipulate induced resistance and create appropriate controls. A new and appealing method of manipulation is the application of chemical elicitors that cause the plant to produce induced resistance. Because elicitors can be easily applied to the plant, and because their effects are dose-dependent (Farmer and Ryan 1990, Baldwin et al. 1998), elicitors have clear advantages in precision and ease of manipulation. However, there are several drawbacks that might outweigh these advantages for studies of the ecological effects of induced resistance. Some elicitors, jasmonic acid for instance, have widespread effects on plant physiology (Creelman and Mullet 1997), so that the effects of elicitor-induced resistance could be confounded with other physiological changes in the plant that are not normally caused by herbivore feeding. Another potential drawback is that while elicitors can turn plants "on" and keep them on via repeated application (Thaler et al. 1996), there is nothing to prevent untreated plants from also being turned on by herbivores. So while elicitors can create initially different conditions, any experiment that runs long enough to allow "natural" induction might run into the problem of elicitor and non-elicitor treatments converging in induction over time. This problem may not materialize if elicitors produce larger responses than herbivores do (Thaler et al. 1996), although if the induced resistance caused by elicitors is outside the natural range of responses, results of experiments using elicitors may not accurately reflect natural conditions.

It might be possible to correct the problem of induced and non-induced treatments converging during an experiment by combining elicitors with methods of inhibiting induced resistance. These methods include pot-binding, which has been shown in at least one case to block induced resistance (Baldwin 1988b), and chemical inhibitors of induced resistance (Hartley 1988, Baldwin et al. 1990, Stout et al. 1998). Pot-binding has the advantage of potentially remaining effective throughout the experiment, thus maintaining differences between treatment and control plants. Disadvantages of pot-binding are that it has only been shown to work in one system, and it may affect aspects of the plant's physiology that influence herbivores in addition to induced resistance. Like chemical elicitors of induced resistance, chemical inhibitors have the potential drawback of affecting aspects of the plant's physiology not normally associated with induced resistance.

Another method of manipulating induced resistance levels is pre-damaging plants using either artificial damage (e.g., Edwards et al. 1985, Hanhimaki 1989) or application of herbivores (e.g., Hanhimaki 1989, Shanks and Doss 1989, Karban 1993, Agrawal 1998). The dynamics of herbivore populations on pre-induced plants could then be compared with those on plants that start out with no damage, and thus no induced resistance. Artificial damage is relatively precise and easy to produce. However, a number of studies indicate that in some systems artificial damage does not produce the same type or strength of induced resistance as natural damage (Baldwin 1988a, Hanhimaki 1989). Using herbivores to produce the initial damage is logistically more difficult, but has the virtue of being more likely to provoke the appropriate induced response. Both methods of pre-damaging plants are subject to the problem of convergence between induced and non-induced treatments during the experiment.

An alternative approach for creating induced and non-induced treatments is using genotypes of plants that inherently differ in their levels of induced resistance. This technique is currently being used to examine the effects of induced and constitutive resistance on herbivore population dynamics in time (N. Underwood, unpublished data) and in space (W. Morris, unpublished data), and has been used to look for relationships between induced and constitutive resistance (Zangerl and Berenbaum 1991, Brody and Karban 1992, English-Loeb et al. 1998, N. Underwood, unpublished data). Among plant differences in characters other than induced resistance could complicate the interpretation of differences in herbivore dynamics among genotypes differing in induced resistance. In theory, it should be possible through screening of genotypes, breeding or transgenics, to obtain lines that are isogenic except for genes controlling the production of induced resistance. Significant genetic variation for induced resistance has been documented in several systems (e.g., Zangerl and Berenbaum 1991, Brody and Karban 1992, English-Loeb et al. 1998, Agrawal, this volume, N. Underwood, unpublished data). Agricultural systems, where resistance and other characters have already been documented for a variety of genotypes, may provide particularly convenient places to search for appropriate genotypes. Once appropriate genotypes are located, using genotypic variation to create treatments with high and low induced resistance is convenient, and differences among treatments should be maintained throughout experiments. A further advantage of this approach is that it allows consideration of evolutionary as well as population dynamic questions.

Evidence from Mathematical and Simulation Models

Over the past three decades, many authors have presented verbal theory concluding that induced resistance might regulate and drive cycles in herbivore populations (Karban and Baldwin 1997). More recently, both analytical and computer simulation modeling have been used to more rigorously explore the interaction between induced resistance and herbivore populations. Some of these models explicitly focus on effects of induced resistance on long-term population dynamics (Fischlin and Baltensweiler 1979, Edelstein-Keshet and Rausher 1989, Lundberg et al. 1994, Underwood 1999). Other models focusing on issues such as the evolution of induced resistance (Frank 1993, Adler and Karban 1994), and effects of induced resistance on the spatial dynamics of herbivores (Lewis 1994, Morris and Dwyer 1997), also make predictions about temporal dynamics. The predictions of these models vary with the type of plant-herbivore system they model, suggesting that characteristics of the plant-herbivore system, such as the strength and timing of induced resistance, mobility and selectivity of the herbivore, and relative lengths of plant and herbivore generation times, will affect the influence of induced resistance on herbivore population dynamics.

Results of existing models suggest that induced resistance has the potential to regulate herbivore populations (i.e., prevent extinction and growth without bound). Regulation is only possible when induced resistance can reduce herbivore population growth rates to zero, and the probability of regulation tends to increase with the strength of induced resistance, though regulation can also depend on the timing of induced resistance, and the relative lengths of plant and herbivore generations (Edelstein-Keshet and Rausher 1989, Underwood 1999).

The likelihood that induced resistance can produce cycles in herbivore populations varies among models. Some models have found that induced resistance by itself cannot produce persistent cycles, though it may produce cycles that damp over time (Frank 1993, Adler and Karban 1994, Lundberg et al. 1994, Morris and Dwyer 1997). Other models have found that induced resistance drives persistent cycles in herbivore populations under certain conditions (Fischlin and Baltensweiler 1979, Edelstein-Keshet and Rausher 1989, Lewis 1994, Underwood 1999). Both damping and persistent cycles are associated with delays in the timing of induced resistance. The longer the delays relative to herbivore generation time, the more likely cycles are, and the less likely regulation becomes.

Induced resistance has been found in a very wide variety of plant-insect systems, involving annual and perennial plants, and both uni- and multivoltine insects. This raises the question of how the relative lengths of the generation times of the plant and herbivore affect the impact of induced resistance on herbivore dynamics. To examine this question, models need to include plant population dynamics as well as herbivore dynamics. Adding plant dynamics to models might increase the likelihood of cycling simply due to the interaction between equations for the two populations, as occurs in some predator-prey models (Edelstein-Keshet 1988). Several continuous time analytical models have included plant population dynamics, with plant and herbivore generation times equal in all cases (Frank 1993, Adler and Karban 1994, Lundberg et al. 1994). All three of these models have some tendency towards oscillation in plant and herbivore populations, although plant dynamics are not discussed at length for these models.

A Model Including Both Plant and Herbivore Population Dynamics

The simulation model of Underwood (1999) can be modified to examine how variable plant population size and the relative lengths of plant and herbivore generations affect the influence of induced resistance on herbivore population dynamics. The model described in Underwood (1999) follows individual plants and herbivores, and consists of three nested loops: a herbivory loop (describing herbivore movement and feeding, and changes in induced resistance in plants), a herbivore generation loop consisting of 30 herbivory loops, and a plant reproduction loop consisting of variable numbers of herbivore generations. There are two equations in the model. The first describes the level of induced resistance in individual plant i at time t+1 ($I_{i,t+1}$) as a function of damage to the plant, which is equal to the herbivore load on that plant ($h_{i,(t-\tau)}$):

$$I_{i,t+1} = \frac{(-\frac{\hat{\alpha}}{\beta} I_{i,t} + \hat{\alpha}) h_{i,(t-\tau)}}{b + h_{i,(t-\tau)}} + I_{i,t}(1-\delta) \qquad (1)$$

In this equation, β represents a physiological maximum induced resistance in an individual plant, $\hat{\alpha}$ is the maximum increase in induced resistance at one time, b is the half-saturation constant, δ is the decay rate of induced resistance and τ is the lag time between damage and induced resistance. In this model it is assumed that $\hat{\alpha} = \beta$. The second equation describes herbivore population size in one generation (H_{t+g}) as a function of its size in the previous generation (H_t) (where one herbivore generation consists of g herbivory loops), the level of induced resistance herbivores have encountered ($\bar{I}$), and the strength of induced resistance (critical level of induced resistance reducing herbivore reproduction to zero, I_c):

$$H_{t+g} = H_t(1+\gamma(1-\frac{\bar{I}}{I_c})) \qquad (2)$$

where γ is the population rate of increase for herbivores.

To incorporate plant population dynamics, a third equation can be added to the model. This equation describes the size of the plant population in one generation (P_{t+n*g}) as a function of its size in the previous generation (P_t) (where one plant generation consists of $n*g$ herbivory loops), the average damage incurred by plants ($\bar{D}$), and the critical damage level reducing plant reproduction to zero (D_c):

$$P_{t+n*g} = P_t(1+\lambda(1-\frac{\bar{D}}{D_c}-\frac{P}{K})) \qquad (3)$$

This formulation of plant reproduction assumes that there is no cost of induced resistance to the plant. In this equation λ is the plant population growth rate and K is a carrying capacity for the plant population set by factors other than herbivore damage. $\bar{D}$ is calculated as the average damage to individual plants in the population at one time (h_i), averaged over all time steps in a plant generation (n*g):

$$\bar{D} = \frac{\sum_{t=1}^{n*g} \frac{\sum_{i=1}^{P} h_i}{P}}{n*g}$$

Because in this model all herbivores occupy a plant at all times, $\sum_{i=1}^{P} h_i = H$. If it is assumed that $\tau = 0$, an equilibrium solution for this model can be found. At equilibrium, $I_i = \bar{I} = I^*$, $H = H^*$ and $h_i = \frac{H^*}{P^*} = \bar{D}$. Substituting these into equations 1 and 2 and combining the two it can be shown that the equilibrium conditions for H are $\gamma = 0$ and:

$$\frac{H^*}{P^*} = \frac{bI_c\delta}{\hat{\alpha}-I_c} \qquad (4).$$

Equation 4 implies that the ratio of herbivores to plants at equilibrium is a constant (θ) determined by the properties of induced resistance. See Underwood (1999) for a more detailed explanation of this model and its results, and Underwood (1997) for derivation of the equilibrium condition. The equilibrium conditions for the plant population are $\lambda = 0$ and:

$$P^* = K(1 - \frac{\frac{H^*}{P^*}}{D_c}). \qquad (5).$$

Equation 5 implies that the equilibrium number of plants is determined by the plant carrying capacity and the influence of herbivores on plants through damage. Note that without an independent carrying capacity for plants (that is, if the plant population is influenced only by herbivores), the model is unstable at all points except $D_c = \theta$.

The effect of induced resistance on herbivore and plant population dynamics was explored by running simulations of the model. All runs had 30 herbivory loops per herbivore generation, and runs were started with initial plant and herbivore populations of 100 individuals. Starting with different initial population sizes did not change the behavior of the model. The effects of the strength of induced resistance (I_c) and the critical damage level for plants (D_c) on plant and herbivore populations were explored over a range of relative numbers of herbivore to plant generations including 1:1, 10:1 and no plant reproduction. The effects of lags in induced resistance (τ) were also considered. For the runs reported here, all other parameters were held constant, with α and $\beta = 100$, $b = 10$, $\gamma = 2$, $\lambda = 1$, $\delta = .07$, and $K = 100$. These parameter values were chosen largely arbitrarily (see Underwood 1999 for discussion of the consequences of changing them).

All simulations were run for at least 150 herbivore generations, or long enough for the output to converge on the asymptotic dynamics. Runs were not replicated because initial exploration of the model indicated that there was not appreciable variation among runs with the same parameter values. Data on average plant and herbivore population sizes were calculated after the dynamics reached their final state (when the behavior of plant and herbivore populations was consistent over at least 50 generations). The behavior of the plant and herbivore populations was divided into the following 5 categories: herbivore extinction, stable herbivore and plant populations, damping cycles in the herbivore population (with stable plants), cycles or fluctuations in the herbivore population (with stable plants), both plants and herbivores cycling, and plant extinction (caused by the uncontrolled increase of herbivores).

Results

Increasing the strength of induced resistance (decreasing I_c) decreases herbivore population size in this model. Smaller herbivore populations allow the plant population to grow, so that plant population size increases with increasing strength of induced resistance (up to the plant carrying capacity K) (Fig. 1). The effect of the strength of induced resistance (I_c) on herbivore and plant populations is modified by the sensitivity of plant reproduction to herbivore damage (D_c). As plants become more sensitive (D_c decreases), plant populations

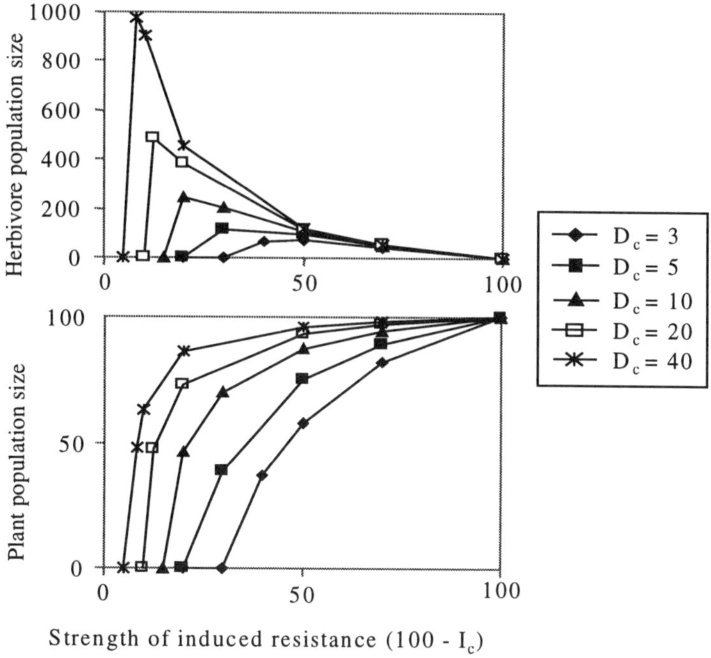

Figure 1: Herbivore and plant population size as a function of the strength of induced resistance. D_c (plant sensitivity to herbivore damage) is the critical damage level reducing plant reproduction to zero. Each point results from a single simulation of the model with one herbivores generation per plant generation and no lag in induced resistance. For these results, plant carrying capacity (K) = 100.

decrease. Small plant populations lead to more herbivores per plant, which causes induced resistance levels in the plants to rise and herbivore populations to fall. The likelihood that herbivore populations are regulated (not extinct or growing without bound) decreases as the number of herbivore generations per plant generation decreases (Fig. 2). Induced resistance is least able to control herbivore populations when herbivore and plant generation times are equal (Fig. 2C). When plants do not reproduce, lags in induced resistance decrease the likelihood that the herbivore population is regulated, and increase the likelihood of persistent cycles in the herbivore population at weaker levels of induced resistance (Fig. 2A). When plants do reproduce, lags still have a destabilizing effect on the herbivore population. As the lag time to induced resistance increases, the likelihood that the herbivore population is unregulated (thus driving plants extinct) or goes extinct (due to induced resistance) increases (Fig. 2B and C).

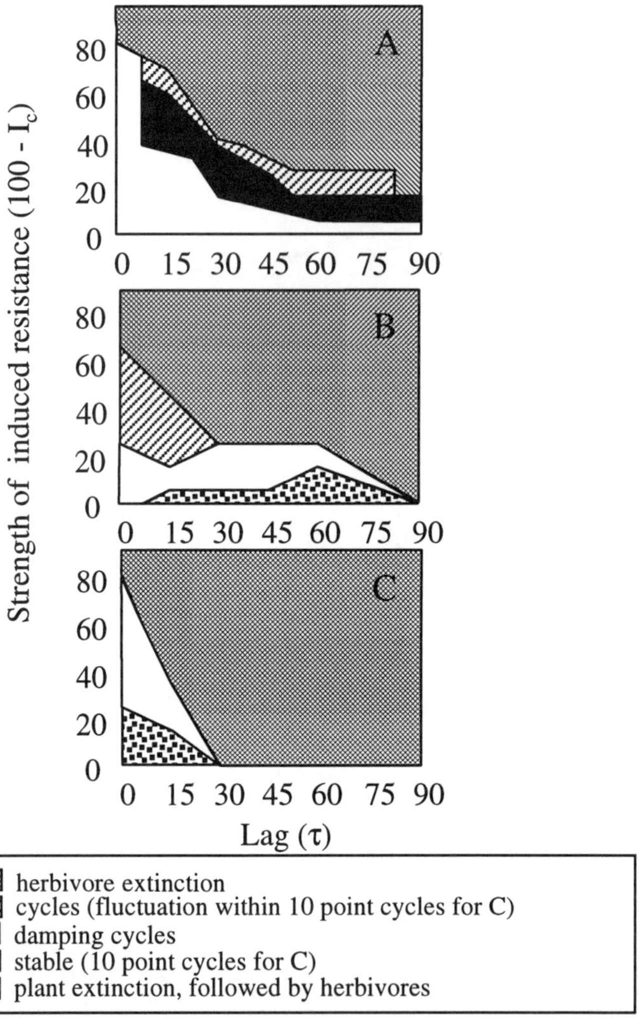

Figure 2: Long-term dynamic behavior of herbivore populations in simulations with A) no plant reproduction, B) ten herbivore generations per plant generation, or C: one herbivore generation per plant generation.

The large majority of the fluctuations exhibited by the model with plant reproduction occur only in the herbivore population, and are forced by the resetting of induced resistance to zero at the beginning of each plant generation. These cycles only occur when herbivores have more than one generation per plant generation (for example, 10 point cycles with 10 herbivore generations per plant generation, Fig. 3A). The reason that cycles driven by lags in induced

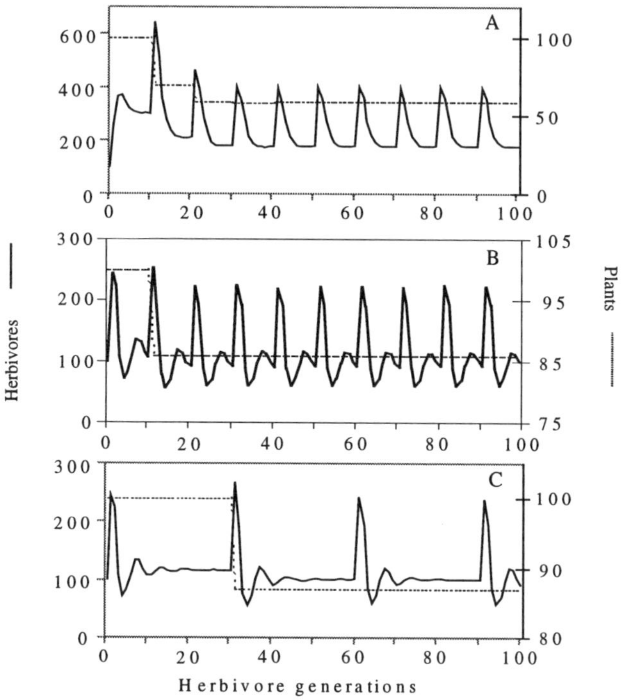

Figure 3: Cycles in the herbivore population forced by plant reproduction, and fluctuations within those cycles, in individual runs of the model. A) and B) 10 herbivore generations per plant generation, c) 30 herbivore generations per plant generation. A) I_c = 80, B) I_c = 60, C) I_c = 60, τ = 15 for all three runs.

resistance rarely occur with plant population dynamics included in the model is that the effect of lags is complicated by the effect of the number of herbivore generations per plant generation. As the number of herbivore generations per plant generation increases, the likelihood of herbivore extinction decreases, and fluctuations within the cycles forced by plant reproduction appear when induced resistance is relatively strong (I_c is relatively low) (Fig. 3B). For example, with 10 herbivore generations per plant generation, there is a region in which the smooth 10-point cycles forced by plant reproduction with relatively weak induced resistance (I_c = 80) (Fig. 3A) are replaced by oscillations with two peaks, one larger and one smaller (Fig. 3B). This second peak is due to a damping oscillation in the herbivore population that is interrupted every 10 generations. These damping oscillations are more apparent if the number of herbivore generations per plant generation is increased (Fig. 3C). These fluctuations within the cycles forced by plant reproduction occur at weaker levels of induced resistance as the lag time gets longer, just as stable and longer damping cycles occur at weaker levels of induced resistance as the lag time gets longer without plant population dynamics.

Initial explorations of this model indicate that it is very rare for the interaction between herbivores and plants in this model to produce joint fluctuations in the plant and herbivore populations. Plant populations remain stable in most configurations of the model because the effect of herbivores on plants is generally either very small, or so large that it drives the plants extinct (Fig. 4A). The effect of herbivores on plants in this model is represented by $\overline{D}$, the amount of damage each plant receives. As shown above, at equilibrium, $\overline{D} = H^*/P^* = \theta$ (the constant describing the characteristics of induced resistance). The relationship between $\overline{D}$ and I_c (the strength of induced resistance) is exponential (Fig. 4A). This means that at most levels of I_c herbivores have relatively little effect on plants, but as I_c increases the effect of herbivores on plants suddenly becomes very large, generally driving the plant population extinct. Persistent cycles occur only in the region between little effect and extinction, and because of the steeply accelerating curve, that region is very small. For example, in Fig. 4B, joint fluctuations of plants and herbivores are found only along a single line dividing stable herbivore and plant populations from plant extinction.

In summary, the following results arise from this model of induced resistance and plant and herbivore population dynamics. First, the interaction between plant and herbivore populations very rarely results in cycling in both populations. Second, the most frequent cycles produced in this model are herbivore population cycles forced by plant reproduction. This result suggests that herbivores with rapid generation times relative to their host plants might be expected to exhibit cyclic dynamics more frequently than herbivores whose generation time is close to the generation time of their host. These forced cycles would, however, only be expected in cases where plant recruitment is relatively synchronous, such as in annual plants, or in perennials whose level of induced resistance is returned to zero when the plant dies back (due to winter or a dry season) and then re-grows. In the field, these cycles would happen within a season, and would be most likely to occur in rapid cycling insects or mites. Such patterns are observed, but are difficult to distinguish from patterns driven by environmental factors such as temperature (see Underwood 1999).

Lags in this model reduce the stability of plant and herbivore populations. However, other than cycles forced by plant reproduction, induced resistance rarely drives cycles in the herbivore or plant populations regardless of lag time, although at intermediate strengths of induced resistance the forced cycles can be complicated by internal fluctuations, giving the appearance of complex behavior.

A clear, though not very surprising, outcome of this model is that the relative number of herbivore to plant generations strongly affects the influence of induced resistance on herbivore population dynamics. A higher number of herbivore generations per plant generation generally increases stability, and also increases the likelihood of stable cycles (both forced and otherwise). This observation suggests that systems with many herbivore generations per plant generation should be more likely to show stable cycles than systems with very

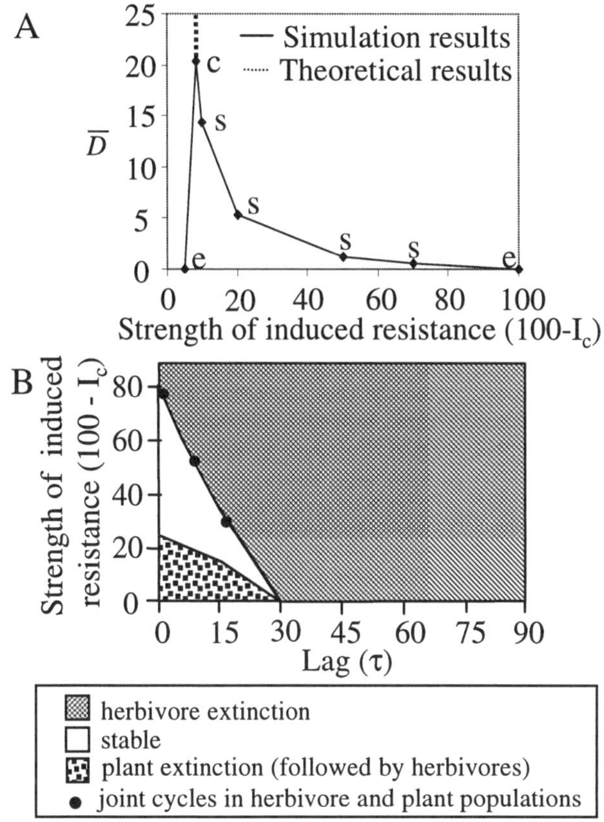

Figure 4: A. Relationship between the strength of induced resistance (I_c) and the influence of herbivores on plants ($\overline{D}$). Points on the line indicate equilibrium values of $\overline{D}$ for single runs of the model. Letters at points indicate dynamic behavior of herbivore and plant populations at each point (s = stable, c = cycles, e = extinct), and the dotted line indicates the relationship predicted by the equations in the model. B. Dynamic behavior of herbivore and plant populations over a range of strengths of induced resistance and lag times, with one herbivore generation to each plant generation. Dots indicate joint cycles in the plant and herbivore population for lags of 0, 8, and 15 herbivory loops.

few herbivore generations per plant generation. In fact, all observations of long-term cycles in insect herbivore populations that have been attributed (tentatively) to induced resistance occur in systems with long-lived trees and short-lived herbivores (Karban and Baldwin 1997). It is important to emphasize that it is the relative number of herbivore and plant generations that should be considered in predicting the effects of induced resistance on population dynamics, rather than whether the herbivore is univoltine versus multivoltine, or the plant is annual versus perennial.

Synthesis

The ecological literature includes many discussions, both theoretical and empirical, of how induced resistance should affect short and long-term herbivore population dynamics. There is, however, very little direct evidence that induced resistance can either regulate or drive cycles in herbivore populations. Although the current lack of data arises largely from the difficulty of approaching this issue experimentally, there is a range of techniques, some of which have only recently been developed, that should allow researchers to empirically address the effects of induced resistance on herbivore population dynamics. Future empirical studies could be guided by existing theoretical explorations of induced resistance and herbivore dynamics, which make some broad predictions for when induced resistance might regulate or drive cycles in herbivore populations. For example, theory predicts that cycles should be more frequent with delays in either the onset or decay of induced resistance. Measuring the timing of induced resistance in the field will be difficult with long-lived plants, but is conceptually straightforward (Underwood 1998), so that the dynamics of herbivore populations in systems with longer versus shorter lag or decay times could be compared. The model presented here provides further predictions of how relative generation times might affect the likelihood of cycles. These predictions remain to be tested. In general, the study of induced resistance and herbivore population dynamics would benefit from a wider use of controlled experiments and model fitting.

Future Directions

Just as induced resistance may affect the population dynamics of herbivores, it may have important effects on the population dynamics of plant pathogens. Because plant pathogens are so much smaller and faster reproducing than their hosts, induced resistance likely affects pathogen population dynamics at two scales: within- and among-hosts. However, most models and studies of plant disease dynamics address only among host dynamics, leaving the dynamics of the disease within the host, as well as plant population dynamics, unknown (Thrall et al. 1997). Results from metapopulation theory suggest that taking into account local (within host) dynamics might change predictions for among host dynamics (e.g., Hastings and Wolin 1989). Because herbivores and pathogens are nearly always both present in plant populations, it would also be valuable to determine how pathogens and herbivores affect each other's population dynamics. The importance of this interaction should depend in part on the specificity of the plant's induced responses (Stout and Bostock, this volume).

The effect of induced resistance on herbivore spatial dynamics also deserves further attention. The few spatially explicit models of induced resistance and herbivore dynamics (Lewis 1994, Morris and Dwyer 1997) suggest that induced resistance may affect herbivore spread differently from constitutive resistance. Few empirical studies have addressed the effect of induced resistance on herbivore movement (e.g., Harrison and Karban 1986, W. Morris, unpublished data) and these have considered relatively small scales. Studies that examine how induced resistance and herbivore characters such as mobility and selectivity interact to determine herbivore spatial distributions could shed light on pest movement as well as the dispersion of native herbivores in natural systems.

Finally, although there is evidence for genetic variation in induced responses from several systems, most discussions of induced resistance and herbivore population dynamics have not explicitly considered variation in induced responses among plants. It would be interesting to determine how variation among plants affects the influence of induced resistance on herbivore population dynamics, especially for systems with selective herbivores. Understanding the effects of population genetic variation for resistance characters such as induced resistance may help evaluate the application of mixed cropping systems in agriculture, as well as improving our understanding of natural systems. Considering variation in induced resistance may also provide an opportunity to link population dynamics and evolution in plant-herbivore systems.

Acknowledgements

I thank A. Agrawal, B. Inouye, R. Karban, M. Rausher, M. Sabelis, P. Tiffin and J. Thaler for thoughtful discussions and helpful editing. This work was supported by NSF DEB grant #9615227 to M. Rausher and by NRI Competitive Grants Program/USDA award #98-35302-6984 to N. Underwood.

Literature Cited

Adler, F. R., and R. Karban. 1994. Defended fortresses or moving targets? Another model of inducible defenses inspired by military metaphors. American Naturalist 144:813-832.

Agrawal, A. A. 1998. Induced responses to herbivory and increased plant performance. Science 279:1201-1202.

Baldwin, I. T. 1988a. The alkaloidal responses of wild tobacco to real and simulated herbivory. Oecologia 77:378-381.

Baldwin, I. T. 1988b. Damage-induced alkaloids in tobacco: pot-bound plants are not inducible. Journal of Chemical Ecology 4:1113-1120.

Baldwin, I. T., D. Gorham, E.A. Schmeltz, C.A. Lewandowski, G.Y. Lynds. 1998. Allocation of nitrogen to an inducible defense and seed production in *Nicotiana attenuata*. Oecologia 115:541-552.

Baldwin, I. T., and E. A. Schmelz. 1994. Constraints on an induced defense: the role of leaf area. Oecologia 97:424-430.

Baldwin, I.T., Sims, C.L. and S.E. Kean. 1990. The reproductive consequences associated with inducible alkaloidal responses in wild tobacco. Ecology 71:252-262.

Baltensweiler, W. 1985. On the extent and the mechanisms of the outbreaks of the larch budmoth (*Zeiraphera diniana* Gn., Lepidoptera, Tortricidae) and its impact on the subalpine larch-cembran pine ecosystem. Pages 215-219 *in* H. Turner and W. Tranquillini, editors. Establishment and tending of subalpine forest: Research and management. Swiss Federal Institute of Forestry Research, Birmensdorf, Switzerland.

Belovsky, G. E., and A. Joern. 1995. The dominance of different regulating factors for rangeland grasshoppers. Pages 359-386 *in* N. Cappuccino and P. W. Price, editors. Population Dynamics: New Approaches and Synthesis. Academic Press, New York.

Benz, G. 1974. Negative feedback by competition for food and space, and by cyclic induced changes in the nutritional base as regulatory principles in the population dynamics of the larch budmoth, *Zeiraphera diniana* (Guenee)(Lep., Tortricidae). Zeitschrift fur Angewandte Entomologie 76:196-228.

Berryman, A. A., N. C. Stenseth, and A. S. Isaev. 1987. Natural regulation of herbivorous forest insect populations. Oecologia 71:174-184.

Berryman, A. and T. Turchin 1997. Detection of delayed density dependence: comment. Ecology 87:318-320.

Brody, A. K., and R. Karban. 1992. Lack of a tradeoff between constitutive and induced defenses among varieties of cotton. Oikos 65:301-306.

Cappuccino, N. 1995. Novel approaches to the study of population dynamics. Pages 3-16 *in* N. Cappuccino and P. W. Price, editors. Population dynamics: New approaches and synthesis. Academic Press, San Diego.

Creelman, R. A., and J. E. Mullet. 1997. Biosynthesis and action of jasmonates in plants. Annual Review of Plant Physiology and Plant Molecular Biology 48:355-381.

Edelstein-Keshet, L. 1988. Mathematical Models in Biology. Random House, New York.

Edelstein-Keshet, L., and M. D. Rausher. 1989. The effects of inducible plant defenses on herbivore populations. 1. Mobile herbivores in continuous time. American Naturalist 133:787-810.

Edwards, P. J., S. D. Wratten, and H. Cox. 1985. Wound-induced changes in the acceptability of tomato to larvae of *Spodoptera littoralis*: a laboratory bioassay. Ecological Entomology 10:155-158.

English-Loeb, G., R. Karban, and M. A. Walker. 1998. Genotypic variation in constitutive and induced resistance in grapes against spider mite (Acari: Tetranychidae) herbivores. Environmental Entomology 27:297-304.

Farmer, E.E. and C.A. Ryan. 1990 Interplant communication: airborne methyl jasmonate induces synthesis of proteinase inhibitors in plant leaves. Proceedings of the National Academy of Science USA 87:7713-7716.

Fischlin, A., and W. Baltensweiler. 1979. Systems analysis of the larch budworm system. Part I. The larch-larch budmoth relationship. Pages 273-289 *in* B. Delucchi and W. Baltensweiler, editors. IUFRO Conference, Zurich, Switzerland.

Frank, S. A. 1993. A model of inducible defense. Evolution 47:325-327.

Hanhimaki, S. 1989. Induced resistance in mountain birch: defence against leaf-chewing insect guild and herbivore competition. Oecologia 81:242-248.

Harrison, S., and N. Cappuccino. 1995. Using density-manipulation experiments to study population regulation. Pages 131-147 *in* N. Cappuccino and P. W. Price, editors. Population dynamics: new approaches and synthesis. Academic Press, San Diego.

Harrison, S., and R. Karban. 1986. Behavioral response of spider mites (*Tetranychus urticae*) to induced resistance of cotton plants. Ecological Entomology 11:181-188.

Hartley, S.E. 1988. The inhibition of phenolic biosynthesis in damaged and undamaged birch foliage and its effect on insect herbivores. Oecologia 76:65-70.

Hassell, M. P. 1975. Density-dependence in single-species populations. Journal of Animal Ecology 44:283-295.

Hastings, A., and C. L. Wolin. 1989. Within-patch dynamics in a metapopulation. Ecology 70:1261-1266.

Haukioja, E. 1980. On the role of plant defences in the fluctuation of herbivore populations. Oikos 35:202-213.

Haukioja, E., and S. Neuvonen. 1987. Insect population dynamics and induction of plant resistance: the testing of hypotheses. Pages 411-432 *in* P. Barbosa and J. C. Schultz, editors. Insect Outbreaks. Academic Press, Inc., San Diego.

Karban, R. 1986. Induced resistance against spider mites in cotton: field verification. Entomologia Experimentalis Et Applicata 42:239-242.

Karban, R. 1991. Inducible resistance in agricultural systems. Pages 403-419 *in* D. W. Tallamy and M. J. Raupp, editors. Phytochemical Induction by Herbivores. Wiley and Sons, New York.

Karban, R. 1993. Induced resistance and plant density of a native shrub, *Gossypium thurberi*, affect its herbivores. Ecology 74:1-8.

Karban, R., and I. T. Baldwin. 1997. Induced Responses to Herbivory. University of Chicago Press, Chicago.

Karban, R., and G. M. English-Loeb. 1988. Effects of herbivory and plant conditioning on the population dynamics of spider mites. Experimental and Applied Acarology 4:225-246.

Karban, R., and J. H. Myers. 1989. Induced plant responses to herbivory. Annual Review of Ecology and Systematics 20:331-348.

Lewis, M. A. 1994. Spatial coupling of plant and herbivore dynamics: the contribution of herbivore dispersal to transient and persistent "waves" of damage. Theoretical Population Biology 45:277-312.

Lundberg, S., J. Järemo, and P. Nilsson. 1994. Herbivory, inducible defence and population oscillations: a preliminary theoretical analysis. Oikos 71:537-539.

Malcolm, S. B., and M. P. Zalucki. 1996. Milkweed latex and cardenolide induction may resolve the lethal plant defence paradox. Entomologia Experimentalis et Applicata 80:193-196.

May, R. M. 1973. Stability and complexity in model ecosystems. Princeton University Press, Princeton.

Morris, W. F., and G. Dwyer. 1997. Population consequences of constitutive and inducible plant resistance: herbivore spatial spread. American Naturalist 149:1071-1090.

Myers, J. H. 1981. Interactions between western tent caterpillars and wild rose: A test of some general plant herbivore hypotheses. Journal of Animal Ecology 50:11-25.

Myers, J. H. 1988. The induced defense hypothesis: does it apply to the population dynamics of insects? Pages 345-365 *in* K.C. Spencer, editor. Chemical Mediation of Coevolution. Academic Press, San Diego.

Pascual, M.A. and P. Kareiva. 1996. Predicting the outcome of competition using experimental data: maximum likelihood and bayesian approaches. Ecology 77:337-349.

Rhoades, D. F. 1985. Offensive-defensive interactions between herbivores and plants: their relevance in herbivore population dynamics and ecological theory. American Naturalist 125:205-238.

Roughgarden, J. 1998. How to manage fisheries. Ecological Applications 8:s160-s164.

Shanks, C. H., and R. P. Doss. 1989. Population fluctuations of twospotted spider mite (Acari: Tetranychidae) on strawberry. Environmental Entomology 18:641-645.

Stout, M.J., K.V. Workman, R.M. Bostock and S.S. Duffey. 1998. Stimulation and attenuation of induced resistance by elicitors and inhibitors of chemical induction in tomato (*Lycopersicon esculentum*) foliage. Entomologia Experimentalis et Applicata 86:267-279.

Thaler, J.S., M.J. Stout, R. Karban, S.S. Duffey. 1996. Exogenous jasmonates simulate insect wounding in tomato plants (*Lycospersicon esculentum*) in the laboratory and field. Journal of Chemical Ecology 22:1767-1781.

Thrall, P.H., J.D. Bever, J.D. Mihail, H.M. Alexander. 1997. The population dynamics of annual plants and soil-borne fungal pathogens. Journal of Ecology 85:313-328.

Turchin, P. 1990. Rarity of density dependence or population regulation with lags? Nature 344:660-663.

Turchin, P. 1997. Population regulation: old arguments and a new synthesis. Pages 19-40 *in* N. Cappuccino and P.W. Price, editors. Population Dynamics: New Approaches and Synthesis. Academic Press, New York.

Underwood, N. 1997. The interaction of plant quality and herbivore population dynamics. Ph.D. dissertation, Duke University, Durham NC.

Underwood, N. C. 1998. The timing of induced resistance and induced susceptibility in the soybean Mexican bean beetle system. Oecologia 114:376-381.

Underwood, N. C. 1999. The influence of plant and herbivore characteristics on the interactions between induced resistance and herbivore population dynamics. American Naturalist 153:282-294.

Woiwood, I. P., and I. Hanski. 1992. Patterns of density dependence in moths and aphids. Journal of Animal Ecology 61:619-629.

Zangerl, A. R., and M. R. Berenbaum. 1991. Furanocoumarin induction in wild parsnip: genetics and populational variation. Ecology 71:1933-1940.

Locally-Induced Responses in Plants: The Ecology and Evolution of Restrained Defense

Arthur R. Zangerl

Abstract

Plants often exhibit locally- or systemically-induced responses to enemy attack. The circumstances under which a localized response is more efficacious than one that is systemic have been the subject of speculation. Among the factors that might influence the spatial dimension of induced responses are adaptively neutral limitations (constraint hypothesis), the cost of the response relative to its benefit (cost hypothesis), the behavior of the herbivore (damage dispersal hypothesis), and the behavior of predators and parasites of the herbivore (natural enemies hypothesis). These hypotheses are evaluated generally, and more specifically for the locally-induced furanocoumarin defense of the wild parsnip.

Introduction

The typical plant is susceptible to attack by a variety of organisms. Attacks may take the form of invasion though breaches in the plant's defenses, followed by contagious spread (pathogens), small gradual tissue removals (most invertebrates), or massive tissue removals (large grazing vertebrates). No single inducible defense mechanism can effectively defend against all of these modes of attack. Indeed, probably no inducible defense, let alone a slow one, can alter the outcome of an interaction between a small herb and a 150 kg ungulate. It seems logical as well, that a rapidly induced response that raises defense throughout a 100-year old tree must certainly be an over-reaction to attack by a lone leaf miner. While the absurdity of the two hypothetical situations just posed seems obvious, the factors that have shaped the evolution of induced responses with respect to spatial dimensions (localized or systemic) are far from resolved. Each form of response is amply represented in the plant kingdom (Karban and Baldwin 1997), but hypotheses that predict which response is more advantageous are rarely tested. The purpose of this chapter is to review key factors that may influence the relative success of systemic and localized responses. Four hypotheses that can account for the existence of localized induced defenses are discussed: the adaptively neutral constraint hypothesis, the cost hypothesis, the dispersed damage hypothesis (involving herbivore behavior), and the natural enemies hypothesis. Within the context of admittedly limited data, the strengths and weaknesses of each hypothesis are explored, and each is evaluated for its utility in accounting for the locally-induced furanocoumarin defense of the wild parsnip, *Pastinaca sativa*. The hypotheses are not mutually exclusive, and it will become clear that much more work is required before a definitive evaluation can made.

Internal Constraint Hypothesis

Whether a plant deploys a localized or systemic response to invasion may have no adaptive significance; plants may simply be incapable of one or the other response. The property of sectoriality that is pervasive in the more advanced members of the plant kingdom is one explanation for the localization of induced responses. Plants are not fully integrated organisms. As a consequence, a branch, leaf, or root can suffer an environmental insult without affecting other branches, leaves, or roots. The proximate factor that seems to be responsible for this independence of plant sectors is the segregation of the plumbing conduits. Xylem and phloem vessels can run distances of many centimeters with no connections to parallel vessels, located fractions of a millimeter away. Consequently, any chemical signaling that relies on xylem and phloem transport is unlikely to be evenly distributed throughout the plant.

There may be advantages to sectoriality within the context of herbivory; it may limit the spread of infections by pathogens or limit the effects of herbivory to the part of the plant that is damaged (Marquis 1996, Price et al. 1996, Sachs and Hassidim 1996, Murphy and Watson 1996, Vuorisalo and Hutchings 1996). However, sectoriality may also constrain the spread of signals that mediate defense induction. Davis et al. (1991) demonstrated that movement of assimilate through the phloem dictates the pattern of wound-induced gene expression among intact poplar leaves. Jones et al. (1993) further demonstrated that changes in undamaged leaves of *Populus deltoides* in response to mechanical damage of a single leaf depended on whether the undamaged leaves were orthostichous (aligned in the same orientation with regard to the stem) to the damaged leaf. Leaves orthostichous to the damaged leaf displayed increased resistance to the leaf-chewing beetle, *Plagiodera versicolora*, while leaves that were not orthostichous remained susceptible. Analysis of the vascular anatomy confirmed that orthostichous leaves share the same vascular traces, while non-orthostichous leaves do not.

Looking at the problem from another perspective, systemic induction may in certain plants be the constrained response. Nicotine, for example, is synthesized in the roots of tobacco species (Dawson 1941, 1942, Baldwin 1991). Extensive study by Baldwin and coworkers (see review in Karban and Baldwin 1997) has confirmed that the increase in shoot nicotine resulting from leaf damage involves communication with the root via phloem transport and subsequent translocation of nicotine from the root to the shoot via xylem transport. However, there is no apparent selective transport of nicotine to damaged plant parts, instead nicotine levels rise throughout the shoot. This systemic response may be an optimal outcome for the plant or may be constrained by an inability of the plant to direct movement of signals to the root and of nicotine from the root over long distances (Karban and Baldwin 1997).

In what is surely the most ambitious spatial mapping study of injury-induced responses to date, Stout et al. (1996) examined the separate effects of soap immersion, forcep-crushing, and feeding by each of two species of insect on changes in the amounts or activities of four constituents of tomatoes: polyphenol oxidase, lipoxygenase, peroxidase, and proteinase inhibitors. Their results reveal clear differences in the spatial extent of induced responses that depend on the response variable and the source of the injury. Production of proteinase inhibitors was almost always induced systemically while induction of the enzymes was to varying degrees localized. The fact that the degree of systemic induction varied, depending on the inducing agent and response variable, suggests that no absolute constraint on signaling exists in tomatoes; more likely, the plant has differing degrees of sensitivity to signals which may or may not have adaptive value. A similar independence between induced responses was demonstrated in bean plants (Jakobek and Lindgren 1993).

Although there may be constraints on the type of induced response allowed in a plant, it would be difficult to predict with any conviction which plants would be more likely to be so constrained. Patterns vary even among plants of similar habit and environment; vascular architecture seems sufficient to account for spatial patterns of response in *Populus*, but no such relationship exists for *Betula pendula* (Mutikainen et al. 1996). As a predictive tool then, the constraint hypothesis is of little value. Instead, its value is primarily as a residual explanation after others have been discarded.

The Cost Hypothesis

The most often cited advantage to inducible defenses is that the cost of defense is avoided until such time as the defense can yield a benefit, i.e., until the plant is attacked (see Karban and Baldwin 1997 and references therein). This conventional wisdom can be extended to explain the value of localized defenses--locally-induced defenses avoid the cost of defending a plant part until that part is attacked. If defenses are indeed costly, a number of predictions can be made about the relative value of localized and systemic responses. Beginning with the premise that the least costly response is a localized one, there are four possible outcomes to a localized response: 1) the attacker dies, 2) the attacker leaves the plant altogether, 3) the response has no effect on the attacker, or 4) the attacker moves to another location within the plant.

From the plant's perspective, the consequences of these outcomes vary greatly. Based purely on a cost/benefit analysis, a response that causes the attacker to leave the plant or to die from ingestion of a lethal dose of a toxin (outcomes 1 or 2) is doubly successful, in that further damage is prevented, and a minimal cost is incurred. A systemic response might have been equally effective but more costly. The advantage of a localized response over one that is systemic is also evident when the defense is wholly ineffectual, as it might be against a specialist herbivore. The limited response, which is less costly, provides the same zero benefit as a systemic response. In the last outcome, the localized response merely causes the herbivore to move to another location on the plant. In this case, the relative effectiveness of localized and systemic responses is conditional on the speed of the response. Defense reactions vary from very rapid (minutes) to very slow (months) (Karban and Baldwin 1997). If a response is slow, a limited response cannot keep pace with the movements of the herbivore; further damage will occur during the lag period at each relocation of the herbivore. Under such circumstances, the systemic response is more effective because, at the end of a single lag period, the entire plant is defended. In situations in which herbivores are merely deterred from extended feeding in one area of the plant, lag times are likely to govern which response is more effective. Long lag times will favor systemic responses while short ones will favor localized responses.

Three plants, for which time courses and degree of localization are known, provide limited support for a connection between speed of a response and its degree of localization. Damage-induced furanocoumarin production in wild parsnip foliage is both rapid and localized; within 3 hours after damage, furanocoumarin concentrations reach peak levels, and the response is confined to the leaflet that is damaged (Zangerl and Berenbaum 1994/1995). In contrast, the nicotine response in wild tobacco shoots is systemic and is considerably slower due in large part to the fact that the nicotine is synthesized in the roots and must be translocated to above-ground parts (Baldwin 1991). Considering these two cases alone, a connection between speed of response and localization would seem to have merit. However, diterpene resin production in grand fir saplings exhibits the slowest response of the three plants and yet is localized (Gijzen et al. 1991, Lewinsohn et al. 1991, 1993).

Comparisons such as these are complicated by the difficulty of classifying responses. Localized and systemic responses can be considered extremes on a continuous spatial scale. Designation of a response as more or less systemic is feasible in exploring patterns of variation within a species, but defining the degree to which a response is systemic in different species is a far more difficult task. For example, given the same stimulus in the form of a small amount of damage, a localized response within a tree may involve far more biomass than a global response within a small herbaceous plant. Whether the tree response is any less extensive than the herb response is debatable. The scale of herbivore attack, measured as size of herbivore or number, relative to the size of the host must also come into play in such comparisons. Attack by a small herbivore may trigger production of defenses in one branch of a tree. Within the context of the movements possible for the herbivore, the tree's response could be global in impact, but from the tree's perspective, the response was clearly localized. Absent a satisfactory means by which spatial responses can be standardized across species, evaluations of any hypothesis regarding spatial extent of induced response might best be limited to single species. Definitive evaluation of the hypothesized relationship between rapidity of a response and its degree of localization must await studies that take advantage either of naturally occurring variation in responses within populations or of techniques to manipulate the response.

All of the predictions in this section rest on the assumption that defenses are costly. Costs can be very difficult to identify because they take many forms and may not appear in all environments. In the broadest sense possible, the cost of defense is defined as the difference in fitness between well-defended phenotypes and less-well defended phenotypes when consumers are absent (Simms and Rausher 1987; for a review of costs, see Agrawal, this volume). One of the more intriguing prospects for looking at costs involves plants that auto-defoliate as a defense. Premature leaf abscission and, to a lesser degree, the hypersensitive response to pathogens are phenomenon that result in immediate lost of leaf area. Premature leaf abscission may be an effective means of shedding herbivores (Faeth et al. 1981, Williams and Whitham 1986, Simberloff and Stiling 1987), but clearly such a response on a systemic scale, while effective, would be extremely costly.

The virtue of the cost hypothesis for explaining localized induced defense is that it makes clear predictions that can be tested. Its weakness is that costs of defense, however real, may not be detectable or may not be of sufficient magnitude to influence the evolution of a defense.

Dispersed Damage Hypothesis

One obvious consequence of a locally-induced defense is that a herbivore, capable both of sensing changes in the quality of what it is eating and of movement, may abandon its feeding site in favor of a less well defended feeding site nearby. Larvae of *Spodoptera littoralis* exhibited a preference in choice experiments for undamaged tomato leaves over damaged leaves, and on intact plants they were likely to abandon damaged leaves for the lower parts of the plant (Barker et al. 1995). Such behavior on the part of the consumer inevitably leads to over-dispersion of the damage that it inflicts. This effect of localized induction prompted Edwards and Wratten (1983, 1985) to suggest that over-dispersion of damage is symptomatic of locally-induced defenses. Moreover, they hypothesized that such patterns of damage, being fairly commonplace, might somehow be less detrimental to the plant than a comparable amount of damage concentrated in one area (Edwards and Wratten 1983).

The number of studies that support this hypothesis is small but slowly growing (for review see Meyer 1998). Marquis (1992) showed that experimental removal of 10% foliage of *Piper arieianum* on a reproductive branch reduced whole-plant seed production, whereas removal of a similar amount of foliage, but from branches throughout the plant, had no effect on seed production. Other studies that examined patterns of partial leaf removal (e.g., Mauricio et al. 1993) gave similar results.

In recent years, efforts have been made to identify the mechanisms that account for this phenomenon. One explanation is that over-dispersed damage is less damaging than concentrated damage to photosynthetic capacity. Presumably, the greater impact of concentrated damage involves the greater likelihood that supply routes to, from, and within the leaf are severed. Most studies that have examined the effects of herbivory or artificial damage on photosynthesis have involved measurements of leaves other than those that have been damaged. These studies often report an enhancement of photosynthesis per unit of leaf area. Among the earliest studies to demonstrate such a photosynthetic enhancement was that of Alderfer and Eagles (1976), in which upper leaves of *Phaseolus vulgaris* were removed with the result that photosynthesis of new leaves was increased. It seems that defoliating insects, in particular, are prone to boost photosynthetic rates of undamaged leaves (Welter 1989). However, effects of herbivores on photosynthesis are anything but uniform. Mesophyll feeders (e.g., Diptera, Hemiptera, Acari) tend to depress photosynthesis of damaged leaves (Welter 1989). This depression of photosynthesis is, perhaps, less a phenomenon unique to mesophyll feeders than an artifact attributable to the fact that mesophyll feedlers leave behind damaged tissues which can be measured for photosynthetic capacity. Herbivores that cleanly remove entire leaves preclude such measurements. It is important to point out that, irrespective of herbivore feeding mode, whole plant photosynthesis typically suffers from herbivore damage (Welter 1989).

The photosynthetic enhancement effects associated with defoliation can often be explained in terms of altered source/sink relationships (Kaitanmiemi and Honkanen 1996). Herbivores that remove leaf tissue alter the amount of source tissue without affecting the amount of sink tissue, e.g., developing leaves, roots, stems and reproductive parts; photosynthesis in the remaining leaf tissues increases to meet the undiminished demand of the sink tissues.

If photosynthetic rates are to be a viable explanation for the advantages of dispersed damage, more extensively damaged leaves must have lower photosynthetic rates per unit of remaining functional area than leaves that are less extensively damaged. Relatively few studies have attempted to examine the effects of damage pattern on photosynthesis of damaged leaves, fewer still have examined the effects of herbivore damage rather than simulated herbivore damage. Morrison and Reekie (1995) simulated naturally occurring patterns of foliage damage in *Oenothera biennis* and found considerable variation in the impact of damage on remaining tissue. Leaves cut in various ways to remove equivalent amounts of tissue exhibited significant differences in area-specific photosynthetic rates, with the negative effect of damage being roughly proportional to the length of cut edges. In the only experiment in which both photosynthesis and components of fitness were quantified in relation to dispersion of damage, Meyer (1998) compared perennial goldenrods (*Solidago altissima*) with half of each leaf removed to plants that had every other leaf removed. Dispersed damage had less impact on photosynthesis of regrowth leaves, and although plants with dispersed damage had temporarily higher growth rates, damage pattern did not affect fitness as measured by that season's inflorescence mass.

An alternative explanation to photosynthesis for the benefits of dispersed damage involves the sectorial nature of plants. If the impact of damage within an integrated sector of a plant is largely, but not exclusively, limited to that sector, and the amount of damage within a sector necessary to effect an impact must exceed some threshold, then the pattern of damage becomes important. A fixed amount of damage that is dispersed among many sectors may fail to reach the threshold in any one of them, although the same amount of damage confined to any one sector will have an adverse effect on that sector. Marquis's (1992) data seem to support this scenario although photosynthetic rates were not measured in that study. This effect would pertain even in situations in which photosynthetic rates of remaining leaf area were unchanged from pre-damage rates.

Another proximate explanation for possible benefits of locally-induced defenses is that herbivores induced to move about the plant will draw the attention of predators (Schultz 1983). Studies have shown that movement of lepidopteran larvae increases risk of predation from spiders (den Boer 1971), pentatomid bugs (Marston et al. 1978), and ants (Bergelson and Lawton 1988). Recent work by Bernays has shown that the mere act of feeding by *Manduca sexta* and *Uresiphita reversalis* increases their risk of death by predators (Bernays 1997). However, no predation link has been established in the context of localized induction. For such a linkage to be established, it would be necessary to show that plant fitness is enhanced by predation in plants with localized induction and less enhanced when the induction is forced to be systemic. It is doubtful that such a result would be obtained, however, because herbivores on systemically induced plants probably move about as frequently as those on locally-induced plants.

There is one potentially serious drawback to defenses that encourage dispersal of damage. A defense that prompts an herbivore to spread its damage throughout the plant also may increase the spread herbivore-vectored pathogens. In excess of 380 plant viruses are vectored by insects; the vectors are primarily homopterans, but thrips, beetles, and mites may also be involved (Nault 1997). Some insects are such effective vectors that they have been evaluated for use in biological control of weeds and as means of inoculation in pathogen resistance screening programs (Vega et al. 1995, Narayana and Muniyappa 1996, Pico et al. 1998). A proliferation of wound sites might also provide opportunities for infection by pathogens that are not vectored. For over-dispersion of damage to be selectively advantageous in a plant, its benefits must outweigh its liabilities with regard to pathogens. In conclusion, the damage dispersion hypothesis may be unlikely to have broad application to the general phenomenon of locally induced responses.

The Natural Enemy Hypothesis

Locally induced responses might, under certain circumstances, be more effective than systemic responses in directing predators and parasites to the location of their herbivorous prey (see Sabelis et al., Paré et al., this volume). The natural enemies hypothesis for localized induction rests on a number of fairly restrictive conditions. First and foremost, it must be demonstrated that a localized response does indeed provide a more precise means of directing the predator or parasite to its host. A localized release of volatiles may be effective at directing nearby flightless predators (e.g., mites). For more mobile predators, however, the signal provided by a spatially limited volatile may be insufficient to permit effective location of the infested plant. Indeed, in plant-herbivore-parasitoid systems in which the parasitoids were capable of flight, plants were found to release volatiles systemically (Potting et al. 1995, Röse et al. 1996, Souissi et al. 1998). Systemic release of volatiles was also effective at attracting flightless predatory mites to groups of Lima bean plants (Dicke et al. 1990).

The extent to which natural enemies can influence whether responses are systemic or localized depends first on whether there are natural enemies and, if there are such enemies, whether they sufficiently impact their prey to enhance plant fitness. The amount of data bearing on this point is admittedly minuscule. However, to the extent that natural enemies may not always be important to the insect, let alone the plant, this hypothesis is probably insufficient to account for localized responses in many cases.

The Wild Parsnip Model

The wild parsnip, *Pastinaca sativa*, is in many ways a useful plant for the study of localized induction. It contains a series of toxic chemicals called furanocoumarins (Murray et al. 1982). These chemicals are contained in oil tubes, present throughout the plant, and their production is inducible to varying degrees by mechanical damage, insects, bacteria, and fungi (Berenbaum and Zangerl 1999). The distribution of constitutive levels of these compounds in reproductive parts has been shown to conform with the predictions of classical optimal defense theory (McKey 1974). This is to say that the amounts of furanocoumarins allocated to reproductive parts, over the course of their development, increases with the investment in, and therefore the value of, those plant parts (Nitao and Zangerl 1987). Similarly, the pattern of damage-inducibility among roots, leaves, and reproductive parts conforms to the expectation that plant parts with high probabilities of attack exhibit minimal or no inducibility, while plant parts that have low probabilities of attack exhibit high degrees of inducibility (Zangerl and Rutledge 1996). In these respects at least, parsnips conform to theoretical expectations.

Wild parsnip leaves, which are inducible by both mechanical and insect damage, are compound, consisting of three to as many as nine leaflets according to the size of the plant. Damage by either an insect or by mechanical wounding to a single leaflet triggers an increase in furanocoumarin concentration in the damaged leaflet, with little or no effect on leaflets located basally or distally to it on the same side of the midvein. Leaflets residing on the other side of the midvein are not induced, and thus the response is highly localized (Zangerl and Berenbaum 1994/1995). This localization has a distinct advantage from the perspective of experimental design. Leaflets on the opposite side of the midvein to a damaged leaflet have the same genetic makeup and morphology as the damaged leaflet, but do not share the same vascular vessels, and hence do not respond to damage of the adjoining leaflet. This makes opposite leaflets ideal controls for comparison with damaged leaflets.

The response to damage in parsnip foliage is characteristically rapid (Fig. 1). Whether induction involves *de novo* synthesis of furanocoumarins in this plant is unknown; $^{14}CO_2$ incorporation experiments demonstrated that furanocoumarin synthesis does not involve recently fixed carbon (Zangerl et al. 1997). Induction of furanocoumarin production in another apiaceous species, parsley, which also has its furanocoumarins localized within oil tubes, appears to proceed *de novo*. Within two hours, wounding of parsley tissues triggers production of mRNAs that code for phenylalanine ammonia-lyase (PAL), the enzyme that mediates the first step in phenylpropanoid synthesis. Within an additional two hours, an increase in production of mRNAs coding for bergaptol O-methyltransferase (BMT) is observed (Lois and Hahlbrock 1992); BMT is the enzyme that converts bergaptol to the furanocoumarin bergapten.

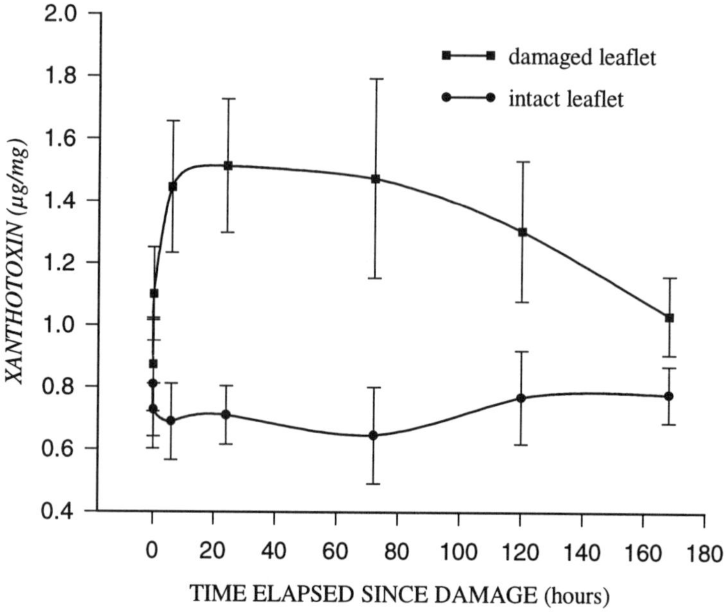

Figure 1: Time course of xanthotoxin induction in damaged leaflets of wild parsnip. Intact leaflets are from the same leaf but located opposite the damaged leaflet (redraw from Zangerl and Berenbaum 1994/1995).

The number of plant chemicals in wild parsnips that is affected by damage is limited. Of 24 primary and secondary plant chemicals studied, only a handful responded to damage. Specifically, furanocoumarins increased, as did soluble proteins and occasionally myristicin, a synergist of furanocoumarin toxicity (Berenbaum and Neal 1985). The only decreases observed have been in the amounts of an uncommon fatty acid (7,10,13-hexadecatrienoic acid) and farnesene (Zangerl et al. 1997, Zangerl and Berenbaum 1998).

Constraints?

The localized response in parsnip foliage can be attributed to the same sort of phenomenon observed by Jones et al. (1993) in *Populus* trees. When dye is applied to the cut surface of a parsnip leaflet, it travels up or down the leaf to adjoining leaflets on the same side of the midvein, but does not cross the midvein (Zangerl and Berenbaum 1994/1995). This suggests a segregation between the vascular traces that serve the two sides of the leaf. However, this constraint does not account for the fact that in most plants, the damage response is limited to the damaged leaflet rather than the entire side of the leaf.

Costs?

That furanocoumarins are costly is evidenced by genetic, phenotypic, and physiological correlations. Negative genetic correlations between the numbers of secondary rays (each of which supports a single flower), an important component of fitness, and concentrations of the furanocoumarins bergapten, xanthotoxin, sphondin, and isopimpinellin were observed in greenhouse-grown plants protected from damage (Berenbaum et al. 1986). The high values of these negative correlations (from -0.632 to -1.0) suggest a genetic trade-off between furanocoumarin production and flower number. This cost was superseded by the benefit of furanocoumarins, as evidenced by a positive correlation between furanocoumarins and seed production in field-grown plants subjected to feeding by the flower- and fruit-feeding specialist parsnip webworm, *Depressaria pastinacella* (Berenbaum et al. 1986). Another indication of a direct cost of furanocoumarin defense takes the form of a negative partial regression of total seed production against seed furanocoumarin concentrations in field-grown plants sprayed with insecticide (Zangerl and Berenbaum 1997). The cost of increasing furanocoumarin concentration by 1mg per gram of seed biomass was equivalent to production of 37 fewer mg of seed.

The cost of furanocoumarin production associated with induction has also been estimated in terms of associated increases in respiration rate, declines in photosynthesis, and declines in biomass accumulation (Zangerl et al. 1997). Leaflets that were administered a uniform and small quantity of damage, in the form of pin pricks (affecting 4% of the area), exhibited temporary reductions in photosynthetic rates--rates that returned to pre-damage rates within three days. There was also a positive association between the increase in furanocoumarin production and the increase in dark respiration rate of leaves two hours after damage ($r = 0.52$, $P = 0.012$). Based on the theoretical energy cost associated with *de novo* furanocoumarin synthesis, the cost of synthesis of the observed increase in furanocoumarin concentrations in damaged leaflets can account for all of the increase in the rate of respiration during the first two hours following damage (Zangerl et al. 1997).

Damage Dispersion?

The benefit of damage dispersion can only be operative if an herbivore responds to induced tissue by relocating to other locations on the plant, and the dispersed damage has less impact on the plant than a comparable amount of damage concentrated in one area. I elected to study two lepidopteran species for which wild parsnips are recorded as hosts. One species, the cabbage looper *Trichoplusia ni*, is a generalist, whose hosts are distributed among 20 plant families (Tietz 1972). Growth rates of cabbage looper larvae fed cut parsnip from previously damaged wild parsnips were significantly reduced compared to those of larvae fed cut foliage that was not previously damaged (Zangerl 1990). In the same study, larvae fed artificial diet that contained the furanocoumarin xanthotoxin, in amounts comparable to the level in induced foliage, also grew significantly more slowly compared to larvae fed a diet containing the furanocoumarin at levels found in intact foliage. These data suggest that the cabbage looper could benefit by engaging in the type of damage avoidance behavior that is prerequisite for dispersed damage and, thereby, enhanced plant fitness. The second herbivore studied was the furanocoumarin-feeding specialist black swallowtail, *Papilio polyxenes*. Black swallowtails feed exclusively on furanocoumarin-containing plants in the family Apiaceae. Compared to the cabbage looper, the black swallowtail metabolizes xanthotoxin 40 times more rapidly (Cohen et al. 1989, Lee and Berenbaum 1989). Despite their high capacity to metabolize furanocoumarins, black swallowtails can be adversely affected by high quantities of these toxins (Berenbaum and Feeny 1981). Unlike cabbage loopers, however, nothing is known about the effects of furanocoumarin induction in parsnips on swallowtail behavior or growth.

The behavioral response of cabbage loopers to induced tissue is instar-dependent. In one experiment, first instar and ultimate fifth instars were placed on either a damaged leaflet of an otherwise intact plant or on an intact leaflet of an intact plant. Damage was inflicted by pressing an array of needles through the leaflet, so that 2% of the treated area was destroyed, an amount sufficient to induce furanocoumarin production (Zangerl and Berenbaum 1994/1995). Whether a larva remained on the leaflet on which it was placed or moved off of it was recorded at intervals up ten hours. The results of this simple experiment revealed that the behavior of cabbage loopers confronted with induced tissues is dependent on their stage of development (Fig. 2). Ultimate instars quickly moved away from target leaflets; within 15 minutes, 90% of the larvae on damaged leaflets had left the leaflet compared to only 55% of the larvae placed on intact leaflets. The rate of abandonment was significantly faster in the first 45 minutes for larvae placed on damaged leaflets compared to larvae on intact leaflets. The results were altogether different for first instars. These larvae did not respond to treatment, and, even after 10 hours, 85% of the larvae remained where they were initially placed (Fig. 2).

The developmental differences in behavior of cabbage loopers confronted with damaged or intact foliage are consistent with the distribution of furanocoumarins in the foliage of this species. Because furanocoumarins are localized in oil tubes within the leaf veins, a small insect, which can forage between leaf veins, can effectively avoid any contact with furanocoumarins. As small larvae with tiny mouth parts, first instar cabbage loopers can and do avoid even the smallest of leaf veins and therefore are not exposed to furanocoumarins. Ultimate instars, having larger mouth parts, are unable to avoid leaf veins; to avoid the negative growth effects of furanocoumarin ingestion, larvae must move

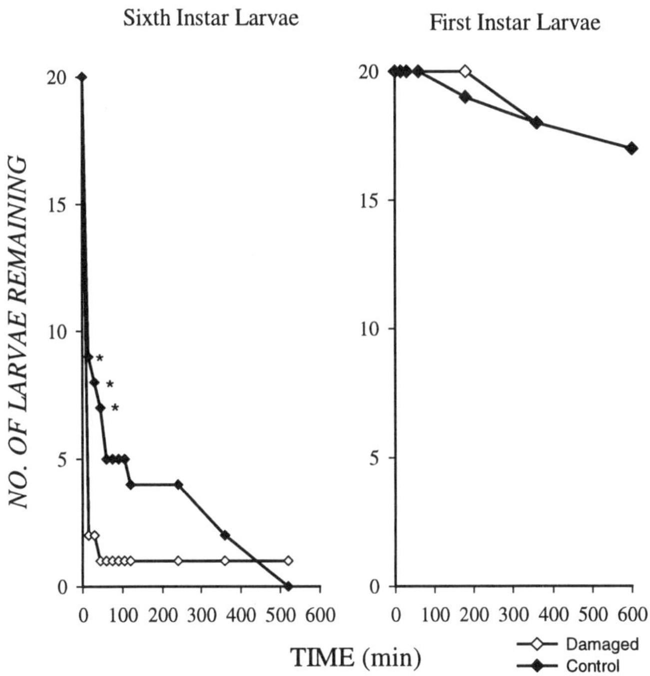

Figure 2: Residency times for first and ultimate instar cabbage loopers placed on intact or damaged wild parsnip leaflets. Sample times with an asterisk indicate significant differences between numbers of larvae on intact and damaged leaflets (chi square test, $P < 0.05$).

off of induced tissue to locate better quality food.

The consequences to the plant of these developmental differences are unclear. The amount of damage that would be inflicted by a first instar cabbage looper is likely to be small and perhaps inconsequential. Behavior of later instars, which consume far more tissue and cannot avoid veins, appears to be consistent with the premise of the damage dispersion hypothesis, that larvae will abandon leaflets on which they have been feeding in order to forage on more suitable leaflets. However, avoidance of damaged tissues alone is not sufficient to satisfy the first requirement of the hypothesis; the behavior must also involve movement to other parts of the plant, resulting in over-dispersion of damage. If larvae are prompted to abandon the plant entirely, the damage will not be dispersed and the benefit of the localized defense cannot be realized.

In nature, larvae, particularly generalist-feeding larvae, may have the option of abandonment. To allow for that option, I performed a second experiment, using plants that either had only one leaflet damaged or had all of their leaflets damaged to simulate a systemic response. With the expectation that they would be too large to avoid feeding on veins, I used third instar larvae in this experiment. All larvae were reared from neonates to the third instar on cut foliage from the same cohort of plants used in the experiment. A single larva was placed on a damaged leaflet of each plant, and disappearance of the larvae was recorded at intervals. Nearly half of the larvae had disappeared from the plants within 2 days and only 20% remained after 9 days (Fig. 3). Of particular interest in these data is the result that treatment had no effect on the period of time that larvae remained on the plant. Thus, it is doubtful that the localized induction is any more or less effective against this herbivore than the systemic induction.

An experiment to assess patterns of movement by ultimate instar black swallowtails on wild parsnip plants revealed that larvae given the option of feeding anywhere on a parsnip plant will feed systematically by defoliating one parsnip leaflet after another until the entire leaf is defoliated. Perhaps these larvae feed so swiftly that they consume tissue before it has had time to respond to damage. In any event, the damage dispersion hypothesis cannot be operative for late instar black swallowtails, which do not disperse their damage.

Earlier instar black swallowtails differ from late instars in that they often partially defoliate leaflets and, therefore, are ideal tools to inflict different concentrations of damage. An experiment was carried out to assess the impact of different concentrations of damage by early instars on photosynthetic rates of

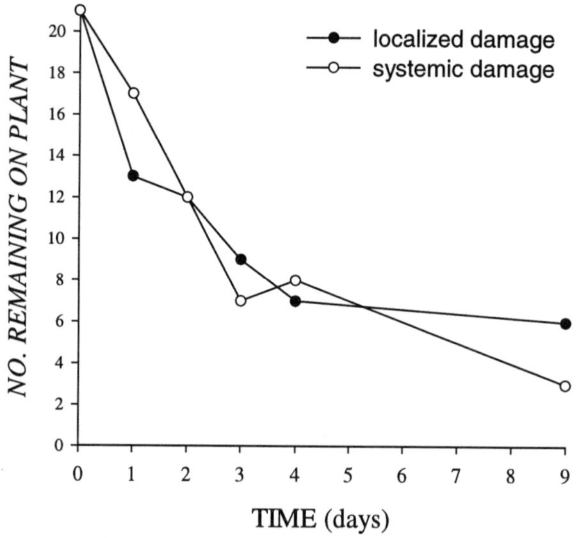

Figure 3: Disappearance of 3rd instar cabbage loopers from wild parsnip plants that had either a single leaf damaged or all of its leaves damaged.

remaining leaf area. A single second instar black swallowtail was placed on each of two leaflets located on either side of the midvein. In one leaflet, the straying larva was repeatedly returned to the leaflet on which it was first placed, while the larva on the adjoining leaflet was permitted to depart from the leaflet. After three days, all of the larvae were removed from the 14 plants included in this experiment. Feeding damage was consistently more concentrated in the leaflet on which the larva was repeatedly replaced. Another three days were allowed to elapse to ensure that damage-induced effects had subsided and that photosynthetic rates had stabilized (Zangerl et al. 1997). Leaflets with the more concentrated damage exhibited rates of photosynthesis that were 28% below those of leaflets with less concentrated damage ($n = 14$ for each treatment, mean $\pm$ s.d. photosynthetic rate, concentrated damage: 0.80 ± 0.28, dispersed damage: 1.11 ± 0.28, paired $t = 2.63$, $P = 0.021$). This result is consistent with the premise that damage dispersion lessens the impact of the damage on photosynthesis.

Synthesis

Although there is some evidence that the localized induced defense in parsnips may be explained in part by the constraint and the dispersed damage hypotheses, the data are far from compelling. First, it is unlikely that anatomical constraints imposed by the vascular system are absolute. At most, vascular system structure can affect the efficiency with which signals are propagated, they do not dictate how much signal is released or the sensitivity of the receptors of those signals. Of course, there may be a genetic constraint on such modifications, in the form of a lack of genetic variation for signal and receptor physiology. In wild parsnips, the degree of localization is not rigidly fixed and is typically less extensive than what is allowed by the vascular system. The response to moderate damage is usually limited to the damaged leaflet, although the vascular system allows extension of the response to all of the leaflets on the same side of a leaf.

The damage-dispersion hypothesis may play some role in the localization of defenses in parsnips, but it is clear that this role depends not only on the behavior of the particular herbivore but also on its developmental stage. These facts, together with the possibility that damage dispersion may be disadvantageous from the perspective of pathogen spread, cast doubt that the damage dispersion hypothesis will have wide application. Similarly, the natural enemies hypothesis, which has yet to be tested extensively in any system, may only be operative in systems in which natural enemies forage at short distances and affect important herbivores.

The only hypothesis among the alternatives that has broad application is the one that invokes a cost of defense. Some costs, such as those associated with lost opportunities or ecological costs (Zangerl and Bazzaz 1992), will vary depending on the plant's physical and biotic environment. Biosynthetic costs, however difficult to measure, are unavoidable and will pertain to every defense system in every plant in which synthesis is required for the induced response. However, the impact of these predictable biosynthetic costs on fitness may well depend on the degree to which plant resources are limiting. In wild parsnips, several kinds of costs have been documented and costs alone may be sufficient to account for the benefit of a localized defense.

Future Directions

The field of study that seeks to explain the evolution of plant defenses is still in its infancy despite the fact that the first coherent theory of defense was proposed some thirty years ago (McKey 1974). If our understanding of the factors shaping defense patterns is to grow, many more tests of specific hypotheses in many different systems are needed. One pitfall of such studies is that the pattern of defense in a plant may not be the product of contemporary selective forces, but of forces exerted over long periods of evolutionary time. This danger is probably overstated. All populations must adapt to the circumstances with which they are currently faced, and plant populations may actually respond quite rapidly to differences in herbivore loads (e.g., Zangerl and Berenbaum 1990, Daehler and Strong 1997).

Specific avenues for research:

1. Are constraints truly constraints or are they adaptive? Constraints that result in localized induction can be evaluated by forcing responses to be systemic (e.g., application of chemical inducers). The fitness of control plants can then be compared to plants with forced systemic responses in the absence and presence of herbivores to assess the cost and benefits of the two responses.

2. Can a classification of responses, regarding lag time and the degree to which a response is systemic, be developed and tested to determine whether there is a relationship between lag time and localization? Comparisons among plants of similar taxonomic and life form could be instructive, as would manipulations of responses within a species.

3. The behavior of herbivores faced with localized and systemic responses is virtually unknown. Do herbivores typically move to other plant parts or leave the plant altogether? If they stay on the plant, do they disperse their damage? Is dispersed damage by actual herbivores less harmful to the plant? Does herbivore movement in response to changing defenses affect their risk of predation? And, does herbivore movement and feeding affect levels of pathogen infection in the plant?

4. What happens when more than one herbivore occupies a plant, are the effects of dispersed damage enhanced or lessened?

Acknowledgments

I thank C. Dubois for assistance with the cabbage looper experiments and NSF DEB 9628977.

Literature cited

Alderfer, R.G. and C.F. Eagles. 1976. The effect of partial defoliation on the growth and photosynthetic efficiency of bean leaves. Botanical Gazette 137:351-355.

Baldwin, I.T. 1991. Damage-induced alkaloids in wild tobacco. Pages 47-69 *in* D. Tallamy and M.J. Raupp, editors. Phytochemical Induction by Herbivores. Wiley, New York.

Barker, A.M., S.D. Wratten, and P.J. Edwards. 1995. Wound-induced changes in tomato leaves and their effects on the feeding patterns of larval lepidoptera. Oecologia 101:251-257.

Berenbaum, M., and P. Feeny. 1981. Toxicity of angular furanocoumarins to swallowtail butterflies: escalation in the coevolutionary arms race? Science 212:927-929.

Berenbaum, M.R. and J.J. Neal. 1985. Synergism between myristicin and xanthotoxin, a naturally cooccurring plant toxicant. Journal of Chemical Ecology 11:1349-1358.

Berenbaum, M.R. and A.R. Zangerl. 1999. Coping with life as a menu option: Inducible defenses of the wild parsnip. Pages 10-32 *in* R. Tolrian and C.D. Harvell, editors. The Ecology and Evolution of Inducible Defenses. Princeton University Press, Princeton, NJ.

Berenbaum, M.R., A.R. Zangerl, and J.K. Nitao. 1986. Constraints on chemical coevolution: wild parsnips and the parsnip webworm. Evolution 40:1215-1228.

Bergelson, J.M. and J.H. Lawton. 1988. Does foliage damage influence predation on the insect herbivores of birch? Ecology 69:434-445.

Bernays, E.A. 1997. Feeding by lepidopteran larvae is dangerous. Ecological Entomology 22:121-123.

Cohen, M.B., M.R. Berenbaum, and M.A. Schuler. 1989. Induction of cytochrome P450-mediated detoxification of xanthotoxin in the black swallowtail. Journal of Chemical Ecology 15:2347-2355.

Daehler, C.C. and D.R. Strong. 1997. Reduced herbivore resistance in introduced smooth cordgrass (*Spartina alterniflora*) after a century of herbivore-free growth. Oecologia 110:99-108.

Davis, J.M., M.P. Gordon, and B.A. Smit. 1991. Assimilate movement dictates remote sites of wound-induced gene expression in poplar leaves. Proceedings of the National Academy of Sciences USA 88:2393-2396.

Dawson, R.F. 1941. The localization of the nicotine synthetic mechanism in the tobacco plant. Science 94:396-397.

Dawson, R.F. 1942. Accumulation of nicotine in reciprocal grafts of tomato and tobacco. American Journal of Botany 29:66-71.

den Boer, M.H. 1971. A colour polymorphism in caterpillars of *Bupalus piniarus* (L.) (Lepidoptera: Geometridae). Netherlands Journal of Zoology 21:61-116.

Dicke, M, M.W. Sabelis, J. Takabayashi, J. Bruin, and M.A. Posthumus. 1990. Plant strategies of manipulating predator-prey interactions through allelochemicals: Prospects for application in pest control. Journal of Chemical Ecology 16:3091-3118.

Edwards, P.J. and S.D. Wratten. 1983. Wound induced defences in plants and their consequences for patterns of insect grazing. Oecologia 59:88-93.

Edwards, P.J. and S.D. Wratten. 1985. Induced plant defences against insect grazing: fact or artifact? Oikos 44:70-74.

Faeth, S.H., E.F. Connor, and D. Simberloff. 1981. Early leaf abscission: a neglected source of mortality for folivores. American Naturalist. 117:409-415.
Gijzen, M., E. Lewinsohn, and R. Croteau. 1991. Characterization of the constitutive and wound-inducible monoterpene cyclases of grand fir (*Abies grandis*). Archives of Biochemistry and Biophysics 289:267-273.
Jakobek, J.L. and P.B. Lindgren. 1993. Generalized induction of defense responses in bean is not correlated with the induction of the hypersensitive reaction. The Plant Cell 5:49-56.
Jones, C.G., R.F. Hopper, J.S. Coleman, and V.A. Krischik. 1993. Control of systemically induced herbivore resistance by plant vascular architecture. Oecologia 93:452-456.
Kaitaniemi, P. and T. Honkanen. 1996. Simulating source-sink control of carbon and nutrient translocation in a modular plant. Ecological Modelling 88:227-240.
Karban, R. and I.T. Baldwin. 1997. Induced Responses to Herbivory. University of Chicago Press. Chicago, IL.
Lee, K, and M.R. Berenbaum. 1989. Action of antioxidant enzymes and cytochrome P450 monooxygenases in the cabbage looper in reponse to plant photoxins. Archives of Insect. Biochemistry and Physiology 10:151-162.
Lewinsohn, E., M. Gijzen, and R. Croteau. 1991. Defense mechanisms of conifers. differences in constitutive and wound-induced monoterpene biosynthesis among species. Plant Physiology 96:44-49.
Lewinsohn, E., M. Gijzen, R.M. Muzika, K. Barton, and R. Croteau. 1993. Oleoresinosis in grand fir (*Abies grandis*) saplings and mature trees. Plant Physiology 101:1021-1028.
Lois, R. and K. Hahlbrock. 1991. Differential wound activation of members of the phenylalanine ammonia-lyase and 4-coumarate:CoA ligase gene families in various organs of parsley plants. Zeitschrift für Naturforschung 47c:90-94.
Marquis, R.J. 1992. A bite is a bite is a bite? Constraints on response to folivory in *Piper arieianum* (Piperaceae). Ecology 73:143-152.
Marquis, R.J. 1996. Plant architecture, sectoriality and plant tolerance to herbivores. Vegetatio 127:85-97.
Marston, N.L., G.T. Schmidt, K.D. Biever, and W.A. Dickerson. 1978. Reaction of five species of soybean caterpillars to attack by the predator, *Podisus maculiventris*. Environmental Entomology 7:53-56.
Mauricio, R., M.D. Bowers, and F.A. Bazzaz. 1993. Pattern of leaf damage affects fitness of the annual plant *Raphanus sativus* (Brassicaceae). Ecology 74:2066-2071.
McKey, D. 1974. Adaptive patterns in alkaloid physiology. American Naturalist 108:305-320.
Meyer, G.A. 1998. Pattern of defoliation and its effect on photosynthesis and growth of goldenrod. Functional Ecology 12:270-279.
Morrison, K.D., and E.G. Reekie. 1995. Pattern of defoliation and its effect on photosynthetic capacity in *Oenothera biennis*. Journal of Ecology 83:759-767.
Murray, R.D.H., J. Mendez, and S.A. Brown. 1982. The Natural Coumarins. John Wiley, Chichester.
Murphy, N. and M.A. Watson. 1996. Sectorial root growth in cuttings *of Coleus rehneltianus* in response to localized aerial defoliation. Vegetatio 127:17-23.
Mutikainen, P., M. Walls, and J. Ovaska. 1996. Herbivore-induced resistance in *Betula pendula*: the role of plant vascular architecture. Oecologia 108:723-727.
Nault, L.R. 1997. Arthropod transmission of plant viruses—a new synthesis. Annals of the Entomological Society of America 90:521-541.
Narayana, Y.D. and V. Muniyappa. 1996. Evaluation of techniques for the efficient transmission of *Sorghum* stripe virus by vector (*Peregrinus maidis*) and screening for disease resistance. Tropical Agriculture 73:119-123.
Nitao, J.K., and A.R. Zangerl. 1987. Floral development and chemical defense allocation in wild parsnip (*Pastinaca sativa*). Ecology 68:521-529.
Pico, B., M. Diez, and F. Nuez. 1998. Evaluation of whitefly-mediated inoculation techniques to screen *Lycopersicon esculentum* and wild relatives for resistance to tomato yellow leaf curl virus. Euphytica 101:259-271.

Potting, R.P.J., L.E.M. Vet, and M. Dicke. 1995. Host microhabitat location by stem-borer parasitoid *Cotesia flavipes*—the role of herbivore volatiles and locally and systemically induced plant volatiles. Journal of Chemical Ecology 21:525-539.

Price, E.A.C., M.J. Hutchings and C. Marshall. 1996. Causes and consequences of sectoriality in the clonal herb *Glechoma hederacea*. Vegetatio 127:41-54.

Röse, U.S.R., A. Manukian, R.R. Heath, and J.H. Tumlinson. 1996. Volatile semiochemicals released from undamaged cotton leaves—a systemic response of living plants to caterpillar damage. Plant Physiology 111:487-495.

Sachs, T, and M. Hassidim. 1996. Mutual support and selection between branches of damaged plants. Vegetatio 127:25-30.

Schultz, J.C. 1983. Impact of variable plant defensive chemistry on susceptibility of insects to natural enemies. Pages 37-54 *in* P.A. Hedin, editor. Plant Resistance to Insects. American Chemical Society, Washington, D.C.

Simberloff, D. and P. Stiling. 1987. Larval dispersion and survivorship in a leaf-mining moth. Ecology 68:1647-1657.

Simms, E.L and M.D. Rausher. 1987. Costs and benefits of plant resistance to herbivory. American Naturalist 130:570-581.

Souissi, R., J.P. Nenon, and B. Leru. 1998. Olfactory responses of parasitoid *Apoanagyrus lopezi* to odor of plants, mealybugs, and plant-mealybug complexes. Journal of Chemical Ecology 24:37-48.

Stout, M.J., K.V. Workman, and S.S. Duffey. 1996. Identity, spatial distribution, and variability of induced chemical responses in tomato plants. Entomologia Experimentalis et Applicata 79:255-271.

Tietz, H.M. 1972. An index to the described life histories, early life stages, and hosts of the Macrolepidoptera of the continental United States and Canada. A.C. Allyn Museum of Entomology, Sarasota, FL.

Vega, F.E., P.F. Dowd, and R.J. Bartelt. 1995. Dissemination of microbial agent using an autoinoculation device and several insect species as vectors. Biological Control 5:545-552.

Vuorisalo, T. and M.J. Hutchings. 1996. On plant sectoriality, or how to combine the benefits of autonomy and integration. Vegetatio 127:3-8.

Welter, S.C. 1989. Arthropod impact on plant gas exchange. Pages 135-151 *in* E.A. Bernays, editor. Insect-plant Interactions. CRC Press, Boca Raton, FL

Williams, A.G. and T.G. Whitham. 1986. Premature leaf abscission: an induced plant defense against gall aphids. Ecology 67:1619-1627.

Zangerl, A.R. 1990. Furanocoumarin induction in wild parsnip: evidence for an induced defense against herbivores. Ecology 71:1933-1940.

Zangerl, A.R., A.M. Arntz, and M.R. Berenbaum. 1997. The physiological price of an induced chemical defense: photosynthesis, respiration, biosynthesis, and growth. Oecologia 109:433-441.

Zangerl, A.R. and F.A. Bazzaz. 1992. Theory and pattern in plant defense allocation. Pages 363-391 *in* Fritz, R. and E. Simms, editors. Ecology and Evolution of Plant Resistance. University of Chicago Press, IL.

Zangerl, A.R., and M.R. Berenbaum. 1990. Furanocoumarin induction in wild parsnip: genetics and populational variation. Ecology 71:1933-1940.

Zangerl, A.R., and M.R. Berenbaum. 1994/95. Spatial, temporal, and environmental limits on xanthotoxin induction in wild parsnip foliage. Chemoecology 5/6:37-42.

Zangerl, A.R. and M.R. Berenbaum. 1997. Costs of chemically defending seeds:furanocoumarins and *Pastinaca sativa*. American Naturalist 150:491-504.

Zangerl, A.R. and M.R. Berenbaum. 1998. Damage-inducibility of primary and secondary metabolites in the wild parsnip (*Pastinaca sativa*). Chemoecology 8:187-193.

Zangerl, A.R., and C.E. Rutledge. 1996. Probability of attack and patterns of constitutive and induced defense: a test of optimal defense theory. American Naturalist 147:599-608.

Induced Plant Defense: Evolution of Induction and Adaptive Phenotypic Plasticity

Anurag A. Agrawal

Abstract

Induced defenses may evolve in natural plant populations where heritable variation in inducibility affects plant fitness. Although this has not been documented for any plant – parasite system, genetic variation in induction has been recently reported for several plant-parasite systems. Correlations between induction and fitness in variable parasite environments suggest that inducibility may indeed be subject to selection, however, these studies focussed on phenotypic (not genetic) correlations. Our current state of knowledge suggests that induced defenses may be an example of adaptive plasticity, where defenses enhance plant fitness when parasites are present, and reduce fitness when parasites are absent. Future work that focuses on consequences of genetic variation in induction and macro-evolutionary trends in inducible versus constitutive defense will help advance our understanding of the evolutionary biology of induced plant defense.

Introduction

Induced plant "responses" to attack consist of any active or passive change in the plant following herbivory or infection. Induced plant "resistance" refers to reduced preference, performance, or pathogenicity of the attacker on induced plants compared to controls. Note that this definition of induced resistance, originally advocated by Karban and Myers (1989), is characterized by the behavior or fitness of the plant attacker, not by attributes of the plant itself. The focus of this chapter is induced plant "defense" rather than resistance. Defense is a term reserved for situations where plants having some resistance character have higher fitness than plants lacking that character. Measuring characteristics that affect plant performance are central to the goals of applied entomology and plant pathology, specifically reducing pest levels below some threshold above which pests cause economic losses. In addition, understanding the factors that influence plant fitness is central to understanding basic ecological and evolutionary principles of plant-parasite interactions. This chapter will focus on two aspects of induced responses and their fitness consequences for plants: 1) Are induced defenses subject to natural selection? and 2) Are their phenotypic benefits and costs associated with induced defenses, indicating that they are an example of adaptive plasticity?

Microevolution of Induced Defenses

Variation in induced resistance (within plants and across populations of plants) can be caused by many biotic and abiotic factors of the environment (Agrawal and Karban 1998). However, induced plant resistance can evolve by natural selection only if there is heritable variation for induction that affects plant fitness. Such genetic variability is detected through a statistical interaction between genetically related families (or clones) of plants and their response to herbivory or pathogen attack. In this review, induced resistance is considered a single trait, in most cases determined by a negative effect on the herbivore or pathogen. Clearly, many genes and pathways are involved in most plant responses to attack. For simplicity, and for the purposes of defining heritability for a suite of traits that may be subject to natural selection, induced resistance is considered a single trait.

In wild plants, 14 plant-herbivore systems have been examined for genetic variation in inducibility (Table 1). Varietal variation in induced resistance against herbivores and pathogens of crop plants has also been reported for several systems. To my knowledge only one wild plant - pathogen system has been examined for genetic variation in induced resistance (Dirzo and Harper 1982). In this case, clover plants exhibited genetic variation in the pre-formed cyanogenic glucoside response to leaf damage.

No study to date has documented fitness consequences for the plant involving genetic variation in induced resistance. Circumstantial evidence that induced resistance may be evolving in response to herbivore attack was provided by Dirzo and Harper (1982), Raffa (1991) and Zangerl and colleagues (Zangerl and Berenbaum 1990, Zangerl and Rutledge 1996, Zangerl, this volume). Although it is likely that induced responses are, indeed, subject to natural selection, we can only infer this from the evidence of current genetic variation in the trait.

Why is there genetic variation in induced plant resistance present in natural populations? The least interesting answer to this question is that genetic variation in induced resistance does not directly affect plant fitness, and induced resistance is a trait that is genetically correlated with other traits that do affect plant fitness. For example, *Brassica rapa* plants with genetically divergent levels of myrosinase (an enzyme involved in hydrolyzing defense products in mustards) showed variation in fitness in a field experiment (Siemens and Mitchell-Olds 1998). However, this study showed that induced levels

Table 1. Systems for which genetic variation in induced resistance has been reported.

Plant	Latin name	Type	Mechanism	Parasite affected	Latin name	Reference
Goldenrod	*Solidago altissima*	Wild	Hypersensitivity	Fly eggs	*Eurosta solidaginis*	Abrahamson and Weis 1997
Wild parsnip	*Pastinaca sativa*	Wild	Furanocoumarins	Moth larvae	*Trichoplusia ni*	Zangerl and Berenbaum 1990
Black mustard	*Brassica nigra*	Wild	Hypersensitivity	Butterfly eggs	*Pieris spp.*	Shapiro and DeVay 1987
Rapid cycling mustard	*Brassica rapa*	Wild	Glucosinolates	Moth larvae	*Plutella xylostella*	Siemens and Mitchell-Olds 1998
Wild radish	*Raphanus sativus*	Wild	?	Aphids	*Myzus persicae*	Agrawal 1998, 1999a
Birch	*Betula pubesens*	Wild	?	Moth larvae	*Epirrita autumnata*	Haukioja and Hanhimaki 1985
Sunflower	*Helianthus annuus*	Wild	Coumarins	Beetles	*Zygogramma exclamationis*	Roseland and Grosz 1997
Quaking aspen	*Populus tremuloides*	Wild	?[1]	Moth larvae	*Lymantria dispar*	Osier and Lindroth, pers. comm.
Narrow leaf plantain	*Plantago lanceolata*	Wild	Iridoid glycosides	N/t	N/t	Bowers and Stamp 1993
Ponderosa pine	*Pinus ponderosa*	Wild	Monoterpenes/ necrosis	Beetles – fungus complex	*Scolytus ventralis, Trichosporium sybioticum*	Raffa 1991
Hounds tongue	*Cynoglossum officinale*	Wild	Pyrrolizidine alkaloids	N/t	N/t	van Dam and Vrieling 1994
Willow	*Salix myrsinifolia*	Wild	Salicortin	N/t	N/t	Julkunen-Tiitto et al. 1995
Turnera	*Turnera ulmifoli*	Wild	Cyanogenesis	N/t	N/t	Schappert and Shore 1995
White clover	*Trifolium repens*	Wild	Cyanogenesis	Several herbivores		Dirzo and Harper 1982
				Rust fungus	*Uromyces trifolii*	
Cotton	*Gossypium hirsutum*	Crop	?	Mites	*Tetranychus urticae*	Brody and Karban 1992
Soybean	*Glycine max*	Crop	?	Beetles	*Epilachna varivestis*	Underwood 1998, unpublished
Grape	*Vitis vinifera*	Crop	?	Mites	*Tetranychus spp.*	English-loeb et al. 1998
Cucumber	*Cucumis sativus*	Crop	Cucurbitacin C	Mites	*Tetranychus urticae*	Agrawal et al. 1999
Wheat	*Triticum aestivum*	Crop	Several enzymes	N/t	N/t	Rybka et al. 1998
			Hydroxamic acids	N/t	N/t	Gianoli et al. 1997

Radish	*Raphanus sativus*	Crop	Necrosis	Leaf spot	*Xanthomonas campestris*	Kamoun et al. 1993
Tomato	*Lycopersicon esculentum*	Crop	PR proteins	Fungal pathogen	*Alternaria solani*	Lawrence et al. 1996
Triticale	*Triticum X Secale*	Crop	Several enzymes	N/t	N/t	Rybka et al. 1998
Cacao	*Theobroma cacao*	Crop	?	Black pod disease	*Phytophthora spp.*	Pires et al. 1997
Pearl millet	*Pennisetum glaucum*	Crop	Hypersensitivity	Downy mildew	*Sclerospora graminicola*	Geetha et al. 1996
Chinese cabbage	*Brassica campestris*	Crop	Glucosinolates	Clubroot disease	*Plasmodiophora brassicae*	Ludwig-Mueller et al. 1997
Tobacco	*Nicotiana tabacum*	Crop	Necrosis	Fungal pathogen	*Phytophthora parasitica*	Bonnet et al. 1996
Barley	*Hordeum vulgare*	Crop	?	Powdery mildew	*Erysiphe graminis*	Newton and Dashwood 1998
Parsnip	*Pastinaca sativa*	Crop	Furanocoumarins	N/t	N/t	Cerkauskas and Chiba 1990

? = unknown
N/t = not tested
[1] not phenolic glycosides, tannins, nitrogen, or water content

of glucosinolates were positively genetically correlated with constitutive levels of glucosinolates. Thus, it is unknown whether induced defenses themselves were under selection, or simply carried along as a correlated trait.

It is likely that induced defense is a trait subject to natural selection in natural populations. However, evolution may very well be acting on induced and constitutive resistance as correlated traits. Theory predicts that there should be a negative genetic correlation between constitutive resistance and inducibility of plants (Brody and Karban 1992, Herms and Mattson 1992). In two studies, levels of constitutive phytochemicals were positively correlated to inducibility of phytochemicals (Zangerl and Berenbaum 1990, Siemens and Mitchell-Olds 1997), while in a third study there was no relationship (Brody and Karban 1992). It is currently unknown whether or not constitutive and induced resistance can be genetically uncoupled. However, many traits with strong phenotypic correlations can be uncoupled with only a few generations of strong selection (e.g., Stanton and Young 1994).

The experimental uncoupling of constitutive and induced resistance will allow us to understand several basic and applied aspects of plant defense. Experimental tests of the benefits of induced resistance over constitutive resistance (and vice versa) in different environments will require this uncoupling. Several non-mutually exclusive alternative benefits of induction have been proposed and are reviewed elsewhere (Karban and Baldwin 1997, Karban et al. 1997, Agrawal and Karban 1998). In addition, understanding how or why induced resistance may be a useful strategy in agriculture will require it to be tested within different backgrounds of constitutive resistance.

As noted above, no studies have documented the fitness consequences for plants with genetically variable induced resistance. However, a few studies have examined the fitness consequences of plants expressing induced versus uninduced phenotypes. In other words, studies have asked if induced plants have higher fitness than uninduced plants when growing in an environment with herbivores; and, if induced plants have lower fitness than uninduced plants when growing in an environment without herbivores. These studies have linked the plant phenotype to plant fitness. This is important for several basic and applied reasons (discussed in the following sections), however, these phenotypic correlations do not directly address the consequences of genetic variation and the evolution of induction. Some authors, have argued and provided evidence that phenotypic correlations are good estimates of genetic correlation (e.g., Cheverud 1988), although this remains controversial.

The Cost-Benefit Framework and Genotype-by-Environment Interaction

The cost-benefit framework is a long standing construct of evolutionary biology that attempts to address why genetic variation in traits that affect fitness persist in natural populations. These ideas were verbally applied to plant defense theory 20 years ago (reviewed by Rhoades 1979). Simms and Rausher quantitatively formalized these arguments for plant defense with theory and experiments in the late 1980's (Simms and Rausher 1987, 1989, Rausher and Simms 1989, Simms 1992a). The basic arguments have stemmed from the observation that plant defenses are beneficial to plants, and yet plant populations are not fixed at maximal levels of defense. Why are not all plants maximally defended? The proposed answer is that although defense is beneficial in some environments, it is also costly. These costs could reduce plant fitness in environments without herbivores, where the benefits of defense cannot be realized (Parker 1992, Simms 1992b, Mole 1994, Agrawal and Karban 1998). This is a classic example of genotype-by-environment interaction, which is thought to be a crucial component in the maintenance of variation in traits affecting fitness (Gillespie and Turelli 1989).

The application of the cost-benefit framework to evolutionary issues of induced defense has been somewhat muddy, in part because variation in defense can exist within a genetic individual (i.e., a single plant can be induced or not). Initial arguments suggested that plants should express high levels of defense when herbivores are present, but not all of the time because of costs. Although this makes perfect sense for an individual plant, as noted above, phenotypic benefits and costs do not necessarily reflect genetic benefits and costs. When under genetic control, such traits may have epistatic, linked, or pleiotropic effects which minimize or exacerbate the phenotypic benefit or cost. Adaptive evolution can only follow from genetic benefits, albeit constrained by genetic costs. However, phenotypic benefits and costs provide valuable information about the adaptive value of phenotypic variation in plant defense.

Induced Plant Defense as Adaptive Plasticity

Plasticity in plant defense reflects biotic or abiotic environmental conditions that affect the expression of resistance characters that can affect plant fitness (Agrawal and Karban 1998). A focus of this chapter is whether or not these induced responses enhance the fitness of plants compared to plants not having the induced responses.

Phenotypic plasticity is thought to evolve as a mechanism for organisms to express adaptive phenotypes in variable environments (Via and Lande 1985, Thompson 1991). Thus, a fundamental prediction of the evolution of adaptive plasticity is that organisms expressing particular phenotypes in particular environments should have higher relative fitness than conspecifics expressing alternative phenotypes (Thomson 1991, Dudley and Schmitt 1996, Kingsolver and Huey 1998). In other words, if an organism changes its phenotype in response to the environment, this is expected to increase its relative fitness compared to organisms that do not alter their phenotypes. An excellent example of adaptive plasticity that has received attention is the stem elongation response of plants growing in crowded conditions. Dudley and Schmitt (1995, 1996) conducted experiments using phenotypic manipulations of plant morphology (elongated or not) by exposing young plants to different red to far red light ratios. Plants of both phenotypes were then placed in both environments (high and low crowding) to evaluate the consequences for relative fitness. Dudley and Schmitt showed that elongated plants in competitive environments have higher relative fitness than non-elongated plants, whereas in non-competitive environments, non-elongated plants have higher relative fitness than elongated plants. It is of value to be tall and spindly if there are many competitors for limiting light; these same traits can be detrimental in the absence of competition.

Inducible plant defenses provide another ideal system to test the adaptive plasticity hypothesis. Do induced plants have higher relative fitness than uninduced plants in environments with plant parasites, and do uninduced control plants have higher relative fitness than induced plants in environments lacking plant parasites? The answer to these questions are not only relevant to understanding whether this type of phenotypic plasticity is adaptive, but also to understanding the potential yield consequences of manipulating induced resistance in agriculture. Is induced defense a viable strategy that will enhance crop yields where pests are present? Under low pest pressure, are induced defenses wasteful and will their costs outweigh their benefits (reducing yield)? Note that viewing induced defenses as adaptive plasticity fits well within the cost-benefit framework outlined above, however, the adaptive plasticity framework is concerned specifically with phenotypic costs and benefits, not genetic costs and benefits (which are potentially responsible for the maintenance of genetic variation).

Recent studies of wild radish plants (reviewed in the next section, Agrawal 1998, 1999a, Agrawal et al. 1999), pepper weed (Agrawal 1999b), and wild tobacco plants (Baldwin 1998) have found support for induced defenses against herbivores as examples of adaptive plasticity by demonstrating fitness benefits and costs of induced resistance. Examples of adaptive plasticity in response to pathogens are fewer. Classic work by Kuć and colleagues in several cucurbit plant – pathogen systems has shown that induced resistance can increase survival and reproduction of challenged plants in the field compared to uninduced control plants (Kuć 1982, 1987, Caruso and Kuć 1977). Similarly, these workers and others have found circumstantial evidence that induction may reduce plant fitness in the absence of pathogens (Kuć 1987, Rasmussen et al. 1991, Hoffland et al.1998). Studies of barley (Smedagaard-Peterson and Stølen 1981) and tobacco (Lagrimini et al. 1997) show that there may be phenotypic costs of induced defense against pathogens, although benefits have not been demonstrated clearly in these systems. Experimental induction of tobacco did protect field-grown plants against blue mold and increased plant growth and yield (Tuzun et al. 1992).

Recent experiments with tomato plants have shown that infection by non-pathogenic *Meloidogyne* spp. nematodes can protect plants against a serious pest nematode, *Meloidogyne hapla* (Ogallo and McClure 1995). In field plots, plants with the initial inducing inoculum were stunted 30% less compared to controls, when both treatments were challenged. A cost assessment has not been made in this system.

Classic work by Raffa and Berryman (1982, reviewed by Raffa 1991) integrates the responses of grand fir trees to insect and pathogen enemies with effects on plant fitness. Fir trees that are attacked by bark beetles are also subject to a fungus which the beetles introduce. The plants respond with localized necrosis and systemic monoterpene emissions. These defensive responses can protect plants against subsequent infection and infestation of trees; inoculated (induced) trees were more likely to survive than uninoculated trees during an outbreak of the pests and pathogens (Raffa and Berryman 1982). Again, a cost assessment of the induced defense has not been made in this system.

Studies of benefits and costs of induced defense have employed various phenotypic manipulations of induced resistance, including: 1) the use of real plant parasites, 2) avirulent plant parasites, and 3) natural and artificial chemical elicitors of induced resistance. In addition, genetic mutants or engineered plants, over- and under-expressing particular inducible gene products can be used to test the adaptive plasticity hypothesis (for theory and a stem elongation example see Schmitt et al. 1995). Plant mutants over- and under-expressing induced defense against herbivores (McConn et al. 1997, Eichenseer et al. 1998) and pathogens (Bowling et al. 1994, Cao et al. 1994, Lagrimini et al. 1997) have also been reported. The technique of using mutants to assess phenotypic benefits and costs is useful because the phenotype does not have to be manipulated and the phenotypes will not change or adjust in different environments through time. However, it is essential to have replicated bred or engineered isogenic lines of over- and under-expressers to properly account for other changes introduced to the plant via the mutagenesis process. Finally, although using engineered plants and mutants are powerful tools for understanding benefits and costs, it is imperative to know where in the biochemical pathway of induction the mutations lie. For example, an under-expresser of induced resistance may have a completely functional pathway, except for one final product. In this case, the under-expresser may not show "cost savings" because most of the expense of the induced defense is already invested.

Case Study: Wild Radish and its Herbivores

In the wild radish system (*Raphanus raphanistrum* and *R. sativus*), induced responses to herbivory include elevated concentrations of indole glucosinolates and increased densities and total numbers of setose trichomes on newly formed leaves of previously damaged plants compared to undamaged controls (Agrawal 1999a, Agrawal et al. 1999). A broad array of herbivores feed on wild radish plants in nature; many of these herbivores are negatively affected by induced resistance, including earwigs, grasshoppers, aphids, flea beetles, and several species of lepidopteran larvae (Agrawal 1998, 1999a, c, unpublished data). Induced resistance in wild radish can be elicited by herbivory by both specialist and generalist herbivores (Agrawal 1999c) and natural elicitors of induced resistance such as jasmonic acid (Agrawal 1999a), but not by some types of mechanical damage (Agrawal 1998, 1999a).

In an experiment to test the first half of the adaptive plasticity hypothesis, that induced plants should have higher relative fitness in the presence of plant parasites than uninduced plants, nearly 500 plants were grown in the field. The plants were randomly distributed among three treatments: unmanipulated control plants, induced plants (plants treated with caterpillar herbivory), and leaf-damage control plants. Leaf-damaged control plants were treated such that an amount of leaf area was removed with a pair of scissors to equal that removed by the caterpillars in the induced resistance treatment. Leaf-damage control plants did not exhibit induced resistance, presumably because there were fewer plant cells damaged by the quick scissors clip and herbivore saliva was lacking (see Felton and Eichenseer, this volume). Thus, leaf damaged control plants assayed for the fitness consequences of losing leaf tissue without the effects of induced resistance.

Herbivores were hand-picked off plants twice daily when the treatments were being imposed. This is important because the experimental procedure did not impose induced resistance on a haphazard set of plants and

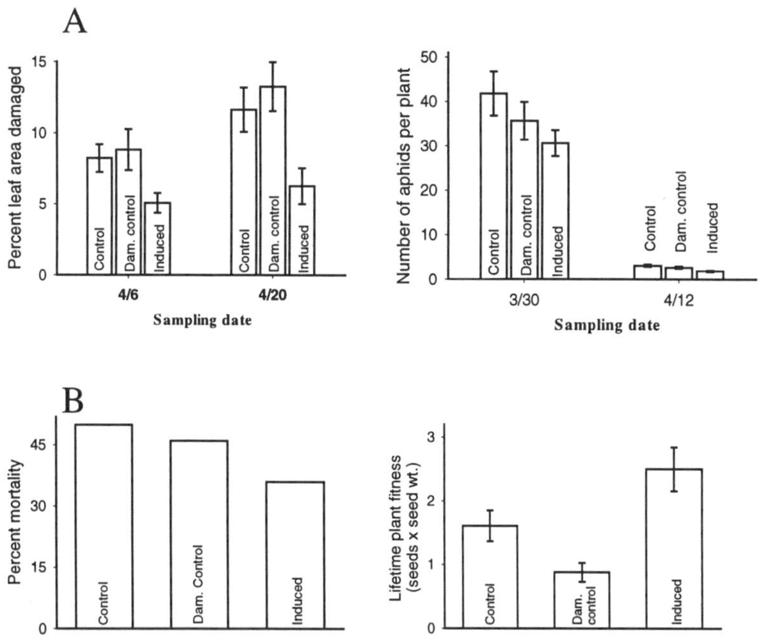

Figure 1: Consequences of induced responses to herbivory in wild radish plants for A) subsequent herbivory by leaf chewers and aphids, and B) components of plant fitness including herbivore imposed mortality and seed production. Dam. control plants received equal leaf area loss to that in the induced treatment but did not have the induced resistance. Error bars represent standard errors. Redrawn from Agrawal (1998).

determine the consequences. Rather, early season induction of resistance was denied in a set of plants (unmanipulated controls, and leaf damage controls) and uniformly imposed on another set of plants (that would have received natural herbivory and induced resistance, if all of the herbivores were not picked off).

The results indicate that plants expressing induced resistance were protected against several herbivores, including aphids, earwigs, and flea beetles (Agrawal 1998, 1999a, Fig 1a). Mortality of plants, due in large part to flea beetles, was differentially seen in the three treatments (Fig. 1b). Finally, induced plants had 60% higher fitness than control plants, calculated per plant as the number of seeds multiplied by mean seed weight (Fig. 1b). Because wild radishes are weedy annual plants, survival and seed production are good estimates of lifetime fitness. Leaf damage controls had the lowest fitness, and thus there was an even greater gap in fitness between leaf damage controls and induced plants. This result demonstrates that leaf loss itself is costly to the plant. However, when plants have induced resistance (coupled with leaf loss) the costs of leaf loss and other costs associated with induced resistance are far outweighed by the benefits (Agrawal 1998, 1999a). The benefits of induction were consistently demonstrated with two species of wild radish (*R. sativus*, and *R. raphanistrum*), conducted over two years and at two field sites (Agrawal 1998, 1999a).

The other half of the adaptive plasticity hypothesis is that plants not expressing defenses should have relatively higher fitness than plants expressing defenses in the absence of plant parasites. In other words, are there phenotypic costs of expressing induced defenses? Costs of induced plant resistance have received relatively more attention than benefits (Brown 1988, Karban 1993, Gianoli and Niemeyer 1997, Zangerl et al. 1997, Baldwin 1998). Among the many studies of phenotypic costs of induced resistance, there have been mixed results, with some studies finding costs (i.e., reduced growth or seed production), while other studies not finding costs. In the studies of costs of induced defenses in wild radish, a broad measure of costs was employed to include male fitness components in addition to traditional female fitness components (i.e., seed set). The number of seeds sired through pollen (male plant fitness) represent half of the genes contributed to the next generation in outcrossing plants.

In the cost studies (Agrawal et al. 1999), we found that induced wild radish plants had reduced fitness compared to uninduced plants in an environment without herbivores. This experiment also controlled for leaf area removal using a factorial design employing plants with and without leaf tissue removal, and plants with and without induced resistance (Fig. 2). This approach is powerful for several reasons, including the fact that the effects of induced defenses and leaf tissue loss can be uncoupled. From a statistical standpoint, the factorial analysis has superior power than non-crossed designs, and has the ability to detect interactions between leaf area removal and induction. Finally, although elicitors of induced plant defenses (and other techniques such as mutagenesis) provide an exciting opportunity to answer ecological and

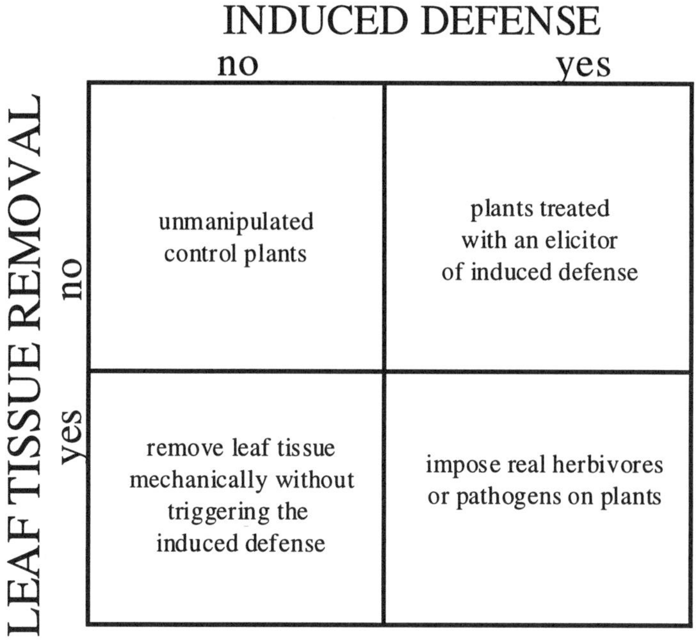

Figure 2: Experimental design recommendation for testing for the phenotypic benefits and costs of induced plant defenses. The factorial design is statistically powerful and can detect an interaction, which is important for calibration of real and artificial eliciting agents.

evolutionary questions, it is important to include real herbivory treatments when using elicitors, so as to calibrate your system and be sure that real herbivory and the elicitors are having the same effect.

The cost of induced defense detected in wild radish was only detected in components of male fitness, pollen grain production and size. These traits have been linked to seed siring ability in wild radish and other systems. Surprisingly, seed number and seed weight were not reduced on induced plants compared to controls. These results point to the need for consideration of other fitness components than seed production when studying fitness in sexually reproducing plants. Physiological studies in plants have directly linked jasmonates to viable pollen production (see Staswick and Leman, this volume), suggesting a mechanism for the cost of induced resistance in terms of male fitness characters.

In summary, induced defenses in wild radish plants are an example of adaptive plasticity because of the phenotypic benefits of induction in the presence of herbivores, and the phenotypic costs of induction in the absence of herbivores. Costs may be manifested in ways other than reduced seed production, including costs in terms of ecological interactions (reviewed by Agrawal and Karban 1998). The key result, however, is that the benefits far outweighed the costs under field conditions.

Predictability: a Requirement for Adaptive Plasticity

In order for a plant parasite to induce an adaptive defense in the plant, the initial attack must provide reliable information, and predict future attack (Karban and Adler 1996, Karban et al. 1999). For example, if plants infected with a pathogen induce production of phytochemicals that protect the plants against subsequent attack, but there is no subsequent attack, then investment in the induced products is presumably wasted (and not adaptive). In many systems, heavy parasite loads early in the season may predict heavy parasite loads later in the season, thereby making induced resistance a good strategy. Likewise, many plant parasites predict future attack by themselves. For example, pathogens that can systemically infect plants may be controlled by induction at the time of initial infection. Herbivores that are relatively immobile, such as leaf miners, small caterpillars, and spider mites all may provide reliable information that predict their own continued presence at the initial time of attack. This predictability is an essential component of most models of induced plant defense (See Table 1 in Karban and Adler 1996). Given that many plant parasites are highly mobile, or seemingly have unpredictable infection dynamics, an important and missing component of evolutionary studies of plasticity in defense is that the parasites are predictable.

Costs of Plasticity Verus Costs of Induced Defenses

For organisms which exhibit phenotypic plasticity, there may be several types of evolutionary constraints that limit expression of maximally beneficial plasticity. Above, phenotypic costs associated with producing defenses were considered. In addition, organisms may be constrained by costs associated with the ability to express alternate phenotypes (plasticity genes, sensory and regulatory mechanisms, etc). Costs of plasticity and their distinction from phenotypic costs of producing an inducible response have been recently reviewed by DeWitt et al. (1998). If quantitative genetic variation in inducibility exists (Table 1), these may be ideal systems to test for costs of plasticity by assessing the relationship between inducibility *per se* and fitness in various environments (DeWitt 1998, Scheiner and Berrigan 1998).

Macroevolution of Induced Plant Defenses

Little is currently known about the macroevolution of induced plant defenses. A phylogenetic perspective could address whether induced defenses are ancestral or derived. Most theoretical models predict that induced defenses are evolutionarily derived from constitutive strategies as a cost saving mechanism (Karban and Adler 1996, Karban and Baldwin 1997, Thaler and Karban 1997). The scenario that has been envisioned is that constitutive defenses were ancestral and that high costs of constitutive defenses were saved as plasticity in defense was favored (Karban and Baldwin 1997). Here costs of plasticity are assumed to be low because plasticity was presumably favored over a fixed defense (i.e., costs of plasticity were smaller than costs of constitutive defense) (Cipollini 1998). An alternative scenario proposed by Karban and Baldwin (1997) and briefly discussed by Cipollini (1998) is that the ancestral state was undefended, not constitutively defended. Here, costs of plasticity are predicted to be higher, and perhaps more easily detectable, because plastic defense was favored as an initial defensive strategy, not as a cheaper alternative to constitutive defense.

Only one study to date has attempted to examine the phylogenetics of induced plant defenses (Thaler and Karban 1997). In this study, the authors examined constitutive and induced resistance of 21 *Gossypium* species to herbivory by spider mites. *Gossypium* spp. are distributed worldwide, and are of economic importance as several species are cultivated for the cotton lint. Constitutive and induced resistance were found to be derived traits in *Gossypium*, suggesting that the ancestral state may have been undefended. Further studies of the macroevolution of induced defenses will be important in understanding large scale patterns in plant defense and constraints on their evolution.

Synthesis

Induced responses against herbivores and pathogens can provide a fitness benefit to plants. Phenotypic benefits and costs of induced resistance have been demonstrated in a few studies; more are needed however, especially those experiments employing a combination of approaches to phenotypic manipulations (i.e., using combinations of real plant parasites, avirulent parasites, elicitors, mutants, etc.). If induced defenses are an example of adaptive plasticity, the next goal should be to understand how they evolve. Although a handful of studies document quantitative genetic variation in inducibility, these studies were largely conducted for other purposes. The evolution of induced defenses and other phenotypically plastic traits is a young field of investigation, and will benefit from field experiments employing phenotypic and genetic manipulations.

Induced defenses add a novel level of complexity to the study of coevolution between plants and their parasites. Specificity in induced resistance (Stout and Bostock, this volume) may promote pairwise coevolution by allowing plants and their enemies to interact on a one-to-one level within the background of other interactions. Inducibilty per se may have been favored to maximize defenses against multiple enemies—allowing plants to fine-tune their defenses upon recognition of their attackers.

Future Directions

Understanding the evolution of induced defense will be advanced by applying our knowledge of quantitative genetic variation in induced resistance to fitness consequences for the plant. Such experiments should be conducted in environments with and without plant attackers. For plant-pathogen interactions, researchers will need to first quantify such genetic variation. It will be useful to impose artificial selection on constitutive and induced defenses, to ask if they can be uncoupled, and to ask how selection might act on them independently and in concert. It is still not clear what biotic and abiotic conditions favor induced plant defenses. Addressing the macroevolution of inducible and constitutive defenses may give us insight into how these traits evolve, and how costs of plasticity may constrain the evolution of plant defense.

Acknowledgements

My work on the evolutionary ecology of inducible plant defenses has been shaped by interactions with Rick Karban. Numerous other colleagues at U.C. Davis, especially committee members Jay Rosenheim and Mau Stanton, and folks in the plant-herbivore discussion group (Strauss and Karban labs) helped me develop the ideas presented in this chapter. I thank Rick Karban, Jennifer Thaler, and Helen Rodd for comments on the chapter. I have been financially supported by the Center for Population Biology at U.C. Davis and also by the National Science Foundation (DEB-9701109).

Literature Cited

Abrahamson, W. G., and A. E. Weis. 1997. Evolutionary ecology across three trophic levels. Princeton University Press, Princeton.

Agrawal, A. A. 1998. Induced responses to herbivory and increased plant performance. Science 279:1201-1202.

Agrawal, A. A. 1999a. Induced responses to herbivory in wild radish: Effects on several herbivores and plant fitness. Ecology (in press).

Agrawal, A. A. 1999b. Induced plant defense as adaptive plasticity: Benefits and costs of induction for *Lepidium virginicum* (Brassicaceae). Submitted to Ecology.

Agrawal, A. A. 1999c. Specificity of induced resistance to herbivores in wild radish: causes and consequences for two specialist and two generalist caterpillars. Manuscript.

Agrawal, A. A., and R. Karban. 1998. Why induced defenses may be favored over constitutive strategies in plants. Pages 45-61 *in* R. Tollrian and C. D. Harvell, editors. The Ecology and Evolution of Inducible Defenses. Princeton University Press, Princeton.

Agrawal, A. A., S. Y. Strauss, and M. J. Stout. 1999. Costs of induced responses and tolerance to herbivory in male and female fitness components of wild radish. Evolution (in press).

Baldwin, I. T. 1998. Jasmonate-induced responses are costly but benefit plants under attack in native populations. Proceedings of the National Academy of Sciences of the United States of America 95:8113-8118.
Bonnet, P., E. Bourdon, M. Ponchet, J. P. Blein, and P. Ricci. 1996. Acquired resistance triggered by elicitins in tobacco and other plants. European Journal of Plant Pathology 102:181-192.
Bowers, M. D., and N. E. Stamp. 1993. Effects of plant age, genotype, and herbivory on *Plantago* performance and chemistry. Ecology 74:1778-1791.
Bowling, S. A., A. Guo, H. Cao, A. S. Gordon, D. F. Klessig, and X. Dong. 1994. A mutation in *Arabidopsis* that leads to constitutive expression of systemic acquired resistance. Plant Cell 6:1845-1857.
Brody, A. K., and R. Karban. 1992. Lack of a tradeoff between constitutive and induced defenses among varieties of cotton. Oikos 65:301-306.
Brown, D. G. 1988. The cost of plant defense: an experimental analysis with inducible proteinase inhibitors in tomato. Oecologia 76:467-470.
Cao, H., S. A. Bowling, A. S. Gordon, and X. Dong. 1994. Characterization of an *Arabidopsis* mutant that is non-responsive to inducers of systemic acquired resistance. Plant Cell 6:1583-1592.
Caruso, F. L., and J. Kuć. 1977. Field protection of cucumber, watermelon, and muskmelon against *Colletotrichum lagenarium* by *Colletotrichum lagenarium*. Phytopathology 67:1290-1292.
Cerkauskas, R. F. and Chiba, M. 1990. Association of phoma canker with photocarcinogenic furocoumarins in parsnip cultivars. Canadian Journal of Plant Pathology 12:349-357.
Cheverud, J. M. 1988. A comparison of genetic and phenotypic correlations. Evolution 42:958-968.
Cipollini, D. 1998. Induced defenses and phenotypic plasticity. Trends in Ecology & Evolution 13:200-200.
DeWitt, T. J. 1998. Costs and limits of phenotypic plasticity: Tests with predator-induced morphology and life history in a freshwater snail. Journal of Evolutionary Biology 11:465-480.
DeWitt, T. J., A. Sih, and D. S. Wilson. 1998. Costs and limits of phenotypic plasticity. Trends in Ecology & Evolution 13:77-81.
Dirzo, R., and J. L. Harper. 1982. Experimental studies on slug-plant interactions IV. The performance of cyanogenic and acyanogenic morphs of *Trifolium repens* in the field. Journal of Ecology 70:119-138.
Dudley, S. A., and J. Schmitt. 1995. Genetic differentiation in morphological responses to simulated foliage shade between populations of *Impatiens capensis* from open and woodland sites. Functional Ecology 9:655-666.
Dudley, S. A., and J. Schmitt. 1996. Testing the adaptive plasticity hypothesis: Density-dependent selection on manipulated stem length in *Impatiens capensis*. American Naturalist 147:445-465.
Eichenseer, H., J. L. Bi, and G. W. Felton. 1998. Indiscrimination of *Manduca sexta* larvae to overexpressed and underexpressed levels of phenylalanine ammonia-lyase in tobacco leaves. Entomologia Experimentalis et Applicata 87:73-78.
English-Loeb, G., R. Karban, and M. A. Walker. 1998. Genotypic variation in constitutive and induced resistance in grapes against spider mite (Acari: Teranychidae) herbivores. Environmental Entomology 27:297-304.
Geetha, S., S. A. Shetty, H. S. Shetty, and H. S. Prakash. 1996. Arachidonic acid-induced hypersensitive cell death as an assay of downy mildew resistance in pearl millet. Annals of Applied Biology 129:91-96.
Gianoli, E., C. M. Caillaud, B. Chaubet, J. P. Di Pietro, and H. M. Niemeyer. 1997. Variability in grain aphid (Homoptera: Aphididae) performance and aphid-induced phytochemical responses in wheat. Environmental Entomology 26:638-641.
Gianoli, E., and H. M. Niemeyer. 1997. Lack of costs of herbivory-induced defenses in a wild wheat: Integration of physiological and ecological approaches. Oikos 80:269-275.
Gillespie, J. H., and M. Turelli. 1989. Genotype-environment interactions and the maintenance of polygenic variation. Genetics 121:129-138.

Haukioja, E., and S. Hanhimaki. 1985. Rapid wound-induced resistance in white birch (*Betula pubescens*) foliage to the geometrid *Epirrita autumnata*: A comparison of trees and moths within and outside the outbreak range of the moth. Oecologia 65:223-228.
Herms, D. A., and W. J. Mattson. 1992. The dilemma of plants: to grow or defend. Quarterly Review of Biology 67:283-335.
Hoffland, E., M. J. Jeger, and M. L. van Beusichem. 1998. Is plant growth rate related to disease resistance? Pages 409-427 *in* H. Lambers, H. Poorter and M. M. I. van Vuuren, editors. Inherent Variation in Plant Growth: Physiological Mechanisms and Ecological Consequences. Backhuys, Leiden, The Netherlands.
Julkunen-Tiitto, R., J. P. Bryant, P. Kuropat, and H. Roininen. 1995. Slight tissue wounding fails to induce consistent chemical defense in three willow (*Salix* spp.) clones. Oecologia 101:467-471.
Kamoun, S., M. Young, C. B. Glascock, and B. M. Tyler. 1993. Extracellular protein elicitors from *Phytophthora* host-specificity and induction of resistance to bacterial and fungal phytopathogens. Molecular Plant-Microbe Interactions 6:15-25.
Karban, R. 1993. Costs and benefits of induced resistance and plant density for a native shrub, *Gossypium thurberi*. Ecology 74:9-19.
Karban, R., and F. R. Adler. 1996. Induced resistance to herbivores and the information content of early season attack. Oecologia 107:379-385.
Karban, R., A. A. Agrawal, and M. Mangel. 1997. The benefits of induced defenses against herbivores. Ecology 78:1351-1355.
Karban, R., A. A. Agrawal, J. S. Thaler, and L. S. Adler. 1999. Induced plant responses and information content about risk of herbivory. Trends in Ecology and Evolution (in press).
Karban, R., and I. T. Baldwin. 1997. Induced Responses to Herbivory. University of Chicago Press, Chicago.
Karban, R., and J. H. Myers. 1989. Induced plant responses to herbivory. Annual Review of Ecology and Systematics 20:331-348.
Kingsolver, J. G., and R. B. Huey. 1998. Evolutionary analyses of morphological and physiological plasticity in thermally variable environments. American Zoologist 38:545-560.
Kuć, J. 1982. Induced immunity to plant disease. BioScience 32:854-860.
Kuć, J. 1987. Plant immunization and its applicability for disease control. Pages 255-274 *in* I. Chet, editor. Innovative Approaches to Plant Disease Control. Wiley, New York.
Lagrimini, L. M., V. Gingas, F. Finger, S. Rothstein, and T. T. Y. Liu. 1997. Characterization of antisense transformed plants deficient in the tobacco anionic peroxidase. Plant Physiology 114:1187-1196.
Lawrence, C. B., M. H. A. J. Joosten, and S. Tuzun. 1996. Differential induction of pathogenesis-related proteins in tomato by *Alternaria solani* and the association of a basic chitinase isozyme with resistance. Physiological and Molecular Plant Pathology 48:361-377.
Ludwig-Mueller, J., B. Schubert, K. Pieper, S. Ihmig, and W. Hilgenberg. 1997. Glucosinolate content in susceptible and resistant Chinese cabbage varieties during development of clubroot disease. Phytochemistry 44:407-414.
McConn, M., R. A. Creelman, E. Bell, J. E. Mullet, and J. Browse. 1997. Jasmonate is essential for insect defense in *Arabidopsis*. Proceedings of the National Academy of Sciences of the United States of America 94:5473-5477.
Mole, S. 1994. Trade-offs and constraints in plant-herbivore defense theory: a life-history perspective. Oikos 71:3-12.
Newton, A. C., and E. P. Dashwood. 1998. The interaction of humidity and resistance elicitors on expression of polygenic resistance of barley to mildew. Journal of Phytopathology 146:123-130.
Ogallo, J. L., and M. A. McClure. 1995. Induced resistance to *Meloidogyne hapla* by other *Meloidogyne* species on tomato and pyrethrum plants. Journal of Nematology 27:441-447.
Parker, M. A. 1992. Constraints on the evolution of resistance to pests and pathogens. Pages 181-197 *in* P.G. Ayres, Editor. Pests and Pathogens: Plant Responses to Foliar Attack. Bios Scientific Publishers Ltd., Oxford.
Pires, J. L., E. D. M. N. Luz, and U. V. Lopes. 1997. Field resistance of cacao clones to black pod disease caused by *Phytophthora* spp. in Bahia, Brazil. Fitopatologia Brasileira 22:375-380.

Raffa, K. F. 1991. Induced defensive reactions in conifer-bark beetle systems. Pages 245-276 *in* D. W. Tallamy and M. J. Raupp, editors. Phytochemical Induction by Herbivores. Wiley, New York.
Raffa, K. F., and A. A. Berryman. 1982. Accumulation of monoterpenes and associated volatiles following inoculation of grand fir with a fungus transmitted by the fir engraver, *Scolytus ventralis* (coleoptera: Scolytidae). Canadian Entomologist 114:797-810.
Rasmussen, J. B., R. Hammerschmidt, and M. N. Zook. 1991. Systemic induction of salicylic acid accumulation in cucumber after inoculation with *Pseudomonas syringae* pv. *syringae*. Plant Physiology 97:1342-1347.
Rausher, M. D., and E. L. Simms. 1989. The evolution of resistance to herbivory in *Ipomoea purpurea*. 1. Attempts to detect selection. Evolution 43:561-572.
Rhoades, D. F. 1979. Evolution of plant chemical defense against herbivores. Pages 4-55 *in* G. A. Rosenthal and D. H. Janzen (Eds.) Herbivores: Their Interaction with Secondary Plant Metabolites. Academic Press, New York.
Roseland, C. R., and T. J. Grosz. 1997. Induced responses of common annual sunflower *Helianthus annuus* L. from geographically diverse populations and deterrence to feeding by sunflower beetle. Journal of Chemical Ecology 23:517-542.
Rybka, K., E. Arseniuk, J. Wisniewska, and K. Raczynska-Bojanowska. 1998. Comparative studies on the activities of chitinase, beta-1,3-glucanase, peroxidase and phenylalanine ammonia lyase in the leaves of triticale and wheat infected with *Stagonospora nodorum*. Acta Physiologiae Plantarum 20:59-66.
Schappert, P. J., and J. S. Shore. 1995. Cyanogenesis in *Turnera ulmifolia* L. (Turneraceae): I. Phenotypic distribution and genetic variation for cyanogenesis on Jamaica. Heredity 74:392-404.
Scheiner, S. M., and D. Berrigan. 1998. The genetics of phenotypic plasticity. VIII. The cost of plasticity in *Daphnia pulex*. Evolution 52:368-378.
Schmitt, J., A. C. McCormac, and H. Smith. 1995. A test of the adaptive plasticity hypothesis using transgenic and mutant plants disabled in phytochrome-mediated elongation responses to neighbors. American Naturalist 146:937-953.
Shapiro, A. M., and J. E. DeVay. 1987. Hypersensitivity reaction of *Brassica nigra* L. (Cruciferae) kills eggs of *Pieris* butterflies (Lepidoptera: Pieridae). Oecologia 71:631-632.
Siemens, D. H., and T. Mitchell-Olds. 1998. Evolution of pest-induced defenses in *Brassica* plants: Tests of theory. Ecology 79:632-646.
Simms, E. L. 1992a. Costs of plant resistance to herbivores. Pages 392-425 *in* R. S. Fritz and E. L. Simms, editors. Plant Resistance to Herbivores and Pathogens. Ecology, Evolution, and Genetics. University of Chicago Press, Chicago.
Simms, E. L. 1992b. The evolution of plant resistance and correlated characters *in* S. B. J. Menken, J. H. Visser and P. Harrewijn, editors. Proceedings of the 8th International Symposium on Insect-Plant Relationships. Kluwer Academic Publishers, Dordrecht.
Simms, E. L., and M. D. Rausher. 1987. Costs and benefits of plant resistance to herbivory. American Naturalist 130:570-581.
Simms, E. L., and M. D. Rausher. 1989. The evolution of resistance to herbivory in *Ipomoea purpurea*. II. Natural selection by insects and costs of resistance. Evolution 43:573-585.
Smedegaard-Petersen, V., and O. Stølen. 1981. Effect of energy-requiring defense on yield and grain quality in a powdery mildew-resistant barley cultivar. Phytopathology 71:396-399.
Stanton, M., and H. J. Young. 1994. Selecting for floral character associations in wild radish, *Raphanus sativus* L. Journal of Evolutionary Biology 7:271-285.
Thaler, J. S., and R. Karban. 1997. A phylogenetic reconstruction of constitutive and induced resistance in *Gossypium*. American Naturalist 149:1139-1146.
Tuzun, S., J. Juarez, W. C. Nesmith, and J. Kuć. 1992. Induction of systemic resistance in tobacco against metalaxyl-tolerant strain of *Peronospora tabacina* and the natural occurrence of the phenomenon in Mexico. Phytopathology 82:425-429.
Thompson, J. D. 1991. Phenotypic plasticity as a component of evolutionary change. Trends in Ecology & Evolution 6:246-249.
Underwood, N. C. 1998. The timing of induced resistance and induced susceptibility in the soybean-Mexican bean beetle system. Oecologia 114:376-381.

van Dam, N. M., and K. Vrieling. 1994. Genetic variation in constitutive and inducible pyrrolizidine alkaloid levels in *Cynoglossum officinale* l. Oecologia 99:374-378.

Via, S., and R. Lande. 1985. Genotype-environment interaction and the evolution of phenotypic plasticity. Evolution 39:505-522.

Zangerl, A. R., A. M. Arntz, and M. R. Berenbaum. 1997. Physiological price of an induced chemical defense: Photosynthesis, respiration, biosynthesis, and growth. Oecologia 109:433-441.

Zangerl, A. R., and M. R. Berenbaum. 1990. Furanocoumarin induction in wild parsnip: Genetics and populational variation. Ecology 71:1933-1940.

Zangerl, A. R., and C. E. Rutledge. 1996. The probability of attack and patterns of constitutive and induced defense: a test of optimal defense theory. American Naturalist 147:599-608.

Behavioral Responses of Predatory and Herbivorous Arthropods to Induced Plant Volatiles: From Evolutionary Ecology to Agricultural Applications

Maurice Sabelis, Arne Janssen, Angelo Pallini, Madelaine Venzon, Jan Bruin, Bas Drukker and Petru Scutareanu

Abstract

Herbivory is known to induce the production of volatiles in plants. These signals are thought to betray herbivores to their predators, which are then attracted or arrested near the plant under attack. Evidence for involvement of herbivore-induced plant volatiles in predator recruitment is largely based on experiments with olfactometers designed to demonstrate a response to odors, not to elucidate the behavioral mechanisms used to locate the source. Since the mechanisms underlying orientation may well operate at a spatial scale beyond that considered in the lab, experiments are required to unravel the tactic and kinetic responses in carefully designed laboratory experiments at a larger scale and to assess the responses under more realistic (greenhouse, field) conditions. We discuss experiments showing the role of odor-conditioned anemotaxis, tactic/kinetic responses to odor gradients, odor-conditioned landing and take-off responses, as well as the role of hunger, associative learning and innate responses in predatory arthropods.

By producing volatiles, plants also betray their presence to arthropods that do not confer a direct benefit to the plant, for instance other herbivores, hyperpredators and omnivores. We review recent results on responses of such non-beneficial arthropods to odors from plants under herbivore attack. Finally, we discuss conditions under which, despite these negative side effects, herbivore-induced synomones have an overall positive effect on plant fitness and we point out possibilities for the practical application of herbivore-induced synomones in plant protection.

Introduction

Plants defend themselves against herbivorous arthropods, not only directly (via toxins, digestion inhibitors, glandular hairs, tough cuticle, etc.), but also indirectly by promoting the effectiveness of the herbivores' predators (Price et al. 1980, Fig. 1a). Indirect plant defense may be achieved by creating protective structures, providing food (pollen, nectar) and by releasing chemical cues which betray the presence of herbivores (Dicke and Sabelis 1988a, Sabelis et al. 1999a, b). These plant-provided facilities can be utilized by predatory arthropods that help the plant to get rid of herbivorous arthropods, but they are also open to other organisms that are not beneficial to the plant. Indeed, mutualistic interactions are never foolproof (Bronstein 1994a) and plant-predator mutualisms are no exception. In this chapter we focus on herbivore-induced plant volatiles (HIPV).

There are two problems to be addressed before we can conclude that attracting predators is indeed one of the functions of HIPV. Firstly, most studies showing attraction of predators by these volatiles are conducted under idealised laboratory conditions, and it remains to be shown that attraction also occurs under natural conditions. Secondly, if plants do attract predators by producing HIPV, it is inevitable that other organisms that are not beneficial to the plant also use these volatiles to their own benefit. Here, we review advances in knowledge on how these volatiles influence the behavior of (1) predatory arthropods and (2) other (possibly non-beneficial) arthropods. By considering the responses to HIPV in the food web of plant-inhabiting arthropods we hope to obtain a more realistic view on the overall effect of producing "alarm calls" on plant fitness.

A food-web perspective on the effects of HIPV is crucial to understand the evolution of plant-predator mutualisms. Hairston et al. (1960) hypothesized that predators control herbivore densities, resulting in the plants being releaved from herbivore damage. This implicitly assumes that direct plant defenses are ineffective. Although these defenses are successful against many herbivores, they are not effective against all (Strong et al. 1984), and this leaves room for the evolution of indirect defenses (plant-predator mutualisms). While the "World is Green" hypothesis (Hairston et al. 1960) may hold for linear tritrophic food chains, one may wonder what happens in more complex food webs (Polis and Strong 1996). When hyperpredation and intraguild predation are prevalent, plant alarm calls may not only benefit the herbivores' predators, but also the predators' predators and competitors. These effects on the fourth trophic level will cascade down the food chain (Fig. 1b): the herbivores' enemies become less abundant, the herbivores increase in number and the plant will suffer more from herbivory. This cascading effect may reduce the plant's benefit in sending

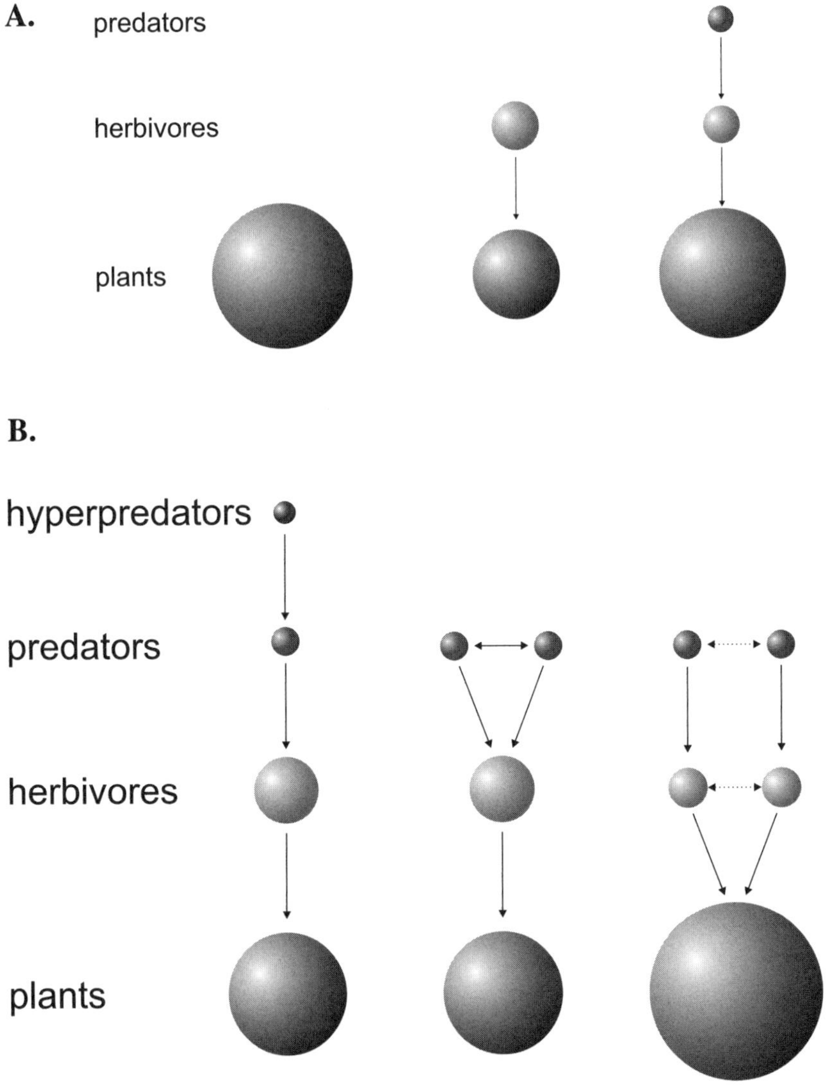

Figure 1: Diagrams of systems with one, two and three trophic levels (1A) or four trophic level and variants thereof (1B). The spheres represent the biomass at each trophic level, assuming 10-20% conversion efficiency for the herbivores and 70-80% for the (hyper) predators. Drawn arrows stand for the trophic interactions, whereas dashed arrows stand for avoidance of (herbivore-herbivore, herbivore-predator, predator-predator) interactions. Defensive interactions are left out to keep the diagram simple.

out alarm calls to the point that it does not outweigh production costs and risks of alerting other herbivores. Thus, it is crucial to determine whether communities of arthropods on plants *behave* like linear tritrophic food chains or not (such as four trophic levels or web of interactions between linear food chains). In the first case, plant investment in indirect defenses is more likely to be favored by selection, since predators will suppress herbivore populations to a greater extent and plants will suffer less from herbivory (Hairston et al. 1960, Strong et al.1984, Sabelis et al. 1999a, b). Consequently, to explain why plants usually retain a green appearance under natural conditions, we should identify the mechanisms that make complex food webs *behave* much like linear tritrophic food chains (Fig. 1b, Hairston et al. 1960). The key mechanisms probably lie in the adaptive behavior of members of the food web. Odor-mediated attraction and avoidance may lead to niche partitioning, thereby decreasing the strength of food web interactions. This is why we focus here on behavioral responses to HIPV and, where possible, also consider responses to other volatiles conveying information on niche quality (Janssen et al. 1998).

Food webs of plant-inhabiting arthropods vary in space and time with respect to their structure and the abundance of their members. Herbivores and their natural enemies make foraging decisions which not only depend on the presence and quality of food but also on the risk of being eaten and the intensity of competition. Thus, these decisions depend on the current and local state of the food web. Although the ultimate evaluation of effects of HIPV on plant fitness should be conducted under natural conditions, there is a danger in evaluating plant fitness effects exclusively in the field, because the temporal and spatial scale of field experiments is bound to be such that only a subset of food web states plays a role. Lab experiments allow a free choice of the composition of the food web, whereas in the field, species composition may vary in space and time. Hence, it seems more wise to develop insight by considering a chain of experiments, starting from controlled conditions in the laboratory, to real, complex conditions in the field.

There is a rich literature on indirect plant defense against herbivores via the attraction of natural enemies by HIPV (Dicke and Sabelis 1988b, 1989, 1992, Dicke et al. 1990b, Dicke 1994, Turlings et al. 1995, Takabayashi and Dicke 1996, Dicke and Vet 1999). In this chapter, we first review literature on the role of HIPV in attracting predatory arthropods. We focus on true predators because they can have a direct impact on plant fitness (e.g., Agrawal and Karban 1997). Predators, unlike many parasitoids, kill the herbivores immediately, thereby preventing further feeding damage to the plant. After reviewing the responses of predators to HIPV, we evaluate the net effect of HIPV production on plant fitness. We close by discussing the practical perspectives of HIPV for crop protection.

Behavioral Responses of Predatory Arthropods to HIPV

The first indication of olfactory responses of arthropods to odors emanating from herbivore-infested plants came from Y-tube olfactometer tests with predatory mites and spider mites on detached bean and apple leaves (Sabelis and Van de Baan 1983). Subsequent experiments showed that the odors do not emanate from the spider mites themselves nor from their silk secretions, but that the odors come, principally from the mite-infested leaves (i.e., after removal of spider mites and their products, Sabelis et al. 1984a) and, to some degree, from mite faeces. Since leaves did not elicit a response until being fed on by spider mites, and infested leaves gradually lost their attractiveness after the removal of spider mites, the hypothesis was formulated that the odors are of plant origin and induced by spider mite feeding (Sabelis and Dicke 1985). Two lines of evidence for the active involvement of plants have been presented. First, headspace analysis revealed that the most abundant chemicals in the blend released upon feeding by two-spotted spider mites, *Tetranychus urticae*, on Lima bean (*e.g.*, methyl salicylate, linalool, (*E*)-β-ocimene, 4,8-dimethyl-1,3(*E*),7-nonatriene, 4,8,12-trimethyl-1,3(*E*),7(*E*),11-tridecatetraene) are known as plant compounds (Dicke 1988, Dicke and Sabelis 1988a, b, 1989, 1992, Dicke et al. 1990a,b). Second, not only the infested leaves release the volatiles, but also uninfested leaves of the same plant (Turlings and Tumlinson 1992, Dicke et al. 1990b, 1993, see also Paré et al., this volume). Thus, there is a local, as well as systemic induction of volatiles upon spider-mite attack. The ability to produce herbivore-induced alarm signals is probably widespread in the plant kingdom, as there is supporting evidence that herbivores elicit such responses across at least 23 plant species in 13 plant families (Dicke 1994, Dicke et al. 1998). This ability is not limited to the interaction of plants with herbivorous mites. Apart from extensive work on the response of parasitoids to plant volatiles induced by caterpillars (Turlings et al. 1995) and aphids (Guerrieri et al. 1997, Du et al. 1998), there is also evidence for responses of anthocorid bugs to odors from bean plants infested by spider mites (Dwumfour 1992), to pear leaves infested by psyllids (Drukker et al. 1995) and to components identified in the odor blends of *Psylla*-infested pear leaves, such as (*E,E*)-α-farnesene and methyl salicylate (Scutareanu et al. 1996, 1997).

There is evidence that the behavioral responses of predatory arthropods to HIPV are based on olfaction. For example, when released in realistic concentrations, single components of the herbivore-induced odor blend (methyl salicylate, linalool, (*E*)-β-ocimene) elicit a positive response of predatory mites in a Y-tube olfactometer (Dicke et al. 1990a). Furthermore, the behavior of chemoreceptor-bearing extremities in response to HIPV has been described for predatory mites (Dicke et al. 1991), chemosensors on tarsi of first legs have been described morphologically (Jagers Op Akkerhuis et al. 1985), and electrophysiological recordings have been obtained from the neurones in these sensors when exposed to one of the blend components (linalool) (De Bruyne et al. 1991). We know little, however, of the role of blend composition, even if it concerns only the major blend components. Can mixes of the major components alone explain all behavioral responses? Or are minor blend components or hitherto undetected volatiles essential? Campbell et al. (1993) compared single-cell recordings of odor-receptor cells with gas chromatograms from odors from hop leaves and found that some spikes in the action potential of the receptor cells did not coincide with detected volatiles, suggesting that other volatiles may also play a role. The only test of the response of a parasitoid to a synthetic mixture of the (eleven) major volatiles from corn fed upon by *Spodoptera exigua* caterpillars failed to reproduce the response to a natural blend of HIPV (Turlings et al. 1991). Moreover, it is well documented that herbivorous mites of different species on the same species of host plant induce the production of blends that differ only in relative amounts of the (major) components. Yet, predatory mites can readily discriminate between leaves infested by either of the spider mite species (Sabelis and Van de Baan 1983, Sabelis and Dicke 1985, Dicke and Groeneveld 1986, Takabayashi et al. 1991). Apparently predatory mites can discriminate between quantitative differences in the composition of major blend constituents, a phenomenon recently also reported for parasitoids (De Moraes et al. 1998). Alternatively, the components that enable the mites to discriminate are present in minor quantities and/or are still to be identified. These volatiles are not necessarily of plant origin as they may well be produced by the herbivores themselves, (pheromones that are only produced in minute quantities) or by associated microorganisms.

Coping With Variability in HIPV

Another salient point arises from the observation that the composition of induced volatiles varies with the species and cultivar of the host plant, even when attacked by the same species of spider mite (Takabayashi et al. 1994). Also, age, tissue, and condition of a host plant influence the composition of plant odor blends (Takabayashi et al. 1991, 1994, Scutareanu et al. 1997). Plants even show diurnal cycles in the production of some of the induced volatiles (Loughrin et al. 1994). One may wonder how predators cope with this bewildering variety of signals.

Margolies et al. (1997) showed that it is possible to select for strength of the response to one of the spider-mite-induced volatiles (linalool) in a culture of the predatory mite, *Phytoseiulus persimilis*. A genetic basis may be expected when there is a cost to flexibility (e.g., the capacity of the sensory and central nervous system to perceive and integrate signals and alter behavioral response) and a cost to dietary change (e.g., due to differences in nutritional quality and secondary plant compounds). Given a sufficiently constant prey availability and signal environment, there will be selection to avoid paying the cost of flexibility, which will then pave the way for genetically fixed preferences. In fact, genetic polymorphisms for prey preference have been found in a predatory mite species associated with roots and bulbs of lilies (Lesna and Sabelis 1998). Thus, there may well be a (partially) genetic determinant of the olfactory response, but this does not imply a rigid prey/plant preference, because the genetic influence may be overruled by plastic responses to environmental stimuli. Indeed, there are indications of effects of experience due to rearing on different host plants attacked by the same herbivore (Dicke et al. 1990c, Takabayashi et al. 1994, Krips et al. 1999). Selection, however, cannot be completely ruled out as a factor modifying the response, because (1) mortality during the rearing period (1 week) may alter the genetic composition and (2) a few hours of rearing on cucumber was not sufficient to alter preference. Recent experiments (Drukker, Jacobs, Bruin, Sabelis, unpublished data), however, have shown that one-day starvation in the presence of one of the HIPV-components (methyl salicylate or linalool) caused a reversal from attraction to avoidance of these odor components. These studies indicate that predators can learn the association between these odors and their state of starvation. Much the same results were obtained with *Anthocoris* females starved in the presence of single HIPV components from pear leaves (methyl salicylate, (*E,E*)-α-farnesene) (Drukker et al.1999). Because there was no mortality in these experiments this is strong evidence that associative learning is implicated in strategies of predatory arthropods to locate herbivorous arthropods. Positive associations may be reversed when volatiles are paired with a hunger stimulus, which leads to dispersal away from the original site. We hypothesize that avoidance responses wane when pairing of the odors and the hunger stimulus is discontinued, which in turn may reinstall the innate responses. When innate responses or coincidental encounters happen to bring the predator in contact with a potential prey on a new host plant, then positive associations with the HIPV in that setting may result (Fig. 2). Learning of positive associations between food and odors has been amply demonstrated in studies on parasitoids of herbivorous insects (Turlings et al. 1993a).

HIPV-Elicited Mechanisms to Locate the Odor Source: Relevance of Spatial Scale

Y-tube olfactometers are suitable to assess whether HIPV elicit a response of the predatory arthropod, but not to analyse the orientation mechanisms. The latter requires a more analytical approach, where each mechanism is tested in a specially designed experimental set-up. Such an approach has so far only been followed in studies on the predatory mite *P. persimilis* (Sabelis and Dicke 1985). This predator showed an ability to orient to odor gradients in an olfactometer with a vertical air flow passing a horizontal fine-mesh gauze which served as a substrate for the predator (Sabelis et al.

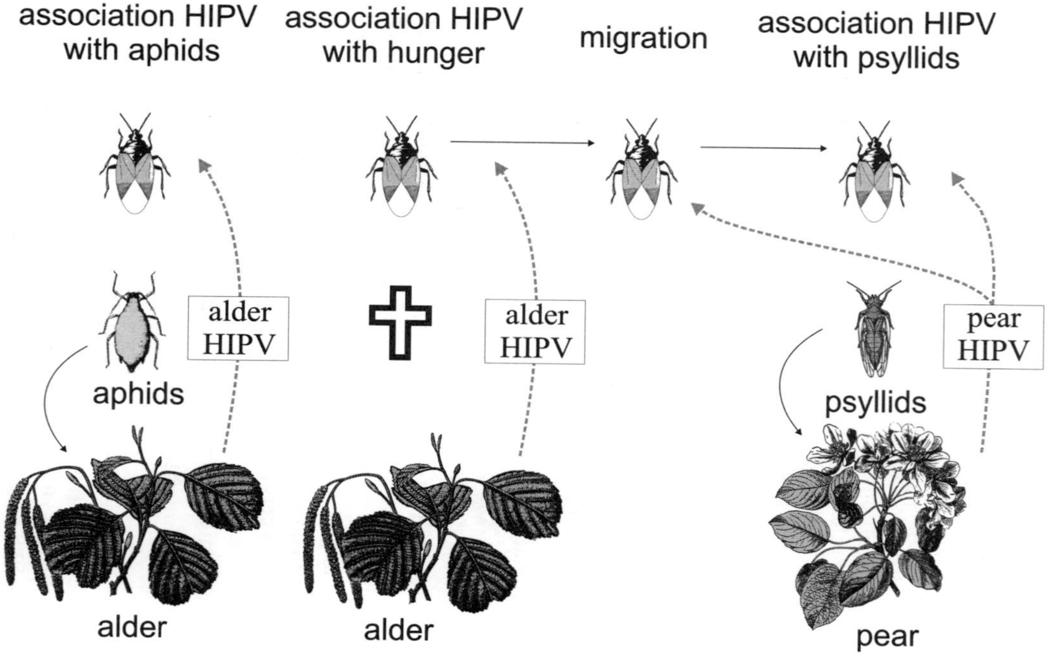

Figure 2: Diagrammatic representation of alternating prey and HIPV preferences by associative learning in anthocorid predators. In May-June the anthocorids feed on aphids where they learn to associate alder HIPV with the presence of aphids as prey. In July aphids migrate or go extinct due to predation. Anthocorids then learn to associate hunger and alder HIPV, which triggers avoidance and emigration. During migration the anthocorid predators search either randomly or respond to an array of odors (other than alder HIPV) according to innate response patterns. Once they land on pear trees and discover psyllid prey, they learn to associate pear HIPV and psyllids as prey, which will elicit arrestment on this new host plant with new prey.

1984b). When the odor source was positioned underneath the center of the gauze, the predatory mites responded to passing the sharp HIPV gradients by performing right-about turns. These turns show that, after passing the gradient, the predator perceived the absence of HIPV and increased the chance of moving back into the area with HIPV. When the odor source was moved in the same direction as the predator's direction of movement, but faster, the predators also responded to the passing gradient by making about turns. Clearly, this behavior would not help them to keep track of a moving odor plume. Hence, orientation to odor gradients is probably only relevant where odor gradients are sufficiently static at the time scale of mite movement. This is probably the case very close to the odor source, i.e., near a spider-mite colony on a leaf, where wind has relatively little impact on the position of the gradient.

The presence of HIPV also suppresses the tendency to disperse aerially in response to herbivore-induced plant odors, as shown by Sabelis and Afman (1994) using a wind tube designed to observe take-off behavior of a single predatory mite on a leaf. Predatory mites are thought to disperse passively in air currents (Sabelis and Dicke 1985), but before becoming airborne, they appear to make decisions on whether to take off or not. These decisions are influenced by their state of food deprivation and dehydration, but also by the presence of HIPV. Even when the predatory mites were severely starved and dehydrated (in the absence of HIPV), the presence of HIPV stimulated them to stay. This response leads to arrestment in the area where HIPV signals the presence of prey to the predatory mites. However, the same response is also relevant to attraction towards the odor source. To understand this, one should realize that airborne predatory mites cannot decide when and where to land, but after landing they would gain by finding clusters of spider mite colonies located not too far from the site of landing. This is done by ambulatory search and the perception of HIPV probably stimulates the predator to continue searching until the source is found.

The ambulatory search to find clusters of spider-mite colonies on plants would be much more effective if HIPV elicits not only take-off suppression but also movement against the wind guided by orientation on the wind direction (positive anemotaxis). This was studied in a wind tunnel with a homogenous air flow and a uniform odor concentration (thus no odor gradients) (Sabelis and Van der Weel 1993). The results showed that starved predators walk upwind, but this positive anemotaxis is also manifested in the absence of HIPV, though in a somewhat less pronounced form. The most surprising result was that in the absence of HIPV, well fed predators move downwind, whereas in the presence of HIPV, this negative anemotaxis disappears or is reversed to a positive anemotaxis depending on the odor concentration. At first sight this result was puzzling: why should satiated predators walk with the wind (= downwind) in absence and against the wind (= upwind) in presence of HIPV? The solution is simple once one realises that predatory mites foraging in clusters of prey colonies are probably satiated and, while moving from one infested leaf to another, may move out of the plant area occupied by the cluster of colonies. Should the predators happen to move out at the upwind (= windward) side of the cluster, then this takes them into HIPV-free air and negative anemotaxis would bring them back into the area with the cluster of colonies. Should they move out at the downwind side of the cluster of colonies, then positive anemotaxis will increase the chance of re-entering the cluster area. The consequence of this behavior is anemotactic arrestment in the prey cluster area.

Whether the above mechanisms are sufficient to explain the orientation responses to HIPV in the field is an open question for future research. So far, effective location of spider mite-infested plants has been demonstrated in wind tunnels (Sabelis and Schippers, unpublished data), in the laboratory (Sabelis and Van der Weel 1993) and under greenhouse conditions (Janssen 1999). Moreover, there is indirect evidence that arrestment of predatory mites in clusters of spider mite-infested leaves plays a decisive role in the rate at which the local populations of spider mites are exploited and ultimately eliminated (Sabelis and Van der Meer 1986, Sabelis 1992, van Baalen and Sabelis, 1995, Pels and Sabelis, 1999). Simulation models of the interaction between *T. urticae* and *P. persimilis* only provided a good fit to data on local predator-prey dynamics when it was assumed that the predatory mites do not emigrate until after all prey are eliminated. Even small rates of predator emigration from the cluster led to drastic effects on the simulated local predator-prey dynamics (much higher prey population peaks, longer periods of predator-prey interaction, much higher overall predator yield). Since predatory mites have to move between spider mite-infested leaves, they have to pass leaf surface, stems and petioles that are devoid of prey. The cloud of HIPV surrounding the cluster of infested leaves may well be instrumental for the predatory mites to stay arrested within the cluster.

The Impact of HIPV in the Field

The first demonstration of the impact of HIPV in the field came from research on biocontrol of psyllids in pear orchards in The Netherlands (Drukker et al. 1995). Pear trees suffer from only a few plant pests, the most important of which are leaf suckers, *Psylla pyri* and *P. pyricola*. These psyllids are absent in spring (March-May), begin to show up in June and may become numerous from July to September. In spring and early summer the predators thought to be most effective against the psyllids, (i.e., heteropteran bugs such as *Anthocoris nemorum*, *A. nemoralis* (Fabricius) and various *Orius* spp.) feed on aphids on other trees (e.g., alder, hawthorn). By the time the psyllid populations start to increase in July, predators begin to immigrate, show up first as adults in June-July, then later in July-September in all developmental stages. The predators build up strong possitive correlations with the population sizes of psyllids

(Scutareanu et al. 1999). It was hypothesized that the immigration of adult predators was triggered by HIPV from *Psylla*-infested pear trees. This was assessed in the field by sampling trees next to cages with trees harboring *Psylla* populations of various sizes. Densities of predatory bugs increased with the density of psyllids in the cages. Since covering the cage with an infested pear tree by a plastic sheet led to a sudden drop in the density of predatory bugs and removal of the sheet was quickly followed by a build-up in predator numbers, there was support for HIPV-triggered immigration of predators (Drukker et al. 1995). This was further substantiated by identification of headspace volatiles from *Psylla*-infested leaves (methyl salicylate and (*E,E*)-α-farnesene) and by testing the behavioral response to *Psylla*-infested vs. clean leaves and to single HIPV-compounds in a Y-tube olfactometer (Scutareanu et al. 1997). Naive predatory bugs showed a weak innate response to HIPV from pear, but once they experienced the association between food and these odors they exhibited a strong positive response (Drukker et al. 1999). When HIPV was paired with a hunger stimulus, the predators exhibited a strong negative response. This ability to learn associations between odors and positive or negative stimuli is probably of great importance in the field, since the predators are faced with a wealth of volatile infochemicals emanating from various plants and plant-herbivore combinations (Fig. 2). We suspect that predatory bugs leave their original prey resources when hungry (a negative association with plant volatiles), subsequently rely on their innate responses until they contact suitable herbivorous prey (such as pear leaf suckers) and experience the associated HIPV (a positive association). These experienced predators then continue to respond positively to the HIPV and end up close to the cages harbouring large *Psylla* populations in the experimental orchard. This may explain the somewhat puzzling observation that predatory bugs aggregate around the cages without being able to contact the prey on the trees inside these cages.

Other examples of predator attraction to infested field-grown plants are still scarce. Shimoda et al. (1997) studied the response of a predatory thrips to odors produced by bean plants infested with spider mites. In the field, thrips were caught on sticky traps baited with infested bean plants but not on traps baited with clean bean plants. However, only a few predatory thrips were recaptured, and only after a long period, and it therefore remains to be seen whether predators arrived in time to benefit the plant.

Limits to Attractiveness: Effects of the Presence of Conspecific Predators

There may well be limitations to the number of predators that a plant can attract (Janssen et al. 1997). When predatory mites are offered a choice between odors from spider mite colonies with and without conspecific predators, they prefer odors from the latter. This probably means that predatory mites are initially attracted towards a plant infested by spider mites, but once the infested

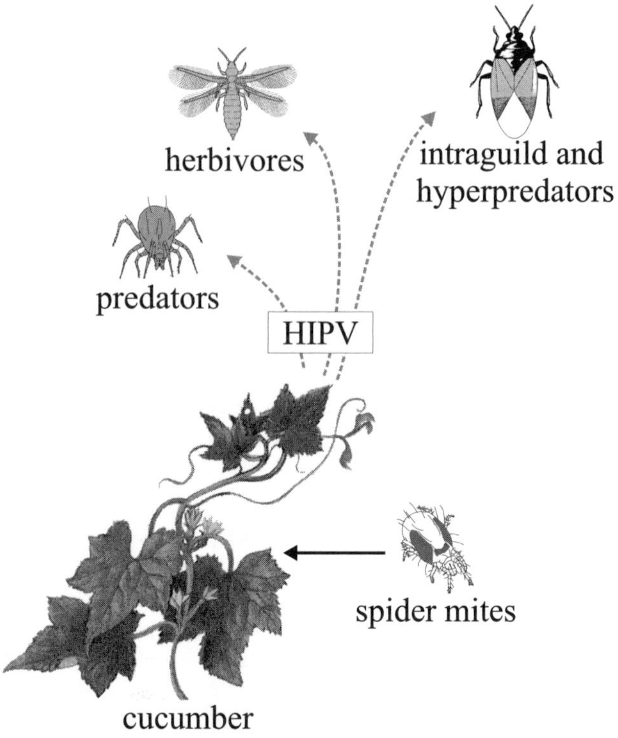

Figure 3: Herbivory (by spider mites) (drawn arrow) induces plant-wide production of volatiles in their host plant (cucumber), thereby providing free information (dashed arrows) for other herbivores (thrips), predators (predatory mites), hyper- and intraguild predators (predatory bugs), and possibly others (e.g., fungivorous or pollen-feeding mites and insects).

leaves are discovered and colonized, others will move on to neighboring plants with spider-mite colonies free of predators. Predatory mites make their own foraging decisions and may well be capable of balancing food gains and competition effects, as predicted by ideal-free foraging theory. They not only use information to locate their prey, but also to avoid intraspecific competition. An open question is where this information on competitors is coming from and whether this information allows predators to discriminate between different competitor species. Janssen et al. (1997) show that production of the odors is associated with the presence of adult spider mites, and they suggest that it is an alarm pheromone of the spider mites. Whether this alarm pheromone conveys information that allows the receiver to specify the enemy (competitor, predator) is not known.

Responses of Non-Beneficial Arthropods to HIPV

Volatile signals may also be picked up by other organisms in the food web, such as herbivores and hyperpredators (Fig. 3), and this can have severe consequences for the net fitness effects of the production of volatiles by plants. Consider, for example, a simple food web occurring on greenhouse cucumber in The Netherlands. One tritrophic food chain studied in this system consists of the plant, two-spotted spider mites and their natural enemy, the predatory mite *P. persimilis*. Another food chain consists of the plant, western flower thrips (*Frankliniella occidentalis*), and its natural enemies. When combining these two food chains in a food web, many more interactions can occur (Janssen et al. 1998, Pallini 1998, Pallini et al. 1997, 1998).

Herbivores could take advantage of the presence of other herbivores on plants, for instance by using the volatiles induced by the other herbivore to acquire information from a distance on the state of the plant. Attack by the first herbivore may induce plant defences, including indirect defense through attraction of natural enemies, making infested plants less suitable for colonization. In this case, the volatiles would signal this defended state to other herbivores that may use them to avoid well-defended plants (e.g., Ohsaki and Sato 1994). Herbivores may also use HIPV to avoid competition or plants that have a high probability of being defended by predators attracted by the volatiles. Under all scenarios, production of HIPV would lead to avoidance of the plant by other herbivores, which will have a positive effect on plant fitness. Alternatively, herbivores attacking may deplete plant constitutive defenses, making the plants a more suitable food source (Tallamy 1986). In this case, production of HIPV could lead to an increase of herbivore numbers and species, which is obviously detrimental to the plant.

Production of HIPV may also attract natural enemies that interfere directly, through interference competition or intraguild predation, with natural enemies already present on the plant (see Janssen et al. 1998 for a review). Effects of these interactions on herbivore densities can be positive or negative (Rosenheim et al. 1995, Holt and Polis 1997, Janssen et al. 1998), and hence will affect the plant's net benefits of HIPV production. Below we review the evidence for the use of HIPV by other members of the food web (see Turlings and Benrey 1998), as well as the evolutionary consequences of this for volatile production.

The Response of Herbivores to HIPV

Effects of HIPV on conspecific herbivores. Dicke (1986) found that two-spotted spider mites dispersed when exposed to odors of bean leaves infested with conspecifics. Although other studies indicate that many of the volatiles emitted by such infested leaves (i.e., general green-leaf volatiles), such as hexenol and specific herbivore-induced volatiles such as terpenoids and the ester methyl salicylate), are indeed produced by the plant, it cannot be ruled out that spider mites themselves also produce volatiles that play a role in eliciting dispersal. Another study of the response of these mites showed a slight attraction to cucumber plants with conspecifics (Pallini et al. 1997). Apart from differences in the plant species, the differences between the two series of experiments may be explained by differences in the spatial scale at which the response occurs; from a distance, spider mites are slightly attracted to plants with conspecifics, but once on the plant, they prefer leaf areas that are unoccupied (see Pallini et al. 1997 for details). This matches with further experiments by Dicke (1986), in which odors from a mixture of clean and infested leaves did not induce dispersal. Hence, from a distance, spider mites perceive a mix of odors from clean and infested leaves to which they are attracted. Once on a plant, they preferentially settle on clean leaf areas.

Landolt (1993) studied the response of adult female cabbage looper moths (*Trichoplusia ni*) to odors of cotton and cabbage plants that were uninfested, mechanically damaged or damaged by conspecifics in a laboratory wind tunnel. In choice experiments, artificially damaged or herbivore-damaged cotton attracted more moths than clean plants, but moths oviposited more on clean plants. In contrast, cabbage with herbivores attracted significantly fewer moths than clean plants did, and there was no difference in attractiveness between clean and artificially damaged cabbage.

Harari et al. (1994) investigated the volatile stimuli that cause aggregation in a polyphagous beetle (the scarabaeid *Maladera matrida*) with a Y-tube olfactometer. Beetles were not attracted to odors from conspecifics only, but were attracted when beetles were feeding on *Duranta repens* leaves. Bolter et al. (1997) studied the response of the Colorado potato beetle (*Leptinotarsa decemlineata*) to clean plants, artificially damaged plants, and plants damaged by conspecifics. More beetles were found to walk upwind when mechanically damaged plants, rather than clean plants, were offered as the odor source, but plants lost their attractiveness soon after being damaged. Plants with older beetle infestations were attractive for a longer period. The volatiles produced were a mix of green-leaf volatiles (fatty acid derivatives) and induced volatiles (terpenoids such as linalool). The responses found in these two studies (Harari et al. 1994, Bolter et al. 1997) may have been elicited by a combination of plant-produced volatiles and beetle odors.

In laboratory olfactometer experiments, Campbell et al. (1993) showed that spring morphs of damson-hop aphids (*Phorodon humuli*) have a significant preference for odors from hop leaves with conspecifics over clean hop leaves. Single-cell recordings of olfactory receptors combined with gas chromatography showed that this preference is likely to be caused by volatiles produced by the plants upon induction by the aphids (hexenal, β-caryophyllene, methyl salicylate and (*E*)-β-farnesene). As mentioned before, some spikes in the action potential of the receptor cells did not coincide with detected volatiles, suggesting that also other volatiles may play a role. The origin of these compounds is unknown, hence, volatiles other than HIPV may play a role. This study also shows some other striking features of arthropod responses to blends of volatiles. First, responses depend on the relative concentration of various components; two components in a natural ratio ((*E*)-2-hexenal and β-caryophyllene) were attractive whereas a 1:1 ratio was unattractive. Second, responses to mixtures differ from responses to individual components; for example, the full blend of HIPV is attractive despite the presence of one component that, by itself functions as a repelling alarm pheromone ((*E*)-β-farnesene). Moreover, the attractive mix of (*E*)-2-hexenal and β-caryophyllene in natural ratios became unattractive when a third compound (methyl salicylate) was added in the natural ratio.

In all of the above examples, it is still possible that minor or even undetected components are responsible for eliciting the responses. A more conclusive example of increased attraction of herbivores by HIPV comes from the work of Loughrin and colleagues on Japanese beetles (*Popillia japonica*: Scarabaeidae). These beetles were shown to induce production of volatiles in crab apple and grapevines (i.e., terpenoids such as ocimene, linalool and farnesene, aliphatic compounds such as hexenyl acetate, and aromatics such as phenylacetonitrile and hexenyl benzoate; Loughrin et al. 1995, 1996). In the field, vines producing these volatiles after overnight feeding by beetles attracted 15 to 30 times as many beetles as plants with non-feeding beetles (Loughrin et al. 1996). Trapping experiments in the field showed that single compounds of the induced volatiles are attractive and that the attractiveness of mixtures of synthetic volatiles increased with the number of compounds added (Loughrin et al. 1998). Some of the most prevalent volatiles produced by plants induced by herbivores (acyclic terpene hydrocarbons such as (*E*)-β-ocimene and farnesene) were not tested however, because they were not commercially available. It is possible that these single compounds would elicit even stronger responses from the herbivores. In conclusion, there are several examples of conspecific herbivores that respond to the odors emanating from plant-herbivore complexes, and there is every reason to assume that heterospecific herbivores use the same volatiles to their own benefit.

Effects of HIPV on heterospecific herbivores. Responses to plant volatiles induced by heterospecific herbivores are less well studied. Some of the already cited studies compare plants infested with heterospecifics to plants with conspecifics. Harari et al. (1994), for example, found that leaves with the desert locust *Schistocerca gregaria* were as attractive to the scarabaeid beetles as leaves with conspecifics, suggesting that common damage-induced volatiles are responsible for the aggregation. The authors further state that in the field, aggregations of beetles were seen close to feeding larvae of *Spodoptera littoralis*. Bolter et al. (1997) found that feeding by beet armyworm larvae

(*Spodoptera exigua*) led to similar responses in the Colorado potato beetle as conspecific feeding. Pallini et al. (1997) found that two-spotted spider mites avoided plants infested with western flower thrips under greenhouse conditions. In contrast, western flower thrips did not show preference when offered clean plants and plants infested with spider mites (Pallini et al. 1999). In all of these studies, it is possible that herbivores responded to odors that were not produced by the plants, such as faeces or body odors of the heterospecific herbivores.

The only study that unambiguously shows that herbivores use plant volatiles induced by heterospecific herbivores when searching for plants is that of Bernasconi et al. (1998). They studied plant selection by the corn leaf aphid (*Rhopalosiphum maidis*). Corn plants were induced to produce volatiles by treating artificially damaged plants with caterpillar (*S. littoralis*) regurgitate. Plants treated in this way are known to produce many typical herbivore-induced volatiles (Turlings et al. 1993b, 1998), amongst others (*E*)-β-farnesene which is an alarm pheromone of aphids. In a Y-tube olfactometer in the lab and in release experiments in the field, it was found that aphids prefer clean plants over induced plants. By using plants that were artificially induced (without herbivores), effects of the presence of odors of herbivores themselves or their faeces were ruled out as a possible cue for avoidance.

Functional explanations. The above shows that there is evidence that herbivores can use volatiles emitted by plant-herbivore complexes to avoid or find plants infested with con- and heterospecific herbivores. Little is known about the adaptive value for such avoidance or attraction. All animals for which odor-mediated attraction towards plants with conspecifics was shown, are known to form aggregations in the field, but it is unclear why they do this, although some reasons have been suggested (Loughrin et al. 1995, 1998, Harari et al. 1994). Bernasconi et al. (1998) suggest that their aphids avoid plants because the volatiles indicate (1) that the plant has started to produce toxic compounds in response to damage, (2) that potential competitors are present on the plant, or (3) that the plant is attractive to natural enemies. Moreover, they suggest that aphids may avoid damaged plants because plants produce an aphid alarm pheromone. This last explanation hinges on the biological function of one isolated compound in aphid ecology (farnesene), and the avoidance is seen as an inevitable by-product of this. However, farnesene was also present in the headspace of infested hops (Campbell et al. 1993) and this did not lead to the whole mixture of volatiles being unattractive to aphids. This last study provides further evidence that mixtures of volatiles may contain unattractive compounds without loosing their overall attractiveness (Campbell et al. 1993). It therefore seems necessary to study and explain responses to blends of volatiles rather than to isolated compounds.

Another study that speculates on the functional explanation for the response of herbivores to HIPV is that of Pallini et al. (1997), who state that two-spotted spider mites avoid plants with thrips to avoid interspecific competition and intraguild predation. Indeed, greenhouse experiments show that the population growth rate of spider mites is lower in the presence than in the absence of thrips (Brødsgaard and Enkegaard 1995).

From the plant's perspective, it is clear that production of induced volatiles may keep other herbivores away in some cases, but can have devastating effects in others. Aggregation of Japanese beetles on certain plants, for example, causes complete defoliation while other hosts in the vicinity are hardly attacked (Loughrin et al. 1996). Since Japanese beetles are an imported pest at the study site, plants may not have adapted to their occurrence. It is expected that plants that have coevolved with this pest would have been selected not to produce such volatiles.

Attraction of Hyperpredators and Intraguild Predators by HIPV

Hyperpredators and intraguild predators may also be attracted by HIPV. To our knowledge, there are no examples of this in the literature. We recently investigated the response of an omnivorous predatory bug, *Orius laevigatus*, to odors of plants attacked by two herbivores, the western flower thrips, the target pest of the predator in greenhouses in The Netherlands, and the two-spotted spider mite, a non-target pest (Venzon et al. in prep.). The bug was attracted to plants infested with either herbivore. Predatory bugs not only attack herbivores, but can also prey on other predators such as predatory mites, the natural enemies of spider mites (Cloutier and Johnson 1993, Brødsgaard and Enkegaard 1995). For this intraguild predation to occur, it is not sufficient that predators are attracted to plants with the same prey, but also that they do not avoid plants with the other predator present (Janssen et al. 1995a,b, 1997). When offered a choice between plants with spider mite prey plus spider mite predators and plants with spider mites alone, the predatory bug showed no preference for either of the two, indicating that it does not avoid plants with the other predator present (Venzon et al.unpublished data). Likewise, the natural enemy of spider mites did not avoid plants with prey and the predatory bug (Janssen et al.unpublished data). This indicates that predators of both species may interact on plants with spider mites. That this can have negative effects on the control of spider mites is indicated by experiments by Brødsgaard and Enkegaard (1995), who compared the dynamics of two-spotted spider mites and *P. persimilis* in the presence or absence of another *Orius* species, *O. majusculus* (Reuter), on gerbera. Their results clearly show that the number of spider mites remained higher in presence of *O. majusculus*. Unfortunately, no data were given on the densities of *P. persimilis*, but it suggests that intraguild predation of *Orius* on *P. persimilis* initially has a positive effect on spider mite densities. Hence, attraction of a generalist predator by odors produced by the plant had positive effects on herbivore density, and may therefore have negative effects on plant fitness. The effects of HIPV on the attraction of hyper- and intraguild predators clearly needs further study.

Synthesis: Overall Impact of HIPV on Plant Fitness

Most studies reviewed above do not enable distinction between volatiles emanating from the herbivores and plant-produced volatiles. Moreover, volatiles may also be produced by microorganisms associated with the plant or with herbivores. The distinction between odors of these various sources and their effects on other herbivores is essential from an evolutionary perspective, since plants may be unable to affect production of volatiles by herbivores or microorganisms. We therefore stress that further identification of the volatiles and determination of their source is essential for determining costs and benefits of HIPV-production by plants.

The concept of plants producing volatiles to attract members of the third trophic level hinges on the assumption that hyperpredators and intraguild predators are unimportant in determining herbivore density and plant fitness. In other words, the effects of hyperpredators should not cascade down to the first trophic level. This is in agreement with the once widely accepted view that herbivore densities are mainly determined by predators, and predator densities are determined by competition rather than hyperpredation (the "World is green" hypothesis, Hairston et al. 1960). This idea however, has lost some of its credibility since it became clear that competition among herbivores occurs frequently (Sih et al. 1985, Denno et al. 1995), and that intraguild predation is widespread and can have important effects on herbivore dynamics (Rosenheim et al. 1995, Holt and Polis 1997). Moreover, the idea that species can be assigned to one trophic level is also the subject of discussion (Polis and Strong 1996, Janssen et al. 1998). Many species are omnivorous and attack both plant and herbivore, or herbivore and other predators. A remarkable example of this is the western flower thrips. It attacks both plants and spider mites, but is mainly viewed as a pest. However, in cotton in California and Texas it is regarded as a predator of spider mites and it is mostly not controlled, although it may occasionally cause substantial damage (Trichilo and Leigh 1986). The thrips can also act as hyperpredator; it kills eggs of predatory mites that prey on spider mites or on the thrips itself (F. Faraji, pers. obs., A. Pallini, pers. obs.).

Viewed from the perspective of such food web complexities, one is inclined to think that conditions favouring the evolution of production of HIPV can only occur under a restrictive set of conditions, i.e., when the overall effect of hyper-predators, intraguild-predators, and simple predators on herbivore densities is negative, or when HIPV-production does not lead to attraction of other herbivores that cause more damage. Moreover, when benefits of volatile production are relatively small, as will be the case when a plant is attacked by a minor pest, the costs may well outweigh the benefits. Hence, assuming that plants can distinguish between herbivore species, we hypothesize that attack by herbivores that cause small negative effects on plant fitness will not lead to induction of volatile production. Yet, we are aware of only one example of a herbivore-damaged plant species that did not produce any volatiles: maize plants infested with aphids (*Rhopalosiphum maidis*) did not show any production of induced volatiles (Turlings et al. 1998). This species can, at times, be a pest in the US and China but is less important than stem-borers that do elicit HIPV production (T.C.J. Turlings, pers. comm.) It is possible that volatile production would be detrimental to the plant because it attracts more severe pests. Alternatively, it is possible that volatiles were produced, but were not detectable by the techniques used. Study of the response of natural enemies of aphids, as well as stem-borers, to plants with aphids will give more insight.

A reason for the apparent generality of HIPV production among plants may be that negative results, i.e., the absence of attraction of natural enemies or production of volatiles, may not be published as readily as positive results. Negative results, however, would enhance our understanding of the evolution of induced volatile production. Another reason might be that most herbivores studied so far are serious pests of crops, hence have enormous effects on plant fitness. In these cases, any defense is better than patiently awaiting death or defoliation. This topic raises further questions on the specificity of volatile signals produced; besides being able to produce specific signals for each herbivore (De Moraes et al. 1998), are plants also capable of shutting down this line of defense when they are attacked by a minor pest?

A third reason for the generality of production of herbivore-induced volatiles may be that complex food webs behave as simple tritrophic food chains (Fig. 4). In other words, the negative effects of hyperpredators and generalist predators on plant fitness are minor compared to the positive effects of predators, and it therefore always pays for a plant to advertise the presence of herbivores. This seems to contradict the notion that interactions such as intraguild predation are widespread (Rosenheim et al. 1995, Holt and Polis 1997). However, many of the empirical studies underpinning this conclusion used confinements to study food web interactions and this leads to inevitable interactions among species (Janssen et al. 1998). Antipredator behavior (Lima and Dill 1990, Lima 1998, Kats and Dill 1998, Pallini et al. 1998) and avoidance of other adverse interactions (Janssen et al. 1995a,b, 1997) may substantially reduce the direct interactions between predators, hyperpredators and generalist intraguild predators (Fig. 4). Still, such avoidance always comes with a cost, so even if predators escape from their adversaries, the act of escaping may reduce their efficiency at reducing herbivore populations.

A last important point to make is that costs and benefits of indirect plant defenses are likely to vary greatly in space and time, as is generally the case in mutualistic interactions (Bronstein 1994a,b). For example, the number of predatory arthropods, but also the number of herbivores and hyperpredators in

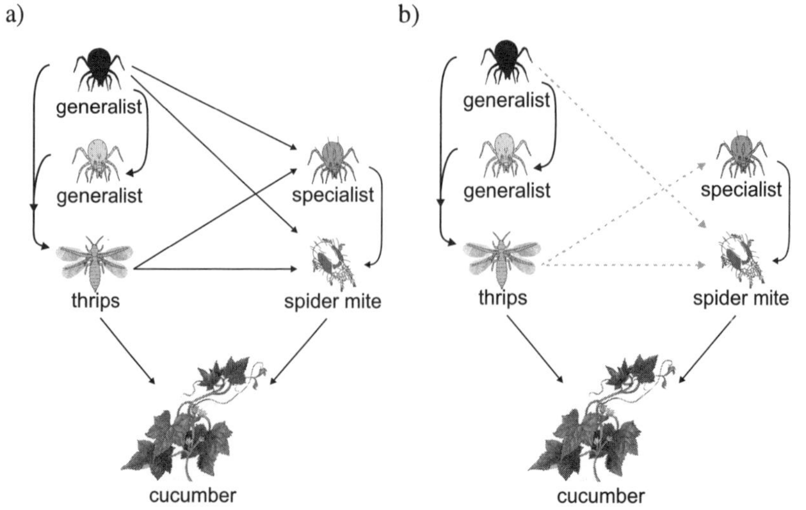

Figure 4: Complex interactions in an artificial food web on cucumber plants consisting of spider mites (the two-spotted spider mite, *Tetranychus urticae*) and their enemy (the predatory mite *Phytoseiulus persimilis*), as well as thrips (the western flower thrips, *Frankliniella occidentalis*) and its enemies (the predatory mites, *Amblyseius cucumeris* and *Iphiseius degenerans*, the predatory bug, *Orius laevigatus*) (4a). The complexities involve herbivore-herbivore, herbivore-predator and predator-predator interactions, as well as the avoidance of these interactions. When the avoidance responses are of overriding importance, the food web behaves like two linear food chains (4b). Dashed arrows indicate weak interactions.

the environment of a plant may vary, thereby causing differential net effects of indirect plant defenses.

In conclusion, we hope to have convinced the reader that, to assess the impact of HIPV on plant fitness, it is essential to consider (1) interactions with other members of the food web, (2) a spatial scale that enables manifestation of the full behavioral repertoire of all organisms in the food web, such as antipredator and avoidance behavior, and (3) a temporal scale that allows the expression of the full range of possible outcomes of interspecific interactions (Bronstein 1994b).

Future Directions: Perspectives of HIPV for Crop Protection

The active role of plants in promoting efficiency of the third trophic level seems an appealing idea for improving crop protection, but it is essential to realize that these plant traits have evolved under natural conditions. One should be cautious when using these traits to improve pest control under agricultural conditions. Clearly, plant fitness and crop yield are not synonymous, although they may overlap in many cases. Moreover, food web structure of arthropods on agricultural crops may differ dramatically from those under natural conditions due to, for example, pesticide use, the introduction of (alien) natural enemies, and reduced plant diversity. Nevertheless, bearing these restrictions in mind, one may speculate on the possibilities for manipulating HIPV production to improve crop protection (Bottrell et al. 1998). Below, we give five applications and discuss their advantages and disadvantages.

Development of predator lures and herbivore deterrents. Components or mixtures of several compounds of HIPV can be used to monitor and/or lure naturally occurring predators to crop fields, or to arrest released predators in the crop where they have been released. Attractive as this might seem, predators may be confused within the crop by the omnipresence of cues signalling prey. This may eventually alter their responsiveness to the compounds either via selection or via learning to avoid them. In fact, when the compounds are offered dissociated from the prey, the application of lures may even become counterproductive because these signals then become associated with hunger (Drukker and Sabelis 1999). Also, the supposed predator lures may additionally attract herbivores. This is nicely illustrated by the field experiments of Molleman et al. (1997). They attempted to attract anthocorid predators to pear orchards by placing delta and funnel traps releasing methyl salicylate, one of the HIPV components from *Psylla*-infested pear leaves that was shown to be attractive in Y-tube experiments. Rather than attracting these predators, they found significant attraction of herbivores, such as silver-Y and apple clearwing moths. Hoverflies were the only natural enemies attracted.

Conditioning of mass-reared predators on the appropriate cues. It is a widely recognised problem that mass-reared natural enemies do not end up on the target crop after release. One possible cause may be that the food/prey used for mass-rearing lacks the association with cues relevant to searching in the field. The simple solution may be that before release the predators are conditioned by offering food associated with appropriate HIPV. For example, the number of *Orius laevigatus* recovered on thrips-infested and spider mite-infested cucumber plants in the greenhouse increased by 30% when the predators had experienced the association of odors and spider mites on cucumber prior to the release experiment (Venzon and Janssen, unpublished data).

Plant breeding for improved HIPV production. Plant species and cultivars differ in the composition and amounts of green leaf volatiles and HIPV, as shown for example in cotton (Elzen et al. 1985), apple (Takabayashi et al. 1991) and gerbera (Krips et al., unpublished data). The responsiveness of natural enemies is not necessarily related to the total amounts of volatiles, but rather to the extent to which the blends stand out in the context of the environment. Conspicuousness may well prove a selectable trait, but the impact on biocontrol is probably context-dependent. In addition, dose-response curves usually exhibit a plateau above which an increase in stimulus is not matched by an increased response. Nevertheless, selecting more attractive varieties due to HIPV release upon herbivore attack has the major advantage that target and signal coincide. Plant varieties attractive due to green leaf volatiles may attract more predators from the environment surrounding the crop, but share the disadvantage with predator lures that the volatiles are not associated with prey and predators may be confused by the omnipresence of cues within the crop.

Elicitors to induce HIPV production. Exogenous application of jasmonic acid and release of volatile methyl jasmonate activates the octadecanoid pathway in plants which in turn triggers the production of proteinase inhibitors, oxidative enzymes (Farmer and Ryan 1990, Thaler et al. 1996, Adviushko et al. 1997, McConn et al. 1997) and plant volatiles known to attract natural enemies of herbivores (Hopke et al. 1994). Thus, these treatments have multiple effects involving both direct and indirect defenses against herbivores (Thaler, this volume). Treating plants with an exogenous elicitor (jasmonic acid) is the most elegant way of producing predator lures; the plant itself produces most of the right ingredients! Whether such application to crops can yield less herbivore damage remains to be seen (See Thaler, this volume). For one thing herbivory also triggers these defenses on site and, for another, application away from the site of herbivore attack may cause the plant to invest in defenses when herbivore densities are low, thus reducing crop yield. Again, there is the danger of attracting predators to places without herbivores, in fact, the bouquet produced by plants consists of major and minor compounds and is thus a realistic mimic of HIPV. This is an extra reason for caution since predators may have difficulty learning to discriminate between JA-induced plant volatiles and plant volatiles induced by herbivory.

Intercropping with plants attractive to predatory arthropods. The use of this technique has recently been illustrated by Khan et al. (1997). They intercropped maize in Africa with molasses grass, a plant that was not colonized by stem-borers, a major pest of maize in this area. When undamaged, this grass normally produces several volatiles that are otherwise known as HIPV, such as β-caryophyllene, humulene and 4,8-dimethyl-1,3(*E*),7-nonatriene. The entire grass plant, as well as one of these volatiles in pure form (nonatriene), was attractive to stem-borer parasitoids, whereas plant volatiles extracted by hydrodestillation deterred oviposition by the stem-borers. Indeed, intercropping resulted in an increase in rate of parasitism and a decrease of stem-borer attacks. This study suggests that undamaged plants releasing volatiles that are otherwise only produced upon herbivore attack may improve parasitoid efficiency, but the role of the volatiles is by no means proven. Moreover, the use of such plants as natural lures faces some of the same problems as described in the previous section. Although fraught with many problems, we think that applications of HIPV hold promise for the future.

Acknowledgements

We thank Anurag Agrawal and Jennifer Thaler for their comments.

Literature Cited

Adviushko, S.A., G.C. Brown, D.L. Dahlman and D.F. Hildebrand. 1997. Methyl jasmonate exposure induces insect resistance in cabbage and tobacco. Environmental Entomology 26:642-854.

Agrawal, A.A., and R. Karban. 1997. Domatia mediate plant-arthropod mutualism. Nature 387:562-563.

Bernasconi, M.L., T.C.J. Turlings, L. Ambrosetti, P. Bassetti and S. Dorn. 1998. Herbivore-induced emissions of maize volatiles repel the corn leaf aphid, *Rhopalosiphum maidis*. Entomologia Experimentalis et Applicata 87:133-142.

Bolter, C.J., M. Dicke, J.J.A. van Loon, J.H. Visser and M.A. Posthumus. 1997. Attraction of Colorado potato beetle to herbivore-damaged plants during herbivory and after its termination. Journal of Chemical Ecology 23:1003-1023.

Bottrell, D.G., P. Barbosa and F. Gould 1998. Manipulating natural enemies by plant variety selection and modification: A realistic strategy? Annual Review of Entomology 43:347-367.

Brødsgaard, H.F. and A. Enkegaard. 1995. Interactions among polyphagous anthocorid bugs used for thrips control and other beneficials in multi-species biological pest management systems. Mededelingen der Faculteit Landbouwwetenschappen, Universiteit Gent 60:893-900.

Bronstein, J.L. 1994a. Our current understanding of mutualism. Quarterly Review of Biology 69:31-51.

Bronstein, J.L. 1994b. Conditional outcomes in mutualistic interactions. Trends in Ecology and Evolution 9:214-217.

Campbell, C.A.M., J. Pettersson, J.A. Pickett, L.J. Wadhams and C.M. Woodcock. 1993. Spring migration of damson-hop aphid, *Phorodon humuli* (Homoptera, Aphididae), and summer host plant-derived semiochemicals released on feeding. Journal of Chemical Ecology 81:1569-1576.

Cloutier, C. and S.G. Johnson. 1993. Predation by *Orius tristicolor* (Hemiptera, Anthocoridae) on *Phytoseiulus persimilis* (Acarina, Phytoseiidae) - testing for compatibility between biocontrol agents. Environmental Entomology 22:477-482.

De Bruyne, M., M. Dicke, W.F. Tjallingii 1991. Receptor cell responses in the anterior tarsi of *Phytoseiulus persimilis* to volatile kairomone components. Experimental and Applied Acarology 13:53-58.

De Moraes, C.M., Lewis, W.J., Paré, P.W., Alborn, H.T., Tumlinson, J.H. 1998. Herbivore-infested plants selectively attract parasitoids. Nature 393:570-573.

Denno, R.F., M.S. McClure and J.R. Ott. 1995. Interspecific interactions in phytophagous insects: competition re-examined and resurrected. Annual Review of Entomology 40:297-331.

Dicke, M. 1986. Volatile spider-mite pheromone and host-plant kairomone, involved in spaced-out gregariousness in the spider mite *Tetranychus urticae*. Physiological Entomology 11:251-262.

Dicke, M. 1988. Infochemicals in tritrophic interactions: origin and function in a system consisting of predatory mites, phytophagous mites and their host plants. Ph. D. Thesis, Agricultural University Wageningen.

Dicke, M. 1994. Local and systemic production of volatile herbivore-induced terpenoids. Journal of Plant Physiology 143:465-472.

Dicke, M. and A. Groeneveld. 1986. Hierarchical structure in kairomone preference of the predatory mite *Amblyseius potentillae*: dietary component indispensable for diapause induction affects prey location behaviour. Ecological Entomology 11:131-138.

Dicke, M. and M.W. Sabelis. 1988a. How plants obtain predatory mites as bodyguards. Netherlands Journal of Zoology 38:148-165.

Dicke, M. and M.W. Sabelis 1988b. Origin and function of semiochemicals in a tritrophic system of host plant, spider mites and predatory mites. Pages 17-18 *in* Les Colloques de l'INRA no. 48. Proceedings of Parasitoid Insects, European Workshop held in Lyon, Sept. 7-10 1987.

Dicke, M. and M.W. Sabelis. 1989. Does it pay plants to advertize for bodyguards? Towards a cost-benefit analysis of induced synomone production. Pages 341-358 *in* H. Lambers, M.L. Cambridge, H. Konings and T.L. Pons, editors. Variation in Growth Rate and Productivity of Higher Plants. SPB Academic Publishing BV, The Hague, The Netherlands.

Dicke, M. and M.W. Sabelis. 1992. Costs and benefits of chemical information conveyance: proximate and ultimate factors. Pages 122-155 *in* B. Roitberg and M. Isman, editors. Insect Chemical Ecology, an Evolutionary Approach. Chapman & Hall, Hants, UK.

Dicke, M., J. Takabayashi, M.A. Posthumus, C. Schütte and O. Krips. 1998. Plant-phytoseiid interactions mediated by herbivore-induced plant volatiles: variation in production of cues and in responses of predatory mites. Experimental and Applied Acarology 22:311-334

Dicke, M., T.A. Van Beek, M.A. Posthumus, N Ben Dom, H. Van Bokhoven and Æ. De Groot. 1990a. Isolation and identification of volatile kairomone that affects acarine predator-prey interactions: involvement of host plant in its production. Journal of Chemical Ecology 16:381-396.

Dicke, M., M.W. Sabelis, R.J.F. Bogaers, M.P.T. Alers and I. van Halder. 1991. Kairomone perception by a predatory mite: behavioural analysis of chemoreceptor-carrying extremities. Proceedings of the section Experimental and Applied Entomology of the Netherlands Entomological Society (NEV), 2:179-184.

Dicke, M., M.W. Sabelis, J. Takabayashi, J. Bruin and M.A. Posthumus. 1990b. Plant strategies of manipulating predator-prey interactions through allelochemicals: prospects for application in pest control. Journal of Chemical Ecology 16:3091-3118.

Dicke, M., P. Van Baarlen, R. Wessels and H. Dijkman. 1993. Herbivory induces systemic production of volatiles that attract herbivore predators: extraction of an endogenous elicitor. Journal of Chemical Ecology 19:581-599.

Dicke, M., K.J. van der Maas, J. Takabayashi and L.E.M. Vet. 1990c. Learning affects response to volatile allelochemicals by predatory mites. Proceedings of the section Experimental and Applied Entomology of the Netherlands Entomological Society (NEV) 1:31-36.

Dicke, M. and L.E.M. Vet. 1999. Plant-carnivore interactions: evolutionary and ecological consequences for plant, herbivore and carnivore. Pages 483-520 *in* H. Olff, V.K. Brown and R.H. Drent, editors. Herbivores between Plants and Predators. Blackwell Science, Oxford.

Drukker, B., J. Bruin, and M.W. Sabelis. 1999. Anthocorid predators learn to associate herbivore-induced volatiles with presence or absence of prey. Entomologia Experimentalis et Applicata (In press).

Drukker, B., P. Scutareanu and M.W. Sabelis. 1995. Do anthocorid predators respond to synomones from *Psylla*-infested pear trees under field conditions? Entomologia Experimentalis et Applicata 77:193-203.

Du, Y., G.M. Poppy, W. Powell, J.A. Pickett, L.J. Wadhams and C.M. Woodcock. 1998. Identification of semiochemicals released during aphid feeding that attract parasitoid *Aphidius ervi*. Journal of Chemical Ecology 24:1355-1368.

Dwumfour, E.F. 1992. Volatile substances evoking orientation in the predatory flower bug *Anthocoris nemorum* (Heteroptera: Anthocoridae). Bulletin of Entomological Research 82:465-469.

Elzen, G.W.H., H.J. Williams, A.A. Bell, R.D. Stipanovic and S.B. Vinson. 1985. Quantification of volatile terpenes of glanded and glandless *Gossypium hirsutum* L. cultivars and lines by gas chromatography. Journal of Agricultural Food Chemistry 33:1079-1082.

Farmer, E.E. and C.A. Ryan. 1990. Interplant communication: Airborne methyl jasmonate induces synthesis of proteinase inhibitors in plant leaves. Proceedings of the National Academy of Sciences USA 87:7713-7716.

Guerrieri, E., F. Pennacchio and E. Tremblay. 1997. Effect of adult experience on in-flight orientation to plant and plant-host complex volatiles in *Aphidius ervi* Haliday (Hymenoptera: Braconidae). Biological Control 10:159-165.

Hairston, N.G., F.E. Smith, L.B. Slobodkin. 1960. Community structure population control and competition. American Naturalist 94:421-425.

Harari, A.R., D. Ben-Yakir and D. Rosen. 1994. Mechanism of aggregation behavior in *Maladera matrida* Argaman (Coleoptera: Scarabaeidae). Journal of Chemical Ecology 20:361-371.

Holt, R.D. and G.A. Polis. 1997. A theoretical framework for intraguild predation. American Naturalist 149:745-764.

Hopke, J., J. Donath, S. Blechert and W. Boland. 1994. Herbivore-induced volatiles: the emission of acyclic homoterpenes from leaves of *Phaseolus lunatus* and *Zea mays* can be triggered by a β-glucosidase and jasmonic acid. FEBS Letters 352:146-150.

Jagers Op Akkerhuis, G., M.W. Sabelis and W.F. Tjallingii. 1985. Ultrastructure of chemoreceptors of the pedipalps and first tarsi of *Phytoseiulus persimilis*. Experimental and Applied Acarology 1:235-251.

Janssen, A. 1999. Plants with spider-mite prey attract more predatory mites than clean plants under greenhouse conditions. Entomologia Experimentalis et Applicata (in press).

Janssen, A., J. Bruin, G. Jacobs, R. Schraag and M.W. Sabelis. 1997. Predators use volatiles to avoid prey patches with conspecifics. Journal of Animal Ecology 66:223-232.

Janssen, A., A. Pallini, M. Venzon and M.W. Sabelis. 1998. Behaviour and indirect interactions in food webs of plant-inhabiting arthropods. Experimental and Applied Acarology 22:497-521.

Janssen, A., J.J.M. van Alphen, M.W. Sabelis and K. Bakker. 1995a. Odour-mediated avoidance of competition in *Drosophila* parasitoids: the ghost of competition. Oikos 73:356-366.

Janssen, A., J.J.M. van Alphen, M.W. Sabelis and K. Bakker. 1995b. Specificity of odour mediated avoidance of competition in *Drosophila* parasitoids. Behavioral Ecology and Sociobiology 36:229-235.

Kats, L.B. and L.M. Dill. 1998. The scent of death: Chemosensory assessment of predation risk by prey animals. Ecoscience 5:361-394.

Khan, Z.R., K. Ampong-Nyarko, P. Chiliswa, A. Hassanali, S. Kimani, W. Lwande, W.A. Overholt, J.A. Pickett, L.E. Smart, L.J. Wadhams and C.M. Woodcock. 1997. Intercropping increases parasitism of pests. Nature 388:631-632.

Krips, O., P.E.L. Willems, R. Gols, M.A. Posthumus, M. Dicke. 1999. The response of *Phytoseiulus persimilis* to spider mite-induced volatiles from gerbera: influence of starvation and experience. Journal of Chemical Ecology (in press).

Landolt, P.J. 1993. Effects of host plant leaf damage on cabbage looper moth attraction and oviposition. Entomologia Experimentalis et Applicata 67:79-85.

Lesna, I. and M.W. Sabelis. 1998. Genetic polymorphism in prey preference at a small spatial scale: a case study of soil predatory mites (*Hypoaspis aculeifer*) and two species of astigmatic mites as prey. Pages 75-96 *in* Bulb Mite Biocontrol: Evolutionary Genetics of Prey Choice in Soil Predators. Ph.D. Thesis, University of Amsterdam, The Netherlands.

Lima, S.L. 1998. Nonlethal effects in the ecology of predator-prey interactions - What are the ecological effects of anti-predator decision-making? BioScience 48:25-34.

Lima, S.L. and L.M. Dill. 1990. Behavioral decisions made under the risk of predation: a review and prospectus. Canadian Journal of Zoology 68:619-640.

Loughrin, J.H., A. Manukian, R.R. Heath, T.C.J. Turlings and J.H. Tumlinson 1994. Diurnal cycle of emission of induced volatile terpenoids by herbivore-injured cotton plants. Proceedings of the National Academy of Sciences USA 91:11836-11840.

Loughrin, J.H., D.A. Potter and T.R. Hamilton-Kemp. 1995. Volatile compounds induced by herbivory act as aggregation kairomones for the Japanese beetle (*Popillia japonica* Newman). Journal of Chemical Ecology 21:1457-1467.

Loughrin, J.H., D.A. Potter and T.R. Hamilton-Kemp. 1998. Attraction of Japanese beetles (Coleoptera: Scarabaeidae) to host plant volatiles in field trapping experiments. Environmental Entomology 27:395-400.

Loughrin, J.H., D.A. Potter, T.R. Hamilton-Kemp and M.E. Byers. 1996. Volatile compounds from crab apple (*Malus* spp.) cultivars differing in susceptibility to the Japanese beetle (*Popillia japonica* Newman). Journal of Chemical Ecology 22:1295-1305.

Margolies, D.C., M.W. Sabelis and J.E. Boyer, Jr. 1997. Response of a phytoseiid predator to herbivore-induced plant volatiles: selection on attraction and effect on prey exploitation. Journal of Insect Behavior 10:695-709.

McConn, M., R.A. Creellman, E. Bell, J.E. Mullet and J. Browse. 1997. Jasmonate is essential for insect defense in *Arabidopsis*. Proceedings of the National Academy of Sciences USA 93:5473-5477.

Molleman, F., B. Drukker and L. Blommers. 1997. A trap for monitoring pear psylla predators using dispensers with the synomone methyl-salicylate. Proceedings of the section Experimental and Applied Entomology of the Netherlands Entomological Society (NEV), Amsterdam 8:177-182.

Ohsaki, N. and Y. Sato. 1994. Food plant choice of *Pieris* butterflies as a trade-off between parasitoid avoidance and quality of plants. Ecology 75:59-68.

Pallini, A. 1998. Odour-mediated indirect interactions in an arthropod food web. PhD thesis, University of Amsterdam, The Netherlands.

Pallini, A., A. Janssen and M.W. Sabelis. 1997. Odour-mediated responses of phytophagous mites to conspecific and heterospecific competitors. Oecologia 110:179-185.

Pallini, A., A. Janssen and M.W. Sabelis. 1998. Predators induce interspecific herbivore competition for food in refuge space. Ecology Letters 1:171-176.

Pallini, A., A. Janssen and M.W. Sabelis. 1999. Do western flower thrips avoid plants infested with spider mites? Interactions between potential competitors. Pages 375-380 *in* J. Bruin, L.P.S. van der Geest and M.W. Sabelis, editors. Ecology and Evolution of the Acari. Kluwer Academic Publishers, Dordrecht, The Netherlands.

Pels, B. and M.W. Sabelis. 1999. Local dynamics, overexploitation and predator dispersal in an acarine predator-prey system. Oikos 86 (in press).

Polis, G.A. and D.R. Strong. 1996. Food web complexity and community dynamics. American Naturalist 147:813-846.

Price, P.W., C.E. Bouton, P. Gross, B.A. McPheron, J.N. Thompson and A.E. Weiss. 1980. Interactions among three trophic levels: Influence of plants on interactions between insect herbivores and natural enemies. Annual Review of Ecology and Systematics 11:41-65.

Rosenheim, J.A., H.K. Kaya, L.E. Ehler, J.J. Marois and B.A. Jaffee. 1995. Intraguild predation among biological-control agents: theory and evidence. Biological Control 5:303-335.

Sabelis, M.W. 1992. Arthropod predators. Pages 225-264 *in* M.J. Crawley, editor. Natural Enemies – The Population Biology of Predators, Parasites and Diseases. Blackwell, Oxford, UK.

Sabelis, M.W. and B.P. Afman. 1994. Synomone-induced suppression of take-off in the phytoseiid mite *Phytoseiulus persimilis* Athias-Henriot. Experimental and Applied Acarology 18:711-721.

Sabelis, M.W., B.P. Afman and P.J. Slim. 1984a. Location of distant spider-mite colonies by *Phytoseiulus persimilis*: Localization and extraction of a kairomone. Pages 431-440 *in* D.A. Griffiths and C.E. Bowman, editors. Acarology VI, Vol. 1. Halsted Press, New York.

Sabelis, M.W. and M. Dicke. 1985. Long-range dispersal and searching behaviour. Pages 141-160 *in* W. Helle and M.W. Sabelis, editors. Spider Mites – Their Biology, Natural Enemies and Control. Elsevier Science Publishers, Amsterdam, The Netherlands, Vol. 1B.

Sabelis, M.W., M. van Baalen, J. Bruin, M. Egas, V.A.A. Jansen, A. Janssen and B. Pels. 1999a. The evolution of overexploitation and mutualism in plant-herbivore-predator interactions and its impact on population dynamics. Pages 259-282 *in* B.A. Hawkins & H.V. Cornell, editors. Theoretical Approaches to Biological Control. Cambridge University Press, Cambridge.

Sabelis, M.W., M. van Baalen. F.M. Bakker, J. Bruin, B. Drukker, M. Egas, A. Janssen, I. Lesna, B. Pels, P.C.J. van Rijn and P. Scutareanu. 1999b. Evolution of direct and indirect plant defence against herbivorous arthropods. Pages 109-166 *in* H. Olff, V.K. Brown and R.H. Drent, editors. Herbivores: Between Plants and Predators. Blackwell Science, Oxford.

Sabelis, M.W. and H.E. Van de Baan. 1983. Location of distant spider-mite colonies by phytoseiid predators: Demonstration of specific kairomones emitted by *Tetranychus urticae* and *Panonychus ulmi* (Acari: Phytoseiidae, Tetranychidae). Entomologia Experimentalis et Applicata 33:303-314.

Sabelis, M.W. and J. Van der Meer. 1986. Local dynamics of the interaction between predatory mites and two-spotted spider mites. Pages 322-344 *in* J.A.J. Metz and O. Diekmann, editors. Dynamics of Physiologically Structured Populations. Lecture Notes in Biomathematics, 68. Springer-Verlag, Berlin, Germany.

Sabelis, M.W. and J.J. Van der Weel. 1993. Anemotactic responses of the predatory mite, *Phytoseiulus persimilis* Athias-Henriot, and their role in prey finding. Experimental and Applied Acarology 17:1-9.

Sabelis, M.W., J.E. Vermaat and A. Groeneveld. 1984b. Arrestment responses of the predatory mite, *Phytoseiulus persimilis*, to steep odour gradients of a kairomone. Physiological Entomology 9:437-446.

Scutareanu, P., B. Drukker, J. Bruin, M.A. Posthumus and M.W. Sabelis. 1996. Leaf volatiles and polyphenols in pear trees infested by *Psylla pyricola*. Evidence for simultaneously induced responses. Chemoecology 7:34-38.

Scutareanu, P., B. Drukker, J. Bruin, M.A. Posthumus and M.W. Sabelis. 1997. Isolation and identification of volatile synomones involved in the interaction between *Psylla*-infested pear trees and two anthocorid predators. Journal of Chemical Ecology 23:2241-2260.

Scutareanu, P., R. Lingeman, B. Drukker and M W. Sabelis. 1999. Cross-correlation analysis of fluctuations in local populations of pear psyllids and anthocorid bugs. Ecological Entomology (in press).

Shimoda, T., J. Takabayashi, W. Ashihara, and A. Takafuji. 1997. Response of predatory insect *Scolothrips takahashii* toward herbivore-induced plant volatiles under laboratory and field conditions. Journal of Chemical Ecology 23:2033-2048.

Sih, A., P. Crowley, M. McPeek, J. Petranka and K. Strohmeier. 1985. Predation, competition and prey communities: a review of field experiments. Annual Review of Ecology and Systematics 16:269-311.

Strong, D.R., J.H. Lawton and T.R.E. Southwood. 1984. Insects on Plants. Blackwell Scientific Publications, Oxford.

Takabayashi, J. and M. Dicke. 1996. Plant-carnivore mutualism through herbivore-induced carnivore attractants. Trends in Plant Science 1:109-113.

Takabayashi, J., M. Dicke and M.A. Posthumus. 1991. Variation in composition of predator-attracting allelochemicals emitted by herbivore-infested plants: relative influence of plant and herbivore. Chemoecology 2:1-6.

Takabayashi, J., M. Dicke and M.A. Posthumus. 1994. Volatile herbivore-induced terpenoids in plant-mite interactions: variation caused by biotic and abiotic factors. Journal of Chemical Ecology 20:1329-1354.

Tallamy, D.W. 1986. Behavioral adaptations in insects to plant allelochemicals. Pages 273-300 *in* L. B. Brattsten and S. Ahmad, editors. Molecular Aspects of Insect-Plant Associations. Plenum, New York.

Thaler, J.S., M.J. Stout, R. Karban and S.S. Duffey. 1996. Exogenous jasmonates simulate insect wounding in tomato plants, *Lycopersicon esculentum*, in the laboratory and field. Journal of Chemical Ecology 22:1767-1781.

Trichilo, P.J. and T.F. Leigh. 1986. Predation on spider mite eggs by the Western Flower Thrips, *Frankliniella occidentalis* (Thysanoptera: Thripidae), an opportunist in a cotton agroecosystem. Environmental Entomology 15:821-825.

Turlings, T.C.J., J.H. Tumlinson, R.R. Heath, A.T. Proveaux and R.E. Doolittle. 1991. Isolation and identification of allelochemicals that attract the larval parasitoid, *Cotesia marginiventris* (Cresson), to the microhabitat of one of its hosts. Journal of Chemical Ecology 17:1251-1253.

Turlings, T.C.J. and B. Benrey. 1998. Effects of plant metabolites on the behavior and development of parasitic wasps. Ecoscience 5:321-333.

Turlings, T.C.J., M. Bernasconi, R. Bertossa, F. Bigler, G. Caloz and S. Dorn. 1998. The induction of volatile emissions in maize by three herbivore species with different feeding habits: Possible consequences for their natural enemies. Biological Control 11:122-129.

Turlings, T.C.J and J.H. Tumlinson. 1992. Systemic release of chemical signals by herbivore-injured corn. Proceedings of the National Academy of Sciences USA 89:8399-8402.

Turlings, T.C.J., F.L. Wäckers, L.E.M. Vet, W.J. Lewis and J.H. Tumlinson. 1993a. Learning of host-finding cues by hymenopterous parasitoids. Pages 51-78 *in* D.R. Papaj and W.J. Lewis, editors. Insect Learning. Chapman and Hall, New York.

Turlings, T.C.J., P.J. McCall, H.T. Alborn and J.H. Tumlinson. 1993b. An elicitor in caterpillar oral secretions that induces corn seedlings to emit chemical signals attractive to parasitic wasps. Journal of Chemical Ecology 19:411-425.

Turlings, T.C.J., J.H. Loughrin, P.J. McCall, U.S.R. Röse, W.J. Lewis and J.H. Tumlinson. 1995. How caterpillar-damaged plants protect themselves by attracting parasitic wasps. Proceedings of the National Academy of Sciences USA 92:4169-4174.

Van Baalen, M. and M.W. Sabelis. 1995. The milker-killer dilemma in spatially structured predator-prey interactions. Oikos 74:391-413.

Part III

Agriculture and Applications

Implementation of Elicitor Mediated Induced Resistance in Agriculture

Gary D. Lyon and Adrian C. Newton

Abstract

We discuss the concept of spray applications of chemicals to enhance resistance of plants to subsequent infection by plant pathogens. We have used the term "elicitor" to refer to a wide range of compounds which are able to induce resistance to subsequent inoculation with a potential pathogen, or which are able to induce host responses normally associated with the resistance response. Knowledge has been derived from fundamental studies designed to better understand the molecular basis of resistance to infection. Such studies have identified compounds involved in the primary stimulation of resistance as well as molecules associated with intracellular signal transduction. With different levels of efficiency these molecules, when applied to the external surface of a plant, are able to enhance or stimulate selected cascades leading to an enhanced level of resistance to subsequent infection. Despite early fears that there would be a yield penalty for up-regulation of any resistance response, or that there would be toxic effects (to the plant) of stimulating some secondary metabolic processes, such fears have proved to be largely unfounded. Thus, there appears to be great potential to further exploit induced disease resistance as an additional tool to control plant diseases within an integrated disease control strategy.

Introduction

Complete immunity has traditionally been the ultimate objective of breeding for disease resistance. This has rarely been achieved due to the difficulty of integrating high levels of disease resistance against a wide range of plant pathogens and pests together with other desirable traits such as high yield and quality. Furthermore, the constant "adaptation" of plant pathogens results in the steady degradation of the efficacy of introduced resistance genes. There is, therefore, a constant requirement to improve the efficacy of existing disease control measures and to develop new approaches with improved efficacy. The concept of plant "immunization" demands close scrutiny in this context, and appears to offer great potential, especially considering its recent origins, at least from an applied perspective. There are many reports of compounds inducing local or systemic resistance in plants in laboratory tests, but relatively little effort has been invested in understanding and overcoming the problems encountered when applying the concept to field environments.

Müller and Börger (1940) demonstrated that inoculation of potato tuber slices with an incompatible isolate of *Phytophthora infestans* was able to protect the tuber from subsequent infection by a compatible isolate of *P. infestans*. This concept of cross protection was later shown to be a general phenomenon applicable to a wide range of plant-pathogen interactions (Matta 1971). Subsequent research, aimed at discovering the molecules leading to this phenomenon of induced resistance, uncovered a wide array of compounds which influenced plant resistance. Thus there are many examples of biotic or abiotic compounds which induced phytoalexin accumulation in plant tissues. The term "elicitors" has been used to describe pathogen produced factors that initiate active defense systems (Keen 1993). However, in this chapter we have used the term "elicitor" to refer to a wider range of compounds which may be derived from a plant pathogen or other microorganism, from the plant, or may be synthetic, and which are able to induce resistance to subsequent inoculation with a pathogen, or which are able to induce some or many host responses normally associated with the resistant response.

While resistance may be induced by a single *avr* gene product in a resistant reaction, many host responses can also be induced by non-race-specific molecules released by the pathogen (Buell, this volume, Fig. 1). There are many different plant receptors which respond to a number of chemical stimuli from the pathogen. It is these non-specific elicitors which have the potential to be exploited by applying them to plants to induce responses and increase resistance to subsequent infection. The rate or efficiency with which these molecules can enter the plant cell, together with the number of receptors, the efficiency of the signal cascade, and the general physiology of the plant, will all have an impact on the speed and magnitude of the plants response. These observations have led to several suggestions that it may be possible to develop new forms of disease control through a form of "immunization" by spray application of chemical resistance elicitors (Sequeira 1984, Tuzun and Kuć 1991).

Other chapters in this book review fundamental aspects of resistance elicitors and their mode of action. We aim to assess the potential for the practical application of elicitors as crop protectants in agriculture and horticulture. We consider factors such as the cost of production of elicitors from natural compounds or by synthetic means, and their environmental acceptability.

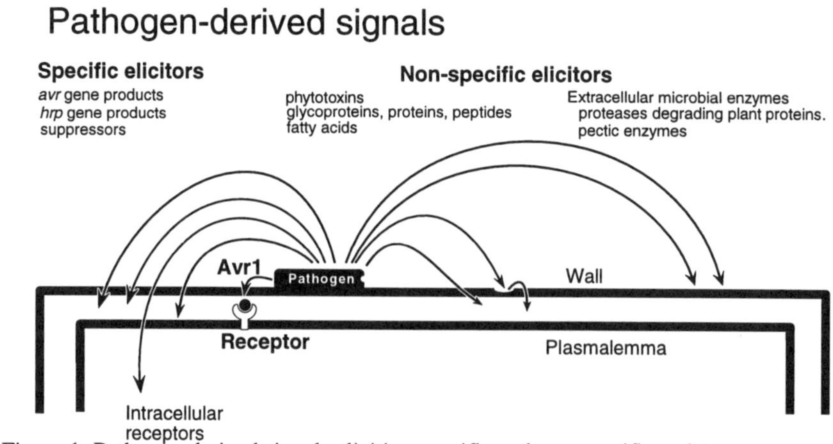

Figure 1: Pathogen-derived signals eliciting specific and non-specific resistance responses.

These factors should be considered within the context of an integrated crop management framework, which is particularly pertinent for resistance elicitors, which must not be seen as replacements to fungicides. Rather they should be considered as an additional option within an Integrated Crop Management (ICM) strategy to be used in conjunction with, for example, fungicides or biological control methods as appropriate.

Types of Elicitors

Various types of elicitor molecules have been reviewed previously (Lyon et al. 1995b, Lyon et al. 1996) and include biotic elicitors such as complex carbohydrates, salicylic acid, fatty acids, jasmonates, amino acids, ethylene, glycoproteins, yeast-derived elicitors, and microbial metabolites, as well as abiotic elicitors and semi-synthetic elicitors.

While many of these elicitors are highly effective at inducing components of plant resistance mechanisms under laboratory conditions, most are unlikely to form the basis of commercial products used for application to crops to control diseases. The reasons for this are numerous and varied: the level of disease control achieved by many elicitors is generally low compared to the levels which can be achieved using traditional antifungal fungicides, complex structures may be too expensive to manufacture, elicitors may not be taken up into plant cells, etc. Solutions exist to overcome some of these problems. Existing elicitors can be used as a model compound to indicate to organic chemists the general properties of more effective elicitors that can be synthesized. This has been the basis for the development of benzothiadiazole (BTH) that was based upon a modification of salicylic acid (see Tally et al., this volume). Complex carbohydrates will remain, in the foreseeable future, too complex to synthesize at a price which is competitive with less complex structures. Similarly, purification of complex carbohydrates from complex starting material (for example, a fermentation process) would again be too expensive to consider. Many protein or peptide elicitors have been described in the literature, e.g. syringolin from *Pseudomonas syringae* pv. *syringae* (Waspi et al. 1998), or elicitin from *Phytophthora infestans* (Kamoun et al. 1998), but unless such elicitors are considered in combination with novel application technologies such as linkage to penetratin, a class of peptides able to translocate hydrophilic compounds such as peptides across the plasma membrane (Derossi et al. 1998), they only have a potential application via plant genetic transformation. In addition, it is likely that commercial production of a peptide elicitor would be uncompetitively expensive in comparison with other forms of disease control techniques.

Extracts of spinach and rhubarb induced resistance to *Colletotrichum lagenarium*, with the active component being identified as oxalic acid (Doubrava et al. 1988). Application of oxalate to cucumber leaves induced systemic resistance to *C. lagenarium* (Doubrava et al. 1988). Dimethyl and diethyl esters of oxalic acid, malonate, formate, glycolate, acetate, phthalate, fumarate and maleate applied as 50mM solutions (adjusted to pH 6.0 with KOH) were inactive, whereas ascorbate and glyoxylate showed some activity. It was noted by the authors that of the compounds tested, oxalic acid is the most active at inducing resistance and is the strongest complexing agent for Ca^{2+}. The importance of Ca^{2+} in signal transduction is well recorded (Trewavas and Knight 1994).

Spraying the abaxial surface of leaf 1 of cucumber plants with 50mM K_2HPO_4, induced systemic resistance to *C. lagenarium* (Irving and Kuć 1990) 2 to 7 days later. Within 16 h of application peroxidase and chitinase activity increased in the treated leaf and they continued to increase over the next 7 days. A single phosphate spray of 0.1 M solution induced a systemic protection against powdery mildew in cucumber caused by *Sphaerotheca fuliginea* and against common rust in maize caused by *Puccinia sorghi*, and northern leaf blight (NLB) caused by *Exserohilum turcicum* (Reuveni and Reuveni 1998). This systemic protection against powdery mildew in cucumber and common rust or NLB in maize was obtained on upper leaves after NPK fertilizer application to the lower leaves.

Root pre-treatment of rice plants with monoepoxy linoleic and linolenic acids induces resistance of leaves to *Pyricularia oryzae*, the imperfect form of *Magnaporthe grisea* (Namai et al. 1993). A topical spray of cutin monomers on rice and barley leaves enhanced resistance to *M. grisea* and *Blumeria* (*Erysiphe*) *graminis* respectively although some genes normally associated with plant defense were not activated (Schweizer et al. 1994).

Specificity of Elicitors

There is some evidence that there are differences in the response of plant cultivars to elicitors (see Agrawal, this volume). For example, Tepper et al. (1989) showed that a galacto-glucomannan isolated from *C. lindemuthianum* was cultivar specific. The elicitor induced phytoalexin accumulation in a bean cultivar resistant to the pathogen and did not induce phytoalexins in the susceptible cultivar. Keen et al. (1983) purified a glucomannan from *P. megasperma* f. sp. *glycinea* which induced less phytoalexin in the compatible cultivar than in the incompatible cultivar. Basidiospore germlings of race 1 of the cowpea rust fungus (*Uromyces vignae*) secrete a cultivar-specific elicitor of plant necrosis (Chen and Heath 1990). Parker et al. (1988) isolated a glycoprotein from *P. megasperma* f. sp. *glycinea* which elicited phytoalexins in parsley (non host) but not in soybean (host) although after pronase treatment the elicitor elicited phytoalexins in soybean and not in parsley. Additionally, deglycosylated protein elicited in parsley but not in soybean. In total, these experiments suggest that there are different structural components of the fungal cell wall eliciting defense responses in different plants, that some plant species may respond differently to different elicitors, and that some cultivars respond differently than others.

Elicitors can also be used to up-regulate genes controlled by elicitor-responsive elements in transgenic plants (Gatz and Lenk 1990). The best example is the use of the *PR-1a* promoter which is responsive to BTH. The PR-1a promoter has already been used to regulate the expression of the *Bacillus thuringiensis* endotoxin in transgenic plants. Many other biotic stress induced genes are known, and some are induced by both biotic and abiotic stress, indicating many common pathways in stress response. Expression of these genes in response to pathogen challenge has been studied much in recent years, and many genes are now being expressed in transgenic plants with a view to providing broad spectrum disease resistance (Oldroyd and Staskawicz 1998). Genes expressed in response to elicitors will give a valuable insight into the mechanisms responsible for effective resistance expression, and can be used in discovery programs (see below).

Elicitors are assumed to induce a general resistance response. However, elicitors are often apparently much more effective against some pathogens than others. This may be only apparent due to the different epidemiology of different pathogens or differences in the crop protection delivery program. Different elicitors presumably trigger different response genes but whether the nature of the resistance mechanism induced is different is largely unknown. One great advantage elicitors have is the potential to control viruses. There are few specific anti-viral compounds for use even in medicine, and none for use against plant diseases, so it is not surprising that, like animal systems, the most effective treatment is through stimulation of the plants own defenses. The analogy of "immunization" is perhaps most appropriate here.

While most elicitors work to some degree on most plants, some are far more effective on certain plants. For example, glucans extracted from fungal mycelium induce very little resistance-related responses in potato whereas DL-β-amino-n-butyric acid (BABA) does induce resistance responses in tomato (Cohen 1994, Cohen et al. 1994) and potato (Miller 1997) against *P. infestans*. The duration of induced resistance tends to be different in dicotyledonous plants compared with monocotyledons. As the genetic and physiological differences between these two groups are likely to be far greater than within, this is not unexpected.

There is a limited amount of evidence that there can be an additive or even a synergistic effect of using different elicitors in combination. Preisig and Kuć (1985) showed that glucans from *P. infestans*, inactive themselves as elicitors of the hypersensitive response on potato, enhanced sesquiterpene phytoalexin accumulation elicited by arachidonic acid. This suggests that while there are some differences in the signal transduction cascades from different elicitors, there must also be some interacting components. Further analysis of such signaling cascades offers further potential to increase the responses of plants to resistance elicitors.

Some elicitors act systemically while others induce local responses. For instance, systemic responses are induced by DL-3-amino butyric acids in tomato (Cohen and Gisi 1994), jasmonates in potato and tomato (Cohen et al. 1993), arachidonic acid in potato (Coquoz et al. 1995), while others, such as yeast-derived elicitors, induce little if any systemic response (Lyon and Newton, unpublished).

Elicitor Discovery Programs

Standard fungicide screening programs are designed to detect substances with high levels of direct antimicrobial activity. The crude formulation may not be effective for delivering elicitor molecules to plant receptor sites, and the environmental conditions of the tests may not be conducive to resistance induction processes. An elicitor molecule may undergo some chemical modification as it enters plant cells as the active molecule(s) may not be the compounds applied. Modification rates will be dependent upon time and environmental factors. While it is reasonable to expect such screens to detect a degree of pathogen control, it is likely that this will fall below acceptable thresholds for "conventional" crop protection chemicals.

As resistance elicitors act by affecting plant resistance expression, not only will the responses be cultivar specific but other factors should also be taken into account when assessing genetic resistance in breeding programs. For example, resistance expression may be growth stage specific (seedlings vs. mature plant) or tissue specific (leaves vs. roots). Thus, different cultivars might express resistance at different levels and times. We might also expect resistance elicitors to be more effective against some pathogens than others, and this too will be cultivar or host specific. Clearly, all the vagaries of the environment, host genetics and physiology, together with their interactions, are compounded in resistance elicitor expression. In the field we must also consider the effects of other pathogens and saprophytes on leaves as these can induce either resistance or susceptibility / accessibility in plants. Thus, the status of the epidemic of these other pathogens must be considered too.

The amount of inducible resistance is also dependent upon the variability of the environment (Falkholf et al. 1988). This makes sense when pathogen attack is seen as biotic stress, just one of the potential inducers of stress responses in plants, the major group being abiotic stress. Stress is an essential and normal component of the developmental program of plants and a "normal" pattern of seasonal and diurnal stress is required for "normal" plant response mechanisms to be fully functional. While "abnormal" conditions can be tolerated for conventional fungicide screens, such conditions may give misleading data when screening for resistance elicitors.

Sources of Elicitors

While phytoalexins themselves have never attracted serious attention as possible fungicides, suggestions have been made that they may serve as models from which to synthesize structurally related analogues (Carter et al. 1978, Arnoldi et al. 1986). Similarly, naturally occurring elicitors are being considered as starting points for developing commercially useful elicitors. Acceleration of the plant's response to infection by pathogens can be achieved under laboratory conditions by application of resistance elicitors derived from plant pathogens. For example, glucans from fungi such as *P. megasperma* var. *sojae* were tested for their ability to induce resistance but would have been rejected as too expensive to synthesize commercial quantities. However, other microorganisms can give similarly effective elicitor-active extracts such as yeast (*Saccharomyces cerevisiae*), a *Mortierella* sp., and a *Mucor* sp. (Newton et al. 1995, Reglinski et al. 1994b). Microorganisms have thus provided a source of potentially novel resistance elicitors from which future commercial products could be derived. Complex molecules are always going to be too expensive to develop for agriculturally-related use although an exception could be the isolation of complex molecules from a fermentation.

Various plant extracts from neem (Jayaraj and Rabindra 1993) to compost have been used to control pathogens. The only one to be commercialized in any significant way is a giant knotweed (*Reynoutria sachalinenis*) extract sold as "Milsana^{TM}" by BASF Corporation. This is claimed to give good protection against powdery mildews in particular (Daayf et al. 1995, 1998).

Many culture filtrates have been recorded as controlling disease in laboratory experiments but the composition of the filtrates is probably both very complex and variable, containing both anti-microbial substances and elicitors.

Filtrates from *Bacillus subtilis* cultures have been shown to control disease under field conditions (Schönbeck et al. 1980, 1982, Schönbeck and Dehne 1986).

Fatty acids such arachidonic and eicosapentaenoic acids are very effective inducers of systemic resistance in potatoes. Fatty acids such as linoleic, linolenic and oleic acids (Cohen et al. 1991) are also able to induce resistance in potatoes, suggesting that there is further potential to develop related compounds for use as elicitors. Interestingly, volicitin, a component of the oral secretion of beet army worms which contains a molecule of linolenic acid, induces corn seedlings to emit volatiles which attract parasitic wasps (Alborn et al. 1997). Thus, fatty acids may also have potential applications in modifying plant/pest interactions. Salicylic acid, an inducer of systemic acquired resistance (SAR), was used as a starting structure from which to develop Bion™ (see Tally et al., this volume).

Strobilurin was isolated from the pine cone fungus (*Strobilurus tenacellus*) and is an antifungal compound used by the fungus to prevent growth of other fungi on infected pine cones (Anke et al. 1977). While not an elicitor it illustrates some important points. The natural/original compound showed only weak activity in glasshouse tests but using the natural strobilurin as model, over 10,000 new compounds were screened for activity from which kresoxim-methyl was selected as the most useful (Ypema and Gold 1999).

Formulation, Application and Registration

The efficacy of any agrochemical can be substantially affected by its formulation (Holloway et al. 1994). In recent years the growth of the adjuvants market is evidence of this. By comparison with conventional fungicides, the formulation problems associated with delivery of elicitors to their receptors are far greater and more complex. Elicitors such as heptaglucans have a highly specific structure and the length of the polymers containing these structures is not only important but also unstable, affected by excreted enzymes from both the plant, pathogen and other surface microbes. The order of magnitude of the elicited resistance response will vary, depending upon the many environmental factors which affect plant metabolic processes, and thus the duration of the resistance effect will also vary.

The activity of resistance elicitors can be considerably increased by the addition of non-eliciting components such as sodium carbonate in both soybean cotyledon phytoalexin assays and against mildew on detached barley leaves (Lyon et al. 1995a). Addition of sodium chloride and calcium chloride in soybean cotyledon assays also boosted phytoalexin production, and while some of this effect could be related to pH changes, it may have also affected receptor efficiency and uptake of the elicitor. Adjuvants such as Agral and LI-700 have been shown to increase the efficacy of yeast-derived elicitors in controlling mildew on barley (Reglinski et al. 1994a). These formulation components and other attributes of adjuvants are very important for assisting the entry of the bioactive molecules into the plant cells by aiding sticking (e.g., rainfastness) and spreading around the tissue surfaces. Adjuvants may also have a direct antimicrobial activity.

Another trait common to elicitors which distinguishes them from fungicides is the lack of a clear dose-response curve. Elicitors are often shown to have very high response from low levels and may or may not show increased response with increased dose. This may be attributable to saturation of receptors, feedback responses, or in the case of complex extracts, the presence of inhibitors or pathogen stimulants. All these factors will influence optimum formulation, dose and timing. There are likely to be more factors which need critical tailoring to individual crops than for fungicides, and thus the importance of decision support systems increases if resistance elicitors are to be successful in practice.

Agrochemical registration systems are geared to conventional chemicals. Some of the criteria to be fulfilled may appear inappropriate, for example yeast-derived elicitors are non-toxic and complex and therefore residue analyses are difficult to conceive. However, the problems are more likely to be with reliable efficacy data as registration authorities will consider reasoned scientific argument for all other criteria. Authorities are aware of the need for such crop protectants as restrictions on the use of toxic products increase and both the public and farming community demand effective, environmentally benign alternatives. It costs around US$160 million to develop a conventional marketable crop protection product. A significant proportion of this is spent after the active ingredient is defined, on formulation as a powder, emulsion, suspension or granular product.

Field Application

Many resistance elicitors have been shown to control plant pathogens by spray application under laboratory conditions, including oxalates (Doubrava et al. 1988), phosphates (Gottstein and Kuć 1989) or ethylenediaminetetra-acetic acid (EDTA) (Walters and Murray 1992). Many elicitors have not been tested under field conditions, but when they have the level of disease control is less than under laboratory conditions or has been considered "poor" (Schönbeck et al. 1982). Such elicitors are, therefore, unlikely to form the basis of commercial products without a substantial improvement in formulation. However, there are a small number of exceptions which do show good disease control in the field. For example, Novartis Crop Protection has developed CGA 245704 (BTH) as a "Plant Activator" and is marketed in some countries under the trade name Bion™. CGA 245704 was obtained through screening compounds structurally-related to salicylic acid and is able to induce resistance in a wide range of plants (Ruess 1998, Tally et al., this volume). Ruess (1998) reported that CGA 245704 induces resistance in all crop plants so far tested, although the degree of induction was variable regarding spectrum, level and durability of resistance. The type of crop and even the variety are, therefore, important parameters affecting the level of induced resistance. In wheat, rates of 30 g ai/ha of CGA 245704 increase disease resistance for a period of approximately 50 days. In general, resistance activation lasts longer in monocots than in dicots, where applications have to be repeated at 7-14 day intervals. Virus protection of potato was improved by weekly applications of 30 g ai/ha of CGA 245704 with good control of PVY and PVS. On tobacco, rates of 12-25 g ai/ha induce resistance against fungi (e.g., *Peronospora tabacina, Rhizoctonia* sp.) bacteria (e.g. *Pseudomonas* sp.) and virus (e.g. PVY).

Bion™ applied as a spray treatment to wheat at growth stages 25, 28 and 32 significantly reduced the severity of powdery mildew (*Blumeria graminis* f. sp. *tritici,* syn. *Erysiphe graminis* f. sp. *tritici*) and to a lesser extent *Septoria tritici* in later growth stages (Buchenauer et al. 1998). CGA 245704 is also being tested (Paulin et al. 1998) against other plant pathogens such as fire blight of apple caused by *Erwinia amylovora.* While encouraging results have been obtained using apple seedlings in the greenhouse, further work is required to be done in orchards and a spray schedule specific to fire blight needs to be proposed (Paulin et al. 1998).

BASF Corporation"s "Milsana™" has been used predominantly on ornamentals and glasshouse crops against powdery mildews (Daayf et al. 1995, Daayf et al 1998, Herger and Klingauf 1990), but it has been shown to have some effect against *S. tritici* in wheat (Metcalf and Wale 1997) and *Botrytis cinerea* on pepper plants (Schmitt et al. 1996).

In a field study, 412 g/ha methyl jasmonate controlled powdery mildew on barley, and although it reduced plant height, grain yield was significantly increased (Mitchell and Walters 1995).

In New Zealand, Reglinski (Personal communication) has been investigating the potential of elicitors to control powdery mildew and *Botrytis cinerea* on grapes. In potted vine experiments, elicitors controlled powdery mildew on Mendoa Chardonnay as effectively as the sulphur fungicide "Super Six". Further work indicated that good *Botrytis* control was achieved with a combination of elicitor and biocontrol agent.

Application of fungal-derived elicitors to cell cultures of trees has demonstrated that resistance-related responses can be induced (Yang et al. 1989, Alami et al. 1998). Reglinski et al. (1998) demonstrated that phenylalanine ammonia-lyase activity increased rapidly in *Pinus radiata* seedlings after treatment with 2mM 5-chlorosalicylic acid (5CSA). Seedlings treated with 5CSA showed an increased resistance to infection by *Sphaeropsis sapinea.* Although 5CSA has some direct antifungal activity the increased resistance of the seedlings was considered to be due, in part, to the involvement of inducible host resistance mechanisms. There is clearly a large potential market for controlling diseases of trees but there is no published information on the efficacy of elicitor application in field trials.

Durability of Induced Resistance

It is argued that induced resistance should be durable in the same way as the *mlo* resistance gene in barley against mildew. The *mlo* gene is not a recognition gene and so the pathogen cannot simply change the gene product being recognized by mutation in the avirulence gene (Büschges et al. 1997). However, it apparently works in a similar way to elicitor treated plants by speeding up recognition of pathogens; the resistance mechanisms could be regarded as partially de-suppressed. Recognition must be taking place and there is some evidence of specificity even against *mlo*. The specificity is clearly not against the *mlo* gene itself as even in mildew isolates with some specificity a high level of resistance is still expressed. Therefore, it may be that the *mlo* gene is responding to signals from multiple recognition genes and that isolates with a degree of virulence are simply not recognized by some of the recognition genes resulting in a quantitative but specific virulence. We could envisage similar durability of elicitor treated plants such that only under very extensive use of elicitors might some low level of specificity towards them occur. Even then, if different elicitors do trigger different genes then durability may be even greater when combinations are used.

Combined Application of Elicitors with Fungicides

Legislation is increasingly restricting overall permitted levels of fungicide active ingredient usage, and in particular use near residential areas or water supplies. While elicitors cannot necessarily be used as replacements for fungicides, as they work in different ways, and have no erradicative properties, they may enable fungicides to retain high efficacy at much reduced doses. Low levels of fungicide active ingredient may also help avoid build up of fungicide resistance depending upon the genetic basis of the resistance. It is unlikely to be less costly even if the elicitor is less expensive to buy, as much of the cost is in the spray operation.

The compound 2,2-dichloro-3,3-dimethyl cyclopropane carboxylic acid, originally isolated as a possible fungicide, was later shown to exert its antifungal activity on rice by "sensitizing" the host such that subsequent inoculation with the rice blast fungus resulted in a resistant reaction (Cartwright et al. 1977). Later, Cahill and Ward (1989) suggested that in infected plants treated with fungicides such as metalaxyl elicitors released from fungal hyphae may stimulate host-defense responses. Interestingly, recent work on NahG and *nim1* mutants of *Arabidopsis thaliana,* which exhibit increased susceptibility to pathogens through an impairment of the induced resistance response, demonstrated that the fungicides metalaxyl, fosetyl and copper hydroxide are much less active on these plants than on normal ones, and fail to control *Peronospora parasitica* (Molina et al. 1998). Not only do such experiments provide evidence on the molecular basis of signal transduction pathways in plants but also confirms that the "normal" level of effectiveness of fungicides could include a component due to induced resistance.

Yield and Quality Implications

It has often been assumed that resistance elicitor applications may be toxic and lead to reduced yield because of the energetic costs of resistance induction. Such costs have been demonstrated (Smedegaard-Petersen and Stølen 1981). However, this has not been borne out in practice, and Reglinski et al (1994b) demonstrated that resistance can be induced by elicitors without causing symptoms of phytotoxicity (see Tally et al., this volume). In addition, there is little, if any, evidence of a yield penalty with their use and yield increases have sometimes been noted. Jasmonate induced resistance to herbivores was not costly for field-grown tomato plants (Thaler 1999, this volume). In field experiments, treatment of winter barley with an inducer of resistance produced by a *Bacillus subtilis* strain led to increased yield after infection with *Blumeria graminis* f.sp. *hordei*, which could not be accounted for by the reduction in disease severity alone (Kehlenbeck et al. 1994, Kehlenbeck and Schönbeck 1995). Such results demonstrated that induced resistance and plant cell death are not interdependent and suggested that commercial application of resistance elicitors to control plant diseases would be possible. Novartis Crop Protection refer to their resistance elicitor, Bion™, as a "plant activator". The term suggests an important concept which can be overlooked by pathologists concentrating on disease control alone. As resistance induction is effectively stimulation of metabolic pathways which are energetically demanding, other cellular processes may also be stimulated or activated. There may be many compensatory metabolic processes which are intended to protect the plants reproductive capability which often represents the agriculturally important yield. Some elicitors may be regarded in part as inducers of disease tolerance.

The yield stimulation effect is most obvious when disease is not well controlled in field trials, yet yields as good as the fungicide control are achieved. However, in economic terms, the quality is often at least as important as the overall yield. Whether through the effects of resistance elicitation, yield metabolic compensation, or the effects of residual pathogen development, the yield can be boosted but quality may be differentially affected. For example, in a winter barley trial, the best elicitor treatments were found to give yield as good as the fungicide, but the thousand grain weight was the same as the untreated control (Reglinski et al. 1994b). Total seed production in plants not selected for seed yield can be compromised. For example, treatment of *Nicotiana attenuata* with methyl jasmonate reduced the number of seed produced (Baldwin 1998, see also Agrawal, Tally et al., Thaler, this volume).

Yield may also be the vegetative parts of the plant. If elicitors induce too much lignification either themselves, or upon subsequent pathogen attack, the digestibility of the vegetation could be affected. Treatments which alleviate the consequences of biotic stresses such as pathogen attack are often correlated with alleviating other stresses such as cold, drought and salt stress. There has been little documented evidence of this yet, but this is probably due to lack of readily available products to test experimentally.

While there has been some concern that induced resistance may affect quality or even safety of food crops there is no evidence that toxic metabolites accumulate in elicitor-treated plants (Lyon and Newton 1997).

Decision Support Systems

Elicitors are a tool for use in high management input systems. Risk-averse management systems rely upon routine, full-dose pesticide applications. They are only competitive with more sophisticated systems if there are large savings in management time costs to compensate. As prices and profit margins for agricultural commodities continue to drop, the need for sophisticated management systems to optimize inputs such as those that utilize decision-support systems, is ever increasing. Some of the controllable factors that can be manipulated in such a system include the choice of cultivar, whether fungicides are to be used in conjunction with elicitor applications - whether as mixtures or alternating sprays, with reduced rate applications of fungicides, or alone (perhaps for an organic market), and factors affecting timing of spray application (Fig. 2). Other uncontrollable inputs affect these decisions, but forecasting can improve decision reliability. Under protected crop conditions where more environmental variables are controllable the efficacy of elicitors may be considerably enhanced and thus affect decisions. Elicitors which do not have any antimicrobial activity should be compatible with application of biocontrol agents. This could be especially useful in environments, such as glasshouses, where biocontrol predominates as the preferred form of pest control.

Non-toxic elicitors may have an overriding advantage over fungicides when the yield component is the same as that part of the plant which needs protecting e.g., lettuce leaves. Other factors such as operator hazard and use with biocontrol agents may give similar overriding advantages. In these instances elicitors are potential alternatives to fungicides. There are still many environmental factors which we believe can influence the level of resistance expression and which have not yet been fully appreciated and characterized. For example, light quality is known to influence resistance. Islam et al. (1998) showed that failure of appressorial penetration of *Botrytis cinerea* on broad bean leaves under yellow or red light (575nm to 700nm) was due to light-activated resistance. Similarly, it has been demonstrated that stress (nutrient or water) can also influence *mlo*-based resistance of barley to mildew (Newton and Young 1996). Thus there are likely to be many environmental factors which could have subtle influences on natural and induced resistance. Management support decisions therefore have the potential to become very sophisticated when the

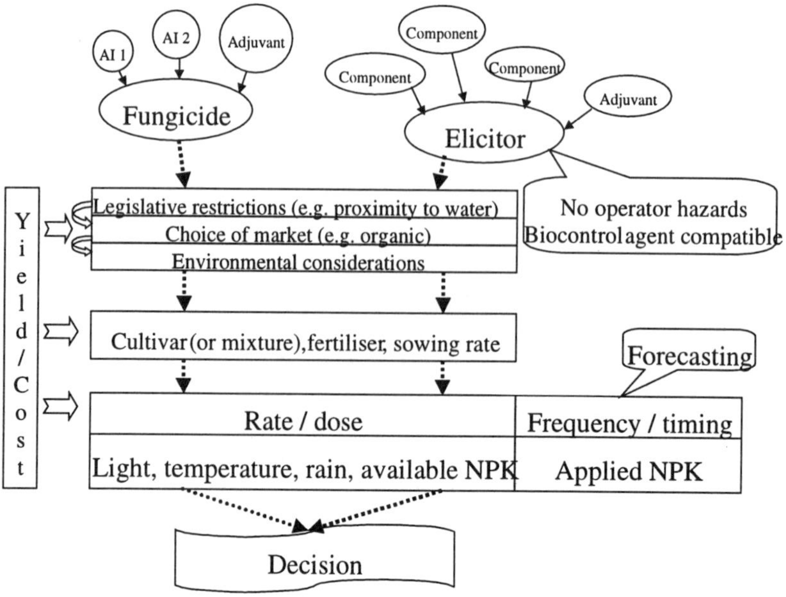

Figure 2: Factors affecting the choice and use of elicitors and / or fungicide applications to crops within an integrated crop management system. The complex nature of these decisions, particularly the loss of direct relationship between disease control and yield loss when using elicitors, dictates that development of decision support systems based on these relationships is necessary for the correct exploitation of elicitor implementation.

nature of inducible resistance is fully understood. In contrast, there are circumstances where application or elicitors could be counter-productive. For example, cultivar or species mixtures rely to a degree upon the effects of induced resistance from avirulent or non-host pathogens (Newton et al. 1997). This component of their efficacy may be simply duplicated by resistance elicitors. It is important that elicitors are applied in ways where they will achieve efficacy comparable with conventional crop protectants. Given the potential problems achieving this, it is essential that they are managed by those who understand their mode of action and therefore make informed decisions about not only how to use them, but also whether to use them in any given circumstances. In most circumstances, the use of a decision support system will be a necessity. Failure to treat elicitors in this way will result in inappropriate use and poor efficacy. Resistance elicitors may then be regarded as unreliable and grouped with "unorthodox" or "fringe" disease control concepts.

Plant Breeding

It has been noted above that cultivars respond differentially to resistance elicitors (See Agrawal, this volume). Implicit in this is that cultivars could be bred for increased response to elicitors. However, selection for greater induction of resistance mechanisms in the absence of pathogens could be detrimental to the plant, or even over expression following subsequent pathogen recognition. In general this has not been the case with current cultivars, and the apparent reduction in yield associated with expression of the analogous *Mlo* resistance to powdery mildew in barley was found to be associated with closely linked heading date and grains per ear loci (Thomas et al. 1998). Thus, there may be considerable as yet unexploited genetic potential for enhancing resistance which could be exploited through resistance elicitors in both spray application and as tools in genetical research.

Synthesis

Crop protection strategies are partly driven by scientific research through the discovery of new forms of disease resistance, or new agrochemicals, but it is increasingly industry driven through the food industry making demands on the quality of food and what type of residues are acceptable. The concept of induced resistance through an "immunization" process would appear, to us, to be particularly compatible with both Integrated Crop Management (ICM) aimed at better disease control and public concerns regarding the safety and quality of food in general.

Synthetic compounds or microbial extracts have been shown to stimulate a number of resistance mechanisms known to be up-regulated after infection by incompatible plant pathogens, and, in turn, these compounds, with varying degrees of efficacy, have been shown to increase resistance to subsequent inoculation by plant pathogens. It is clear that it is not sufficient to simply identify a compound able to enhance resistance and expect that such a compound may be a suitable candidate for a commercial product. There are many other factors to take into consideration. These include formulation, and how the product can be used within an ICM system. There will be many opportunities for exploitation of such a concept but some will be more commercially viable than others, due, either to increased disease control, or to other factors such as reduced pesticide residues.

Future Directions

Through a search for a comprehensive understanding of the molecular basis of disease resistance, new opportunities for exploiting that knowledge will become apparent. New biotic elicitors (derived from pathogens or initially classified as signal transduction molecules within plants) will be tested. Some will be potentially usable in their own right, but others may form a basis for synthesizing structural analogs for commercial exploitation. Some of these molecules may have wide applicability, while others may be suitable for certain plants or for the control of a narrow range of pathogens. It is still too early to know whether induced resistance is going to be more efficient against biotrophs or necrotrophs.

Control of plant diseases through the application of elicitors will undoubtedly become another component of integrated disease management strategies enabling better control of plant diseases to occur through reduced rate application of fungicides, or better control through application of existing fungicides. Additionally they may preserve the useful life of a fungicide by extending the period before pathogens develop intolerance to the active ingredient of the fungicide.

The future for resistance elicitors, as with so many interactions with variable plant genetic responses, is inextricably linked with biotechnological developments. Understanding and manipulating the plant's receptors will be the key to enhancing and targeting elicitor responses. This approach can be extended to any developmental- or recognition-based genetic pathway. Thus elicitors could be used to switch on pathways in order to phase or optimize timing of end product production. Optimization factors might be market timing or targeting for energy or raw material quantitative or qualitative availability and utilization. The acquisition of elicitor-responsive transcription factors for such pathways will enable genetic transformation with genes which could be activated by simple spray application of elicitors (gene switching). This has been suggested for initiation of traits such as flowering (Caddick et al. 1998, Gatz and Lenk 1998).

Acknowledgements

We are grateful for financial support from the Scottish Office of Agriculture, Environment and Fisheries Department.

Literature Cited

Alami, I., S. Mari, and A. Clerivet. 1998. A glycoprotein from *Ceratocystis fimbriata* f. sp. *platani* triggers phytoalexin synthesis in *Platanus* x *acerifolia* cell-suspension cultures. Phytochemistry 48:771-776.

Alborn H.T., T.C.J. Turlings, T.H. Jones, G. Stenhagen, J.H. Loughrin, and J.H. Tumlinson. 1997. An elicitor of plant volatiles from beet armyworm oral secretion. Science 276:945-949.

Anke, T., F. Oberwinkler, W. Steglich, and G. Schramm, 1977. The strobilurins - new antifungal antibiotics from the basidiomycete *Strobilurus tenacellus* (Pers. Ex Fr.) Sing. Journal of Antibiotics 30:806-810.

Arnoldi, A., G. Farina, R. Galli, L. Merlini, and M.G. Parrino. 1986. Analogs of phytoalexins. Synthesis of some 3-phenylcoumarins and their fungicidal activity. Journal of Agricultural and Food Chemistry 34:185-188.

Baldwin I.T. 1998. Jasmonate-induced responses are costly but benefit plants under attack in native populations. Proceedings of the National Academy of Sciences USA 95:8113-8118.

Buchenauer, H., M. Stadnik, J. Chamsai, Y. Ch. Jeun, M. Orober, J. Siegrist, and G. Anfoka, 1998. Induction of resistance in different crops against fungal and viral diseases. SCI Conference on Systemic Acquired Resistance, 10th March 1998, London. Abstract.

Büschges, R., K. Hollricher, R. Panstruga, G. Simons, M. Wolter, A. Frijters, R. van Daelen, T. van der Lee, P. Diergaarde, J. Groenendijk, S. Töpsch, P. Vos, F. Salamini, and P. Schulze-Lefert. 1997. The barley *Mlo* gene: a novel control element of plant pathogen resistance. Cell 88:695-705.

Caddick, M.X., A.J. Greenland, I. Jepson, K-P. Krause, N. Qu, K.V. Riddell, M.G. Salter, W. Schuch, U. Sonnewald, and A.B. Tomsett. 1998. An ethanol inducible gene switch for plants used to manipulate carbon metabolism. Nature Biotechnology 16:177-180.

Cahill D.M., and E.W.B. Ward. 1989. Effects of metalaxyl on elicitor activity, stimulation of glyceollin production and growth of sensitive and tolerant isolates of *Phytophthora megasperma* f.sp. *glycinea*. Physiological and Molecular Plant Pathology 35:97-112.

Carter, G.A., K. Chamberlain, and R.L. Wain. 1978. Investigations on fungicides. XX. The fungitoxicity of analogues of the phytoalexin 2-(2'-methoxy-4'-hydroxyphenyl)-6-methoxy-benzofuran (vignafuran). Annals of Applied Biology 88:57-64.

Cartwright D., P. Langcake, R.J. Pryce, D.P. Leworthy, and J.P. Ride. 1977. Chemical activation of host defense mechanisms as a basis for crop protection. Nature 267:511-513.

Chen, C.Y., and M.C. Heath. 1990. Cultivar-specific induction of necrosis by exudates from basidiospore germlings of the cowpea rust fungus. Physiological and Molecular Plant Pathology 37:169-177.

Cohen, Y. 1994. 3-aminobutyric acid induces systemic resistance against *Peronospora tabacina*. Physiological and Molecular Plant Pathology 44:273-288.

Cohen, Y., and U. Gisi. 1994. Systemic translocation of ^{14}C-DL-3-aminobutyric acid in tomato plants in relation to induced resistance against *Phytophthora infestans*. Physiological and Molecular Plant Pathology 45:441-456.

Cohen, Y., U. Gisi, and E. Mosinger. 1991. Systemic resistance of potato plants against *Phytophthora infestans* induced by unsaturated fatty acids. Physiological and Molecular Plant Pathology 38:255-263

Cohen, Y., U. Gisi, and T. Niderman. 1993. Local and systemic protection against *Phytophthora infestans* induced in potato and tomato plants by jasmonic acid and jasmonic methyl ester. Phytopathology 83:1054-1062.

Cohen, Y., T. Niderman, E. Mösinger, and R. Fluhr. 1994. 8-aminobutyric acid induces the accumulation of pathogenesis-related proteins in tomato (*Lycopersicon esculentum* L.) plants and resistance to late blight infection caused by *Phytophthora infestans*. Plant Physiology 104:59-66.

Coquoz, J.-L., A.J. Buchala, Ph. Meuwly, and J.-P. Métraux. 1995. Arachidonic acid induces local but not systemic synthesis of salicylic acid and confers systemic resistance in potato plants to *Phytophthora infestans* and *Alternaria solani*. Phytopathology 85:1219-1224.

Daayf, F., A. Schmitt, and R.R. Belanger. 1995. The effects of plant extracts of *Reynoutria sachalensis* on powdery mildew development and leaf physiology of long English cucumber. Plant Disease 79:577-580.

Daayf, F., A. Schmitt, and R.R. Belanger. 1998. Evidence of phytoalexins in cucumber leaves infected with powdery mildew following treatment with leaf extracts of *Reynoutria sachalensis*. Plant Physiology 113:719-727.

Derossi, D., G. Chassaing, and A. Prochiantz. 1998. Trojan peptides: the penetratin system for intracellular delivery. Trends in Cell Biology 8:84-87.

Doubrava, N.S., R.A. Dean, and J. Kuć. 1988. Induction of systemic resistance to anthracnose caused by *Colletotrichum lagenarium* in cucumber by oxalate and extracts from spinach and rhubarb leaves. Physiological and Molecular Plant Pathology 33:69-79.

Falkholf, A.G., H.W. Dehne, F. Schönbeck. 1988. Dependence of the effectiveness of induced resistance on experimental conditions. Journal of Phytopathology 123:311-321.

Gatz C., and I. Lenk 1998. Promoters that respond to chemical inducers. Trends in Plant Science 3:352-358.

Gottstein, H.D., and J. Kuć. 1989. Induction of systemic resistance to anthracnose in cucumber by phosphates. Phytopathology 79:176-179.

Herger, G., and F. Klingauf. 1990. Control of powdery mildew fungi with extracts of the giant knotweed (*Reynoutria sachalinensis*). Mededelingen van het Faculteit Landbouwwetenschappen Universiteit Gent 55:1007-1014.

Holloway, P.J., R.T. Rees, and D. Stock, editors. 1994. Interactions between adjuvants, agrochemicals and target organisms. Ernst Schering Research Foundation Workshop 12. Springer-Verlag, Berlin.

Irving, H.R., and J.A. Kuć. 1990. Local and systemic induction of peroxidase,chitinase and resistance in cucumber plants by K_2HPO_4. Physiological and Molecular Plant Pathology 37:355-366.

Islam, S.Z., Y. Honda, and S. Arase. 1998. Light-induced resistance of broad bean against *Botrytis cinerea*. Journal of Phytopathology 146:479-485.

Jayaraj, S., and R.J. Rabindra. 1993. The local view on the role of plant protection in sustainable agriculture in India. Ciba Foundation Symposia 177:168-180.

Kamoun, S., P. van West, V.G. Vleeshouwers, K.E. de Groot, and F. Govers. 1998. Resistance of *Nicotiana benthamiana* to *Phytopthora infestans* is mediated by the recognition of the elicitor protein INF1. Plant Cell 10:1413-1426.

Keen N.T. 1993. An overview of active disease defense in plants. Pages 3-11 *in* B. Fritig and M. Legrand, editors. Mechanisms of Plant Defense Responses. Kluwer Academic Publishers, Dordrecht, The Netherlands.

Keen N.T., M.Yoshikawa and M.C. Wang. 1983. Phytoalexin elicitor activity of carbohydrates from *Phytophthora megasperma* f. sp. *glycinea* and other species. Plant Physiology 71:466-471.

Kehlenbeck, H., C. Krone, E.C. Oerke, and F. Schönbeck.1994. The effectiveness of induced resistance on yield of mildewed barley. Zeitschrift-für-Pflanzenkrankheiten-und-Pflanzenschutz 101:11-21.

Kehlenbeck, H. and F. Schönbeck. 1995. Effects of induced resistance on disease severity/yield relations in mildewed barley. Journal of Phytopathology 143:561-567.

Lyon, G.D., and A.C. Newton. 1997. Do resistance elicitors offer new opportunities in integrated disease control strategies? Plant Pathology 46:636-641.

Lyon, G.D., R.S. Forrest, and A.C. Newton. 1996. SAR - The potential to immunize plants against infection. Brighton Crop Protection Conference 18-21 November 1996. 939-946.

Lyon, G.D., T. Reglinski, R.S. Forrest, and A.C. Newton. 1995a. The use of resistance elicitors to control plant diseases. Aspects of Applied Biology. Physiological responses of plants to pathogens 42:227-234.

Lyon, G.D., T. Reglinski, and A.C. Newton. 1995b. Novel disease control compounds: the potential to "immunize" plants against infection. Plant Pathology 44:407-427.

Matta, A. 1971. Microbial penetration and immunization of uncongenial host plants. Annual Review of Phytopathology 9:387-410.

Metcalfe, R.J. and S.J. Wale. 1997. Evaluation of Milsana for the control of *Septoria tritici* in wheat. Annals of Applied Biology 130 (supplement). Test of Agrochemicals and Cultivars 18:52-53.

Miller, S.K. 1997. Assessment of the potential to control potato diseases by the application of resistance elicitors. PhD thesis, University of Dundee, Scotland, UK.

Mitchell, A.F., and D.R. Walters. 1995. Systemic protection in barley against powdery mildew infection using methyl jasmonate. Aspects of Applied Biology. Physiological responses of plants to pathogens 42:323-326.

Molina, A., M.D. Hunt, and J.A. Ryals. 1998. Impaired fungicide activity in plants blocked in disease resistance signal transduction. Plant Cell 10:1903-1914.

Müller, K.O. and H. Börger. 1940. Experimentelle Untersuchungen uber die *Phytophthora*-Resistenz der Kartoffel zugleich ein Beitrag zum Problem der Aerworbenen Resistenz im Pflanzenreich. Arbeitn der Biologischen Reichsanstalt für Land und Forstwirlschaft (Berlin) 23:189-231.

Namai, T., T. Kato, Y. Yamaguchi, and T. Hirukawa. 1993. Anti-rice blast activity and resistance induction of C-18 oxygenated fatty-acids. Bioscience Biotechnology and Biochemistry 57:611-613.

Newton, A.C., T. Reglinski, and G.D. Lyon. 1995. Resistance elicitors from fungi as crop protectants. Pages 419-422 *in* M. Manka, editor. Environmental Biotic Factors in Integrated Plant Disease Control. Proceedings of the Third European Foundation for Plant Pathology meeting, Poznan, Poland.

Newton, A.C., R.P. Ellis, C.A. Hackett, and D.C. Guy. 1997. The effect of component number on *Rhynchosporium secalis* infection and yield in mixtures of winter barley cultivars. Plant Pathology 45:930-938.

Newton A.C., and I.M.Young. 1996. Temporary partial breakdown of *Mlo*-resistance in spring barley by the sudden relief of soil water stress. Plant Pathology 45:970-974.

Oldroyd, G.E.D., and B.J. Staskawicz. 1998. Genetically engineered broad-spectrum disease resistance in tomato. Proceedings of the National Academy of Science USA 18:10300-10305.

Parker, J.E., K. Hahlbrock, and D. Scheel. 1988. Different cell wall components from *Phytophthora megasperma* f. sp. *glycinea* elicit phytoalexin production in soybean and parsley. Planta 176:75-82.

Paulin J.P., R. Chartier, J. Guillaumes, and M.N. Tanne, 1998. Activity and potential use of CGA 245704 against fire blight of apple (*Erwinia amylovora*). SCI Conference on Systemic Acquired Resistance: 10th March 1998, London. Abstract.

Preisig, C.L., and J.A. Kuć. 1985. Arachidonic acid-related elicitors of the hypersensitive response in potato and enhancement of their activities by glucans from *Phytophthora infestans* (Mont.) deBary. Archives of Biochemistry and Biophysics 236:379-389.

Reglinski, T., A.C. Newton, and G.D. Lyon. 1994a. Induction of resistance mechanisms in barley by yeast-derived elicitors. Annals of Applied Biology 124:509-517.

Reglinski, T., A.C. Newton, and G.D. Lyon. 1994b. Assessment of the ability of yeast-derived resistance elicitors to control barley powdery mildew in the field. Journal of Plant Disease and Protection 101:1-10.

Reglinski, T., F.J.L. Stavely, and J.T. Taylor. 1998. Induction of phenylalanine ammonia-lyase activity and control of *Sphaeropsis sapinea* infection in *Pinus radiata* by 5-chlorosalicylic acid. European Journal of Forest Pathology 28:153-158.

Reuveni R, and M. Reuveni. 1998. Foliar-fertilizer therapy - a concept in integrated pest management. Crop Protection 17:111-118.

Ruess, W. 1998. Field experience with CGA 245704 in a range of crop plants. SCI Conference on Systemic Acquired Resistance, 10th March 1998, London. Abstract.

Schönbeck F, H-W. Dehne, and W. Beicht. 1980. Activation of unspecific resistance mechanisms in plants. Zeitschrift für Pflanzenkrankheiten und Pflanzenschutz 87:654-666.

Schönbeck, F., and H.-W. Dehne. 1986. Use of microbial metabolites inducing resistance against plant pathogens. Pages 363-375 *in* N.J. Fokkema and J. van den Heuvel, editors. Microbiology of the Phyllosphere. Cambridge: Cambridge University Press.

Schönbeck, F., H.W. Dehne and H. Balder. 1982. On the efficiency of induced resistance under practical growing conditions I. Powdery mildew on grapevine, cucumber and wheat. Zeitschrift für Pflanzenkrankheiten und Pflanzenschutz 89:177-184.

Schweizer P., A. Jeanguenat, E. Mösinger, and J-P. Métraux. 1994. Plant protection by free cutin monomers in two cereal pathosystems. Pages 371-374 *in* M.J. Daniels, J.A. Downie and A.E. Osbourn, editors. Advances in Molecular Genetics of Plant-Microbe Interactions, vol. 3. Kluwer Academic Publishers, Dordrecht, The Netherlands.

Sequeira, L. 1984. Cross protection and induced resistance: their potential for plant disease control. Trends in Biotechnology 2:25-29.

Schmitt, A., S. Eisemann, S. Strathmann, K.A Emslie, and B. Seddon. 1996. Mode of action of extracts from giant knotweed (*Reynoutria sachalinensis*) on *Botrytis cinerea*, the causal organism of gray mold. Federal Biology Institute:32:421

Smedegaard-Petersen, V., and. O. Støen. 1981. Effect of energy-requiring defense reactions on yield and grain quality in a powdery mildew-resistant cultivar. Phytopathology 71:396-399.

Tepper, C.S., F.G. Albert, and A.J. Anderson. 1989. Differential mRNA accumulation in three cultivars of bean in response to elicitors from *Colletotrichum lindemuthianum*. Physiological and Molecular Plant Pathology 34:85-98

Thaler, J. S. 1999. Induced resistance in agricultural crops: effects of jasmonic acid on herbivory and yield in tomato plants. Environmental Entomology 28:30-37.

Thomas W.T.B., E. Baird, J.D. Fuller, P. Lawrence, G.R. Young, J. Russell, L. Ramsay, R. Waugh and W. Powell. 1998. Identification of a QTL decreasing yield in barley linked to *Mlo* powdery mildew resistance. Molecular Breeding 4:381-393.

Trewavas, A., and M. Knight. 1994. Mechanical signalling, calcium and plant form. Plant Molecular Biology 26:1329-1341.

Tuzun, S., and J. Kuć. 1991. Plant immunization: an alternative to pesticides for control of plant diseases in the greenhouse and field. Pages 30-39 *in* The Biological Control of Plant Diseases. Food and Fertilizer Technology Centre Book Series. No. 42.

Walters, D.R., and D.C. Murray. 1992. Induction of systemic resistance to rust in *Vicia faba* by phosphate and EDTA: effects of calcium. Plant Pathology 41:444-448.

Waspi, U., D. Blanc, T. Winkler, P. Ruedi, and R. Dudler. 1998. Syringolin, a novel peptide elicitor from *Pseudomonas syringae* pv. *syringae* that induces resistance to *Pyricularia oryzae* in rice. Molecular Plant Microbe Interactions 11:727-733.

Yang, D., R.S. Jeng, and M. Hubbes. 1989. Mansonone accumulation in elm callus induced by elicitors of *Ophiostoma ulmi*, and general properties of elicitors. Canadian Journal of Botany 67:3490-3497.

Ypema, H.L. and R.E. Gold. 1999. Modification of a naturally occurring compound to produce a new fungicide. Plant Disease 83:4-19.

Jasmonic Acid Mediated Interactions Between Plants, Herbivores, Parasitoids, and Pathogens: A Review of Field Experiments in Tomato

Jennifer S. Thaler

Abstract

In this chapter recent experiments are reviewed in which jasmonic acid was sprayed on agriculturally grown tomato plants to manage pests. Induction with jasmonates simulated herbivore-mediated induction, and was associated with high levels of several putative defense proteins. This chemical induction was also associated with induced resistance to a wide variety of the major herbivores of tomato, as well as increased levels of parasitism of caterpillars. However, jasmonate-mediated induction reduced the plants' ability to induce salicylate-mediated resistance against pathogens. Conversely, induction of systemic resistance to pathogens (by a different inducing agent, BTH) reduced the plants' ability to induce jasmonate-mediated resistance against herbivores. The ultimate effects of these multispecies interactions on plant yield is, at this stage, ambiguous, although under high pest conditions jasmonate-mediated induction has the potential to be an effective means of pest control.

Introduction

We are searching for ways to apply our knowledge of chemical ecology to solve pest problems in agriculture with reduced pesticide inputs. There is great potential for using chemical elicitors of plant resistance to protect plants against their insect and pathogen attackers. Work in the area of plant resistance to pathogens has advanced more quickly than studies of plant resistance to herbivores, and has resulted in the production and marketing of elicitors of salicylate-dependent plant resistance to pathogens (Lyon and Newton, Tally et al., this volume). Research on jasmonate-mediated plant resistance to insects has recently gained a lot of momentum from both basic and applied researchers, and hopefully elicitors of plant resistance to insect herbivores will be employed in the near future.

The purpose of this chapter is to review what has been learned about how jasmonate-mediated induced plant resistance works in an agricultural system. In lieu of a literature review, I summarize the recent results of our group of laboratories simply because our effort has been the first systematic program to examine the consequences of jasmonate-mediated induction in a field setting. We have studied the effects of exogenous applications of jasmonic acid on multiple species interactions in the tomato system. Our ultimate goal is to understand how the manipulation of plant resistance influences the community of organisms feeding on the plant. Induction of resistance represents a multitude of changes in the plant, and these changes will affect a broad variety of organisms, many of which have the potential to affect plant yield. The connection between the chemical changes induced by the elicitors and the effects that these changes have on plants and their associated organisms in the field is a critical step in the development of elicitors as pest management tools.

Jasmonates and Induction in the Tomato System

In many plants, localized wounding increases allocation to defensive chemistry in wounded tissues as well as in distant, unwounded portions of the plant (Karban and Baldwin 1997, Constabel, this volume). Systemic responses of this sort utilize signal(s), generated at the site of wounding, to regulate their expression. Several putative signal molecules have been identified, among them is the linolenic acid derivative, jasmonic acid (JA) (Reinbothe et al. 1994, Staswick and Lehman, this volume). In many plant species, wounding increases endogenous levels of JA, and exogenous application of JA stimulates the expression of defensive compounds (Reinbothe et al. 1994, Staswick and Lehman, this volume). The latter fact suggests that it may be possible to use applications of JA to stimulate natural induced plant resistance on a large scale without damaging individual plants and without the use of transgenic plants.

In foliage of the tomato (*Lycopersicon esculentum* Mill), caterpillar feeding initiates an array of defense-related responses which exhibit a characteristic spatial pattern (Stout et al. 1996). The most prominent systemic responses are the increases in the levels and activities of polyphenol oxidase (PPO) and of proteinase inhibitors (PIs). Local responses include increased activity of the oxidative enzyme lipoxygenase (LOX) in addition to the PPO and PI responses. Peroxidase (POD), another oxidative enzyme, is also induced, but less markedly and consistently. All of these proteins are thought to decrease the nutritive value of tomato foliage to herbivores, particularly noctuid larvae (Duffey and Felton 1991, Duffey and Stout 1996). Incorportation of these proteins into artificial diets reduced the performance of noctuids (Duffey and Felton 1991, Duffey and Stout 1996). Evidence indicates that the most prominent protein responses- the systemic increases in the levels of PIs and PPO- are co-regulated at the transcriptional level by an octadecanoid-based signal transduction pathway which includes JA (Farmer et al. 1992, Constabel et al. 1995).

Application of JA or its volatile derivative, jasmonic acid methyl ester (methyl jasmonate, MJ) to tomato foliage causes increased production of PI and PPO proteins (Farmer et al. 1992, Thaler et al. 1996). Whether other chemical responses to caterpillar feeding are controlled by the same signaling pathway as the PPO/PI response, or whether these other responses are controlled by separate signaling pathways, is under investigation. Recent developments demonstrate that the application of jasmonates to tomato foliage also induces LOX and sometimes POD. Jasmonate-induced proteins show the same spatial pattern of induction as that following feeding by the noctuid caterpillar, *Helicoverpa zea* (corn earworm) (Thaler et al. 1996). Exogenous JA also causes a systemic increase in plant resistance to herbivores which is similar to that caused by *H. zea* feeding. These results prompted us to test the effects of exogenous JA on the defensive chemistry and actual resistance of field-grown tomato plants.

General Methods

Over 4 years (1995 - 1998), tomato plants (*L. esculentum* var. Ace) were grown from seed or transplanted using standard agricultural practices in the tomato growing region of Northern California (Davis, CA). Jasmonic acid was dispersed in one ml of acetone per 8 liters of water; the control consisted of one ml of acetone per 8 liters of water. Generally, plants were randomly assigned to control or JA sprayed treatments: control plants received a spray of water and acetone, JA plants received a spray of 0.5 mM JA (circa 0.545 micromoles of JA per plant). These doses are well below those that cause toxic responses in plants (Thaler et al. 1996). This JA treatment simulates the level of induced resistance roughly equivalent to 24 hours of feeding by one fourth instar *H. zea* larva on greenhouse grown plants (Thaler et al. 1996).

Chemical analyses were conducted to examine effects of jasmonate-mediated induction on PIs and PPO (see Thaler et al. 1996 for details). Because application of jasmonic acid stimulates chemical induction in a consistent way as herbivore damage, we can utilize the plants' natural signaling molecules to: 1) induce plants in the field without introducing pests, and 2) manipulate the timing and extent of induction.

Jasmonate Spray Induces Plants in the Field

In the field, plants treated with JA had higher peak activities of both PPO and PIs compared to control plants (Fig. 1, Thaler et al. 1996, 1999a). The increased activities of PPO and PIs in induced plants were obtained three days after the plants were sprayed with JA (Thaler et al. 1996, 1999a). Elevated PPO and PI activities were maintained by a single spray for at least three weeks. By three weeks, the differences in PPO activity between treatments had decreased but remained statistically distinguishable (Fig. 1), while the differences in PI

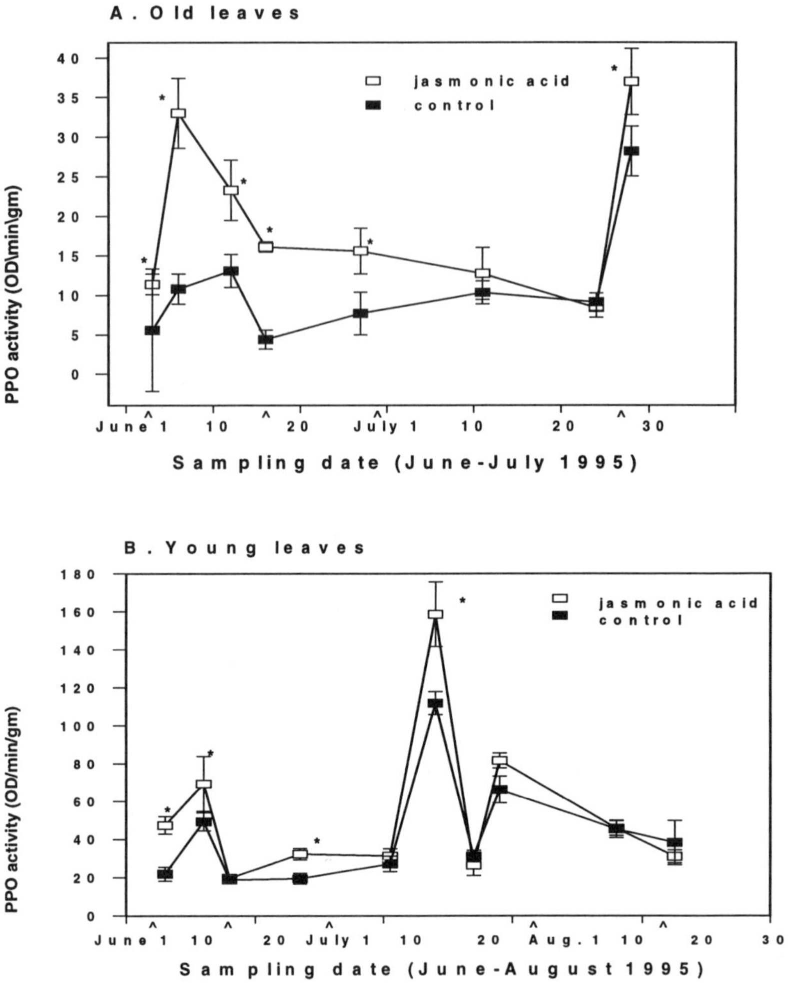

Figure 1: Polyphenol oxidase levels in (A) old leaves and (B) young leaves of field grown plants in summer 1995. Stars indicate significance at the 0.05 level. Arrows along x-axis indicate spray dates. Modified from Thaler et al. (1996).

induction persisted with large differences between the treatments. After three weeks, an additional JA treatment was applied to increase the differences in chemical activity between the control and JA treated plants. Thus, elevated chemical activities were sustained for at least 6 weeks by the two applications of JA. This time included the period of flowering and initial fruit formation. In addition, we have observed dose-dependent induction in the laboratory and field, with higher concentrations of JA inducing higher levels of defensive compounds (Thaler et al. 1996, 1999a).

Effects of Jasmonate-Mediated Induction on a Community of Herbivores

Over 3 years of field monitoring, all of the common herbivores were reduced in abundance by jasmonate-mediated induction, including thrips (*Frankliniella occidentalis*) (cell content feeders), noctuid larvae (*Spodoptera exigua*), flea beetles (*Epitrix hirtipennis*) (leaf chewers), and aphids (*Myzus persicae*) (phloem feeders) (Thaler et al. 1999a). We conducted controlled experiments in order to understand the ecological mechanisms leading to the effects seen during the monitoring. In field cage experiments, survivorship and growth rate of noctuid larvae were reduced 50-80% on induced plants compared to controls. For a more mobile herbivore, adult flea beetles, we measured preference for control vs. induced plants. Beetles preferred uninduced plants four times as often as induced plants. Other herbivores of tomato plants, which we did not observe in our study, have also been reported to be negatively affected by induced responses, including two-spotted spider mites (*Tetranychus urticae*), leafminers (*Lyriomyza trifolii*), corn earworm caterpillars (*H. zea*) (Stout et al. 1994, 1998, Stout and Duffey 1996), and hornworm caterpillars (*Manduca sexta*) (Johnson et al. 1989, Orozco-Cardenas et al. 1993). The generality of this induced resistance may also extend to some, but not all, pathogens (Table 1). For instance, other work from our laboratory demonstrates that the chemical responses induced by herbivores negatively affect the bacterium *Pseudomonas syringae* pv. *tomato* under some conditions, but not the fungus *Phytophthora infestans* (Stout et al. 1998, 1999).

Aphids showed the least consistent responses to plant induction. The effects of induced plant responses on aphids, while strong in one year, were not detected in other years. We have found other variable effects of induction on aphids which may be due to seasonal effects; induced responses had negative effects on aphids more often early in the growing season (Thaler, unpublished data). Resistance of tomato plants to aphids has been found to depend not only on the aphid species involved, but also on biotypes within a species, and plant developmental age (Kaloshian et al. 1997, Rossi et al. 1998).

Our results demonstrate that there is a general negative effect of jasmonate-mediated induced resistance on herbivores, but the magnitude of this effect varies according to species. These results are consistent with the work on generalized resistance that has emerged from several other well-studied systems. For example, many herbivores and some pathogens are affected by damage induced resistance in young cotton plants (Karban 1991). Chemical changes induced in damaged soybean foliage also affected many herbivores and pathogens (Kogan and Paxton 1983, Wheeler and Slansky 1991). Damage to birch foliage reduced its quality to many, though not all, herbivore species (Hartley and Lawton 1987, 1991, Neuvonen and Haukioja 1991). In the birch studies, several types of herbivores were negatively affected by induced

Table 1. Summary of the effects of elicitor-mediated induced resistance in tomato plants on herbivores, natural enemies of herbivores, and pathogens. Descriptions indicate how organisms perform on induced plants compared to control plants. Data from: [1]Thaler et al. 1999a, [2]Thaler et al. 1996, [3] Stout et al. 1998, [4]Inbar et al 1998, [5]Thaler 1999b, [6]Thaler, unpublished data, [7]Fidantsef et al. 1999, [8]Cohen et al. 1993, [9]Benhamou and Bélanger 1998, [10]Attitalla et al. 1998, [11]Thaler et al. 1999b. Blank cells indicate that there is no data. BTH is benzothiadiazole-7-carbothioic acid S-methyl ester and induces salicylate mediated-induced resistance.

Species	Common name/ disease name	Feeding guild	Abundance	Preference	Survivorship	Growth	Greenhouse/ field	Inducing agent
Herbivores								
Trichoplusia ni	Cabbage looper	leaf chewer			negative[1]	negative[1]	field	JA
Spodoptera exigua	Beet armyworm	leaf chewer	negative[1]		negative[1, 2]	negative[1, 2]	field[1] and lab[2]	JA
Spodoptera exigua		leaf chewer			no effect[3] positive[3, 1]		field[1] and greenhouse[3]	BTH
Epitrix hirtipennis	Flea beetle	leaf chewer	negative[1]	negative[1]			field	JA
Frankiliniella occidentalis	Western flower thrips	cell content feeder	negative[1]				field	JA
Myzus persicae	Green peach aphid	phloem feeder	variable[1] negative				field	JA
Liriomyza trifolii	Leafminer	mesophyll	negative[4]	negative[4]			field and greenhouse	BTH
Natural enemies of herbivores								
Hyposoter exiguae	Wasp	parasitoid	positive[5]				field	JA
Syrphidae	Syrphid fly	aphid predator	negative[6]				field	JA
Aphelinidae	Wasp	aphid parasitoid	no effect[6]				field	JA
Plant Pathogens								
Phytophthora infestans	Late blight	hemibiotrophic				no effect[7]	greenhouse	BTH
Phytophthora infestans	Late blight	hemibiotrophic				negative[8]	greenhouse	JA
Alternaria solani	Early blight	necrotroph				negative[4]	field	BTH
Fulvia fulva	Leaf mold	necrotroph				negative[4]	field	BTH
Fusarium oxysporum f. sp. *radicis-lycopersici*	Crown and root rot	necrotroph				negative[9]	greenhouse	BTH
Fusarium oxysporum f. sp. *lycopersici*	Tomato wilt	necrotroph				negative[10]	greenhouse	Salicylic acid
Pseudomonas syringae pv *tomato*	Bacterial speck	necrotroph				no effect[11]	field	JA

responses including leaf chewers and leaf miners. Aphid species, in particular, gave results that differed from other kinds of herbivores, similar to our results in the tomato system (Wratten et al. 1984, Fowler and MacGarvin 1986). That jasmonic acid can be applied to plants in the field to stimulate generalized induced resistance to herbivores (and some pathogens) provides an exciting possibility for a novel means of pest management.

Interactions Between Host Plant Resistance and Natural Enemies

Herbivore infested plants release increased quantities of volatiles and attract natural enemies such as parasitic wasps (Dicke and Sabelis 1988, Turlings et al. 1993, De Moraes et al. 1998, Paré et al., Sabelis et al., this volume). Recently, it has been suggested that the jasmonate pathway may regulate production of these volatile compounds that attract host seeking parasitic wasps (Boland et al. 1995, Alborn et al. 1997). I sought to test whether herbivores feeding on induced plants are parasitized more often than herbivores feeding on uninduced plants. Since plant resistance compounds that reduce preference and performance of herbivores can also have direct negative effects on parasitoids (Campbell and Duffey 1979), the net effects of jasmonate induction on mortality of herbivores by parasitic wasps are unclear. Even if parasitoids are attracted to induced plants, the herbivores on these plants may be of poor quality.

I examined the effects of induction on a plant-herbivore-natural enemy interaction. *Hyposoter exiguae* is an endoparasitic wasp and an important mortality agent for the agronomic pest *S. exigua* (Strand 1998). Three weeks after the field was sprayed, the number of naturally occurring *H. exiguae* pupae, which had developed in naturally occurring *S. exigua* caterpillars, were counted. Twice the number of *H. exiguae* pupae were found on induced plants relative to control plants (details in Thaler 1999b).

To determine whether plant traits *per se* were responsible for this increased parasitism, I conducted an experiment in which "sentinel" caterpillars were placed underneath the canopy of JA- treated and control plants in the field. The sentinel *S. exigua* caterpillars were reared on an artificial diet and placed in cups. Since these sentinel caterpillars never fed on plants, and never experienced induced resistance, they were not influenced by changes in the plant. In the field, they were not allowed to feed on plants, but were still accessible to foraging parasitoids. Parasitism in caterpillars associated with induced plants was 37 percent greater than in the control treatment (Fig. 2). Based on these findings, expression of the jasmonate pathway appears to result in increased attractiveness of plants to parasitoids and increased mortality for injurious herbivorous insects. Irrespective of herbivore quality or quantity, jasmonate-mediated induction in tomato increased natural parasitism of herbivores in the field.

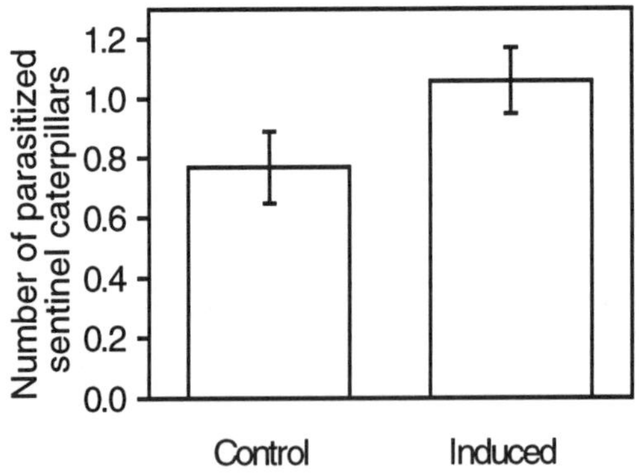

Figure 2: The number of parasitized sentinel *Spodoptera exigua* caterpillars from underneath the canopies of induced and control plants. Bars indicate mean ± SE. Figure modified from Thaler 1999b.

Induced plant resistance can also negatively affect parasitic wasps. In a series of controlled laboratory experiments, wasps were reared in caterpillars to assess the indirect effects of plant resistance on parasitoid success via altered host quality. In the first experiment, caterpillars from induced and control foliage were matched for age at the time of parasitism by naive, mated female wasps. In the second experiment, caterpillars were matched for body mass. In both experiments, parasitoid performance was reduced on caterpillar hosts grown on induced compared to control foliage (Thaler 1999b). While egg to pupa development time increased and pupal weight decreased on induced plants, pupa to adult development time was not different for wasps reared in caterpillars on induced and control foliage. When tested in the absence of plants, the parasitoids were not able to differentiate between equal sized caterpillars grown on induced and control foliage. Wasps are also not directly attracted or repelled by jasmonic acid compared to acetone controls.

How do Plants Coordinate their Responses to Attack From Insects and Pathogens?

Some defenses against insects may be effective against attackers from diverse kingdoms including bacterial and fungal pathogens, while other defenses are specific to particular attackers (Stout and Bostock, this volume). Since defenses effective against all challengers do not exist, plants must coordinate multiple defense strategies which may have different effects on different pests (Linhart 1991). Responses to pathogens and herbivores appear to be highly conserved in plants (Herrmann et al. 1989, Choi et al. 1994, Schneider et al. 1996). Studies of induced resistance to pathogens and induced resistance to insects initially proceeded with little consideration of possible interactions between the two (Karban and Baldwin 1997, Durner et al. 1997, but see Hatcher 1995). An understanding of how pathways controlling induced resistance to pathogens and insects interact will be critical for using chemical inducers to provide effective pest management.

Salicylic acid (SA) is a key compound in a pathway that regulates resistance to fungal, bacterial, and viral pathogens and provides a signal for expression of pathogenesis-related (PR) proteins and other potentially protective factors induced following pathogen challenge (Enyedi et al. 1992, Ryals et al. 1996, Hammerschmidt and Smith-Becker, this volume). As discussed above, jasmonic acid, produced by the octadecanoid pathway via lipoxygenation of linolenic acid, serves as a signal for expression of a number of chemicals that contribute to plant resistance against many insect attackers (Farmer et al. 1992, Staswick and Lehman, Constabel, this volume). Laboratory studies suggest that there can be interference between the salicylate and octadecanoid signaling pathways (Doherty et al. 1988), whereby salicylic acid appears to inhibit jasmonic acid biosynthesis and the subsequent chemical responses (Peña-Cortés et al. 1993, Doares et al. 1995). There is also evidence for inhibition of salicylate action by jasmonic acid (Sano and Ohashi 1995, Niki et al. 1998). This negative interaction between the two response pathways, previously demonstrated at the biochemical level, may compromise the ability of plants to coordinate defense against simultaneous challenge from pathogens and herbivores typically encountered in field settings.

Below I summarize the evidence, from a field experiment, for a tradeoff between jasmonate-mediated induced resistance to *S. exigua* and salicylate-mediated induced resistance to the bacterial speck pathogen, *Pseudomonas syringae* pv. *tomato*. BTH, benzothiadiazole-7-carbothioic acid S-methyl ester, a mimic of SA that elicits salicylate-mediated induced resistance was used to activate the SA pathway (Tally et al., Lyon and Newton, this volume). Plants treated with compounds that engage the two pathways separately (JA and BTH) resulted in the induction of the appropriate chemical responses and biological resistance to challenge insects and pathogens previously reported for each pathway (Fidantsef et al. 1999, Stout et al. 1999).

However, when plants were sprayed to stimulate both response pathways simultaneously, the corresponding biological effects on resistance to the pathogen and the insect were compromised. Levels of the pathogenesis-related (PR) protein P4 were higher in the plants sprayed with BTH and lower in plants sprayed with JA compared to control plants (Fig. 3a). The plants sprayed with both BTH and JA had P4 transcript levels similar to the controls, suggesting that the individual effects of BTH and JA may cancel each other. PPO activity, which was used to estimate the extent of JA-mediated induction, was highest in the plants sprayed with JA and lowest both in the plants sprayed with BTH and the control plants (Fig. 3b). The plants sprayed with both BTH

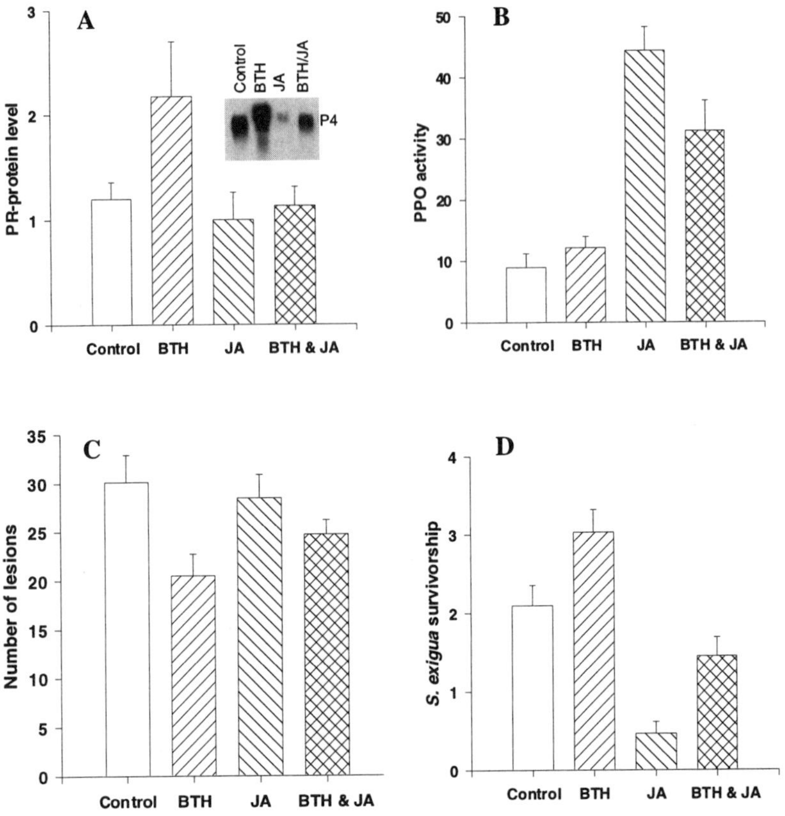

Figure 3: Field grown tomato plants were treated with 1) carrier control, 2) BTH to stimulate induced resistance to pathogens, 3) Jasmonic acid (JA) to stimulate induced resistance to insects, 4) both BTH and JA. Mean ± SE are shown. A) Pathogenesis-related protein P4 mRNA abundance is expressed relative to the maximum level of transcript abundance obtained in the four treatments. Inset: Representative blot of the levels of P4 transcripts detected with the probe in one sample, B) Polyphenol oxidase activity (ΔOD/gm/min) measured in the terminal leaflet of the fourth youngest leaf, C) The number of bacterial speck lesions per leaflet, and D) The number of surviving *S. exigua* larvae per plant in each treatment ten days after five neonates were caged onto each plant. Figure modified from Thaler et al. 1999b.

and JA had intermediate PPO activities compared with the plants treated separately. Differences in PR protein mRNA levels and PPO activity correlated with the subsequent performance of *P. syringae* pv. *tomato* and *S. exigua* on challenged leaves. Plants sprayed with BTH had the fewest bacterial speck lesions, whereas the plants sprayed with JA and controls had the most lesions (Fig. 3c). The plants sprayed with both BTH and JA had an intermediate number of lesions, a result that could have been caused by JA attenuation of BTH action.

JA reduced and BTH enhanced survival of the *S. exigua* larvae (Fig. 3d). Insects feeding on plants in the BTH/JA treatment had intermediate survivorship. The greater survivorship of *S. exigua* on BTH plants compared with control plants may have reflected the greater ability of control plants to activate resistance mechanisms against the assay larvae. In contrast, the BTH plants may not have been able to activate resistance because BTH may interfere with JA synthesis and action.

Thus, there appears to be reciprocal attenuation of signaling pathways that produce induced resistance to insect herbivores and pathogens in tomato plants under field conditions. The tradeoff in mobilization of defense responses operates in both directions. The expression patterns of defense-related indicators suggest that at least a portion of this phenomenon is due to biochemical interference of signal-response pathways. This potential interference between the induction of defenses effective against herbivores and pathogens poses a potential problem in disease and insect management. Since crop plants like tomato are typically challenged by multiple parasites at the same time, it may not be possible to simultaneously maximize defense against all kinds of attackers. Inhibition of jasmonate-mediated signaling by salicylates might also interfere with the attraction of the natural enemies of herbivores (see above). It will be important to consider the tradeoffs that may occur when host plants encounter multiple stresses, typical of field settings, during the development of new strategies that incorporate elicitors of inducible resistance.

How Do Multispecies Interactions Combine to Affect Plant Yield?

The use of chemical inducing agents is an alternate approach that may circumvent the damage caused by resistance-inducing herbivores, and the aversion of growers to introducing a potential pest into agricultural crops. Chemical agents are a practical way to induce plant responses because they can be patented, manufactured, and applied to large numbers of plants by conventional spray technologies. Before the discovery of chemical inducing agents, induced resistance was not a practical tool for use in agriculture because inducing plants with biotic challengers (e.g., herbivores) to protect against future, more damaging attackers was not generally feasible (for an exception, see Karban et al. 1997). However, in order to use induced responses as an effective pest management tool, the effects of induced responses on plant performance and yield in an agricultural setting must be evaluated. The manipulation of plant induced responses will only be effective if 1) the responses benefit the plant by reducing herbivory, 2) the benefits of induced responses outweigh the costs to the plant, 3) the net benefits are realized in agronomic traits, and 4) the increased crop value is greater than the economic cost of the elicitor and cost of comparable methods (i.e., traditional pesticides).

Over the course of the three years, I was not able to detect a cost or a benefit of jasmonate-mediated induction in the tomato system (Thaler 1999a, Table 2). Induction had no effect on the quantity or quality of the fruits

Table 2. Summary of the effects of jasmonate mediated induction on yield of tomato plants compared to yield of control plants in field experiments conducted over three years in Davis, California.

Trait	Affected?
Number of leaves with damage	Decreased
Number of flower buds	No
Number of flowers	Decreased early in the season
Number of fruits	No
Weight of fruits	No
Presence of sunburn	No
Presence of wet wounds	No
Presence of dry wounds	No

produced. This absence of detectable agronomic costs means that the economic costs of induction using jasmonates and other elicitors may be limited to the costs of purchasing and applying the chemical (see also Tally et al., this volume). Other indirect costs to the plant appear to be minimal. Because economically important pests are decreased by induced responses, under conditions where herbivores are important to the plant, induction may benefit the plant.

Synthesis and Future Directions

The work summarized in this chapter elucidates the effects of plant defensive chemistry on the community of interacting organisms which are supported by tomato plants. By combining an understanding of the biochemical basis of resistance with an ecological approach, our goal was to predict how jasmonate-mediated induction will affect how organisms interact in the field.

There are four major conclusions from this work. First, jasmonate-mediated induction simulated herbivore-mediated induction, and the resulting induced resistance had negative effects on many insect herbivores. All of the herbivores encountered in my studies, leaf chewers, cell content feeders, and phloem feeders were negatively affected by the induction of resistance, with the effects on aphids being the most variable. The generality (i.e., lack of specificity) in effects of induced resistance observed in tomatoes, may or may not occur in other plant – parasite systems (see also Stout and Bostock, this volume). Tally et al. (this volume) report that induction of resistance in tomato plants provides broader control of various pathogens than induced resistance in some other plant-pathogen systems. We need to understand the source of these similarities and differences between plants so that we can apply what we learn from tomatoes to other plants. There is greater incentive for industry to develop chemical inducing agents which provide generalized protection, in many plants, because these products can be sold to a larger market.

Second, there may be synergies between direct and indirect plant resistance. The effects of induced responses on natural enemies of herbivores will be a balance of direct and indirect effects, but the importance of these features may vary for different plant-herbivore-natural enemy combinations. The positive effects of increased attraction or retention of natural enemies to induced plants will only be realized if host finding is a limiting process for the natural enemy. Natural enemies that have difficulty finding hosts may benefit from the increased attractiveness of induced plants. For natural enemies for whom host finding is not a limiting step, the positive effects of increased host finding will be outweighed by the negative effects of feeding on poor quality hosts from induced plants. For example, the wasp *H. exiguae* was found in higher abundance on induced plants, but aphelinid parasitoids and syrphid fly larvae (both aphid feeders) were not affected by induced resistance (Thaler, unpublished data). This could be because *H. exiguae* feeds on low density hosts (caterpillars) whereas the other two natural enemies feed on high density hosts (aphids). Finding a patch with a caterpillar may be difficult, but finding a patch with aphids may be relatively easy. Future pest management strategies will rely on the use of combinations of strategies; this research suggests situations where induced plant resistance and biological control may provide compatible and potentially synergistic control of herbivores.

Third, under some conditions there are tradeoffs between plant resistance to herbivores and pathogens. This could be an important constraint on the expression of plant resistance. If this tradeoff occurs with many pairs of insects and pathogens, it may help to explain why plants employ inducible defenses in addition to constitutive defenses. It may benefit the plant to wait for cues regarding which kind of attacker is in the environment, and then induce an appropriate response. Our current work is aimed at understanding whether the timing or amount of inducing agent influences the tradeoff in expression of these two pathways. We know that some pathogens can induce the products of both of these pathways, indicating that either the two systems can be induced at the same time, or that when attacked by this pathogen the plant induces the same products using different biochemical pathways. By using chemical inducers of plant resistance, we may be able to tailor the resistance profile of plants to suit the pest pattern of the particular region in which the crop is being grown. In order to do this effectively, we must understand how expression of plant resistance to one pest influences plant resistance to other pests.

Fourth, costs or benefits of induction were not detected. Herbivores were in low abundance during the experiments, which may explain why a benefit of induction was not detected. Herbivores are thought to negatively affect yield in tomato plants under some conditions (Lange and Bronstein 1981). It is possible that the costs of induction are equal to the benefits of induction, or that only under conditions with damaging levels of herbivores are the benefits detected. Crops where early season damage is important are candidates for the successful use of induction in agriculture, since induction is often stronger in younger plants.

Acknowledgements

I thank Anurag Agrawal and Rick Karban for help with this chapter. I am grateful to the USDA-NRI program for support from grants #96-02065 and #98-02362. The UC-Davis SAREP program provided funding which got this project going.

Literature Cited

Alborn, H. T., T. C. J. Turlings, T. H. Jones, G. Stenhagen, J. H. Loughrin, and J. H. Tumlinson. 1997. An elicitor of plant volatiles from beet armyworm oral secretion. Science 276:945-949.

Attitalla, I. H., P. Quintanilla, and S. Brishammar. 1998. Induced resistance in tomato plants against *Fusarium* wilt invoked by *Fusarium* sp, salicylic acid and *Phytophthora cryptogea*. Acta Phytopathologica et Entomologica Hungarica 33:89-95.

Benhamou, N., and R. R. Bélanger. 1998. Benzothiadiazole-mediated induced resistance to *Fusarium oxysporum* f. sp. *radicis-lycopersici* in tomato. Plant Physiology 118:1203-1212.

Boland, W., J. Hopke, J. Donath, J. Nuske, and F. Bublitz. 1995. Jasmonic acid and coronatin induce odor production in plants. Angewandte Chemie 34:1600-1602.

Campbell, B. C. and S. S. Duffey. 1979. Tomatine and parasitic wasps: potential incompatibility of plant antibiosis with biological control. Science 204:700-702.

Choi, D., R.M. Bostock, S. Avdiushko and D. F. Hildebrand. 1994. Lipid-derived signals that discriminate wound- and pathogen-responsive antimicrobial isoprenoid pathways in plants: methyl jasmonate and the fungal elicitor arachidonic acid induce different 3-hydroxy-3-methylglutaryl-coenzyme A reductase genes in *Solanum tuberosum* L. Proceedings of the National Academy of Sciences USA 91:2329-2333.

Cohen, Y., U. Gisi, and T. Niderman. 1993. Local and systemic protection against *Phytophthora infestans* induced in potato and tomato plants by jasmonic acid and jasmonic methyl ester. Phytopathology 83:1054-1062.

Constabel, C. P., D. R. Bergey and C. A. Ryan. 1995. Systemin activates synthesis of wound-inducible tomato leaf polyphenol oxidase via the octadecanoid defense signalling pathway. Proceedings of the National Academy of Sciences USA 92:407-411.

De Moraes, C. M., W. J. Lewis, P. W. Pare, H. T. Alborn, and J. H. Tumlinson. 1998. Herbivore-infested plants selectively attract parasitoids. Nature 393:570-572.

Dicke, M. and M. W. Sabelis. 1988. How plants obtain predatory mites as bodyguards. Netherlands Journal of Zoology 38:148-165.

Doares, S. H., Narvaez-Vasquez, J., Conconi, A. and Ryan, C. A. 1995. Salicylic acid inhibits synthesis of proteinase inhibitors in tomato leaves induced by systemin and jasmonic acid. Plant Physiology 108:1741-1746.

Doherty, H. M., Selvendran, R. R. and D. J. Bowles. 1988. The wound response of tomato plants can be inhibited by aspirin and related hydroxybenzoic acids. Physiological and Molecular Plant Patholol gy 33:377-384.

Duffey, S. S., and G. W. Felton. 1991. Enzymatic antinutritive defenses of the tomato plant against insects. Pages 167-197 *in* P. A. Hedin, editor. Naturally Occurring Pest Bioregulators. ACS Symposium Series 449. Dallas, Fall 1989. American Chemical Society, Washington, D.C.

Duffey, S. S. and M. J. Stout. 1996. Antinutritive and toxic compounds of plant defense against insects. Archives of Insect Biochemistry and Physiology 32:3-37.

Durner, J., J. Shah and D. Klessig. 1997. Salicylic acid and disease resistance in plants. Trends in Plant Science 2:266-274.

Enyedi, A. J., N. Yalpani, P. Silverman and I. Raskin. 1992. Signal molecules in systemic plant resistance to pathogens and pests. Cell 70:879-886.
Farmer, E. E., R. R. Johnson, and C. A. Ryan. 1992. Regulation of expression of proteinase inhibitor genes by methyl jasmonate and jasmonic acid. Plant Physiology 98:995-1002.
Fowler, S. V. and M. MacGarvin. 1986. The effects of leaf damage on the performance of insect herbivores on birch, *Betula pubescens*. Journal of Animal Ecology 55:565-573.
Fidantsef, A. L., M. J. Stout, J. S. Thaler, S. S. Duffey and R. M. Bostock. 1999. Signal interactions in pathogen and insect attack: expression of lipoxygenase, proteinase inhibitor II, and pathogenesis-related protein P4 in the tomato, *Lycopersicon esculentum*. Physiological and Molecular Plant Pathology (in press).
Hartley, S. E. and J. H. Lawton. 1987. The effects of different types of damage on the chemistry of birch foliage and the responses of birch feeding insects. Oecologia 74:432-437.
Hartley, S. E. and J. H. Lawton. 1991. Biochemical aspects and significance of the rapidly induced accumulation of phenolics in birch foliage. Pages 105-132 *in* D. W. Tallamy and M. J. Raupp, editors. Phytochemical Induction by Herbivores. John Wiley, New York.
Hatcher, P. E. 1995. Three-way interactions between plant pathogenic fungi, herbivorous insects and their host plants. Biological Review 70:639-694.
Herrmann, G., Lehmann, J., Peterson, A., Sembdner, G., Weidhase, R. A. and Parthier, B. 1989. Species and tissue specificity of jasmonate induced abundant proteins. Plant Physiology 134:703-709.
Inbar, M., H. Doostar, R. M. Sonoda, G. L. Leibee and R. T. Mayer. 1998. Elicitors of plant defense systems reduce insect densities and disease incidence. Journal of Chemical Ecology 24:135-150.
Johnson, R., J. Narvaez, G. An, and C. A. Ryan. 1989. Expression of proteinase inhibitors I and II in transgenic tobacco plants: effects on natural defense against *Manduca sexta* larvae. Proceedings of the National Academy of Sciences USA 86:9871-9875.
Kaloshian, I., M. G. Kinsey, D. E. Ullman, and V. M. Williams. 1997. The impact of *Meu*1-mediated resistance in tomato on longevity, fecundity, and behavior of the potato aphid, *Macrosiphum euphorbiae*. Entomologia Experimentalis et Applicata 83:181-187.
Karban, R. 1991. Inducible resistance in agricultural systems. Pages 403-419 *in* D. W. Tallamy and M. J. Raupp, editors. Phytochemical Induction by Herbivores. John Wiley, New York.
Karban, R. and I. T. Baldwin. 1997. Induced Responses to Herbivory. University of Chicago Press, Chicago.
Karban, R., G. English-Loeb, and D. Hougen-Eitzman. 1997. Mite vaccinations for sustainable management of spider mites in vineyards. Ecological Applications 7:183-193.
Kogan, M. and J. Paxton. 1983. Natural inducers of plant resistance to insects. Pages 153-171 *in* P.A. Hedin, editor. Plant Resistance to Insects. American Chemical Society, Washington D. C.
Lange, W. H. and L. Bronson. 1981. Insect pests on tomatoes. Annual Review of Entomology 26:345-371.
Linhart, Y. B. 1991. Disease, parasitism and herbivory: multidimensional challenges in plant evolution. Trends in Ecology and Evolution 6:392-396.
Neuvonen, S. E., and E. Haukioja. 1991. The effect of inducible resistance in host foliage on birch-feeding herbivores. Pages 277-291 *in* D. W. Tallamy and M. J. Raupp, editors. Phytochemical Induction by Herbivores. John Wiley, New York.
Niki, T., I. Mitsuhara, S. Seo, N. Ohtsubo, and Y. Ohashi. 1998. Antagonistic effect of salicylic acid and jasmonic acid on the expression of pathogenesis-related (PR) protein genes in wounded mature tobacco leaves. Plant and Cell Physiology 39:500-507.
Orozco-Cardenas, M., B. McGurl, and C. A. Ryan. 1993. Expression of an antisense prosystemin gene in tomato plats reduces resistance toward *Manduca sexta* larvae. Proceedings of the National Academy of Sciences USA 90:8273-8276.
Peña-Cortés, H., T. Albrecht, S. Prat, E. W. Weiler and L. Willmitzer. 1993. Aspirin prevents wound-induced gene expression in tomato leaves by blocking jasmonic acid biosynthesis. Planta 191:123-128.

Reinbothe, S., B. Mollenhauer and C. Reinbothe. 1994. JIPs and RIPs: The regulation of plant gene expression by jasmonates in response to environmental cues and pathogens. Plant Cell 6:1197-1209.

Rossi, M., F. L. Goggin, S. B. Milligan, I. Kaloshian, D. E. Ullman, and V. M. Williamson. 1998. The nematode resistance gene *Mi* of tomato confers resistance against the potato aphid. Proceedings of the National Academy of Sciences USA 95:9750-9754.

Ryals, J. A., U. H. Neuenschwander, M. G. Willits, A. Molina, H.-Y. Steiner and M. P. Hunt. 1996. Systemic acquired resistance. Plant Cell 8:1809-1819.

Sano, H. and Y. Ohashi. 1995 Involvement of small GTP-binding proteins in defense signal-transduction pathways of higher plants. Proceedings of the National Academy of Sciences USA 92:4138-4144.

Schneider, M., P. Schweizer, P. Meuwly and J. P. Métraux. 1996. Systemic acquired resistance in plants. International Review of Cytology 168:303-339.

Stout, M. J. and S. S. Duffey. 1996. Characterization of induced resistance in tomato plants. Entomologia Experimentalis et Applicata 79:273-283.

Stout, M. J., Fidantsef, A. L., Duffey, S. S. and Bostock, R. M. 1999. Signal interactions in pathogen and insect attack: systemic plant-mediated interactions between pathogens and herbivores of the tomato, *Lycopersicon esculentum.* Physiological and Molecular Plant Pathology (in press).

Stout, M. J., K. V. Workman, R. M. Bostock, and S. S. Duffey. 1998. Specificity of induced resistance in the tomato, *Lycopersicon esculentum.* Oecologia 113:74-91.

Stout, M. J., K. V. Workman, and S. S. Duffey. 1994. Differential induction of tomato foliar proteins by arthropod herbivores. Journal of Chemical Ecology 20:2575-2594.

Stout, M. J., J. Workman, and S. S. Duffey. 1996. Identity, spatial distribution, and variability of induced chemical responses in tomato plants. Entomologia Experimentalis et Applicata 79:255-271.

Strand, L. L. editor. 1998. Integrated Pest Management for Tomatoes, 4th Edition. University of California Statewide Integrated Pest Management Project, Division of Agriculture and Natural Resources. Publication 3274.

Thaler, J. S., M. J. Stout, R. Karban, and S. S. Duffey. 1996. Exogenous jasmonates simulate insect wounding in tomato plants, *Lycopersicon esculentum*, in the laboratory and field. Journal of Chemical Ecology 22:1767-1781.

Thaler, J. S. 1999a. Induction of plant resistance to herbivores and effects on yield of field grown tomato plants. Environmental Entomology 28:30-37.

Thaler, J. S. 1999b. Jasmonate-inducible plant defenses cause increased parasitism of herbivores. Nature (in press).

Thaler, J.S., M. J. Stout, R. Karban, and S. S. Duffey. 1999a. Linking biochemical mechanisms of induced plant resistance and herbivore population dynamics. Submitted.

Thaler, J. S., A. L. Fidantsef, S. S. Duffey, and R. M. Bostock. 1999b. Tradeoffs in plant defense against herbivores and pathogens: a field demonstration. Journal of Chemical Ecology (in press).

Turlings, T. C. J., P. J. McCall, H. T. Alborn, and J. H. Tumlinson. 1993. An elicitor in caterpillar oral secretions that induces corn seedlings to emit chemical signals attractive to parasitic wasps. Journal of Chemical Ecology 19:411-425.

Wheeler, G. S., and F. Slansky. 1991. Effect of constitutive and herbivore-induced extractables from susceptible and resistant soybean foliage on nonpest noctuid caterpillars. Journal of Economic Entomology 84:1068-1079.

Wratten, S. D., P. J. Edwards, and I. Dunn. 1984. Wound-induced changes in the palatability of *Betula pubescens* and *B. pendula*. Oecologia 61:372-375.

Microbe-Induced Resistance Against Pathogens and Herbivores: Evidence of Effectiveness in Agriculture

Geoffrey W. Zehnder, Changbin Yao, John F. Murphy, Edward R. Sikora, Joseph W. Kloepper, David J. Schuster and Jane E. Polston

Abstract

This chapter presents a summary of the results of experiments conducted in Alabama and Florida over a five year period to evaluate strains of plant growth-promoting rhizobacteria (PGPR) for induction of resistance against insect-transmitted diseases on field-grown cucumber and tomato. Experiments with cucumber demonstrated that treatment with PGPR significantly reduced the incidence of wilt symptoms caused by the bacterial pathogen Erwinia tracheiphila, *and also reduced numbers of the cucumber beetle vectors of the bacteria. Cotyledons from PGPR-treated plants contained significantly lower concentrations of cucurbitacin, a secondary plant metabolite and cucumber beetle feeding stimulant, than untreated plants. The PGPR-induced change in cucurbitacin metabolism may be associated with the production of other plant defense compounds during the induction of a systemic plant defense response. Subsequent studies with tomato were conducted to identify PGPR strains for induction of systemic resistance against cucumber mosaic cucumovirus (CMV) and whitefly-transmitted tomato mottle virus (ToMoV). Tomatoes treated with PGPR demonstrated a reduction in the development of disease symptoms, and often a reduction in the incidence of viral infection and increase in tomato yield. While preliminary, the results of these experiments in cucumber and tomato demonstrate that PGPR-mediated induced resistance represents a viable and environmentally-friendly approach to crop disease management, particularly for insect-transmitted diseases which are often difficult or impossible to control with pesticides.*

Introduction

Plant-associated microorganisms, including eubacteria, actinomycetes, and fungi are part of the natural ecosystem of healthy plants, occurring in the major habitats of the rhizosphere, leaf surfaces, and inside plant tissues. Some of these naturally occurring microorganisms develop symbiotic relationships with the plant in which the microbes live within plant tissues in a modified morphological state and contribute to plant growth and development, as in the case of mycorrhizal fungi and the nitrogen-fixing rhizobia on legumes. Much research over the past century has been devoted to using free-living plant-associated microorganisms to benefit plants. One such group of free-living microorganisms which has been extensively investigated are "rhizobacteria" which are the subset of rhizosphere bacteria known to aggressively colonize roots (Schroth and Hancock 1982). When used in relation to microbial inoculants, the term "root colonization" denotes an active process whereby bacteria survive inoculation into seeds or soil, multiply in the spermosphere in response to seed exudates rich in carbohydrates and amino acids (Kloepper et al. 1985), attach to the root surface (Suslow 1982), and colonize the developing root system in soils containing indigenous microorganisms. Therefore, rhizobacteria have been shown to be efficient microbial competitors that can displace native root-colonizing microorganisms (Kloepper and Schroth 1981) and persist throughout some or all of the crop season. Typically, introduced rhizobacteria colonize roots at the mid-stages of host-plant ontogeny at population densities of 10^3 - 10^6 colony-forming units (CFU)/ g root fresh weight (Bahme et al. 1988, Kloepper and Beauchamp 1992). Rhizobacteria are distributed in the rhizosphere in a lognormal pattern (Loper et al. 1984) and are sporadically dispersed along roots in microcolonies (Bahme and Schroth 1987).

The general effects of rhizobacteria on host-plants range from deleterious to neutral to beneficial (Glick 1995, Lazarovits and Nowak 1997). Rhizobacteria that exert beneficial effects on plant development are termed "plant growth-promoting rhizobacteria" (PGPR) (Kloepper and Schroth 1978) because their application is often associated with increased rates of plant growth. PGPR also provide benefits to plants by suppression of soil-borne pathogens (Schippers et al. 1987) and through induction of systemic resistance (see below). Under practical agricultural conditions, in non-sterile field soils, it is not possible to conclusively differentiate mechanistically between growth promotion and biological control, because some soil borne pathogens, such as *Pythium* spp., are present in nearly all soils. However, under controlled laboratory or growth room conditions, one can demonstrate direct plant growth promotion by some PGPR strains (Glick 1995). The ability of PGPR to elicit induced plant resistance varies among bacterial strains, and plants also vary in the expression of resistance upon induction by specific bacterial strains (van Loon et al. 1998).

Pioneering studies beginning in the 1950s by researchers in Russia, China and several western countries showed the potential of bacteria for plant disease management (reviewed in Backman et al. 1997). Building on this early research, more recent studies have demonstrated the biocontrol activity of numerous PGPR strains against many soilborne pathogens, including *Aphanomyces* spp., *Pythium* spp., *Fusarium oxysporum, F. solani, Gaeumannomyces graminis* var. *tritici, Phytophthora* spp., *Sclerotium rolfsii,* and *Thielaviopsis basicola* (reviewed in Weller 1988, Schippers 1988). In 1985, Gustafson, Inc. (Plano, Texas) introduced the first commercial PGPR products in the United States using Broadbent's (Broadbent et al. 1977) *Bacillus subtilis* A-13 strain and related strains GB03 and GB07 (sold under the trade names Quantum®, Kodiak®, and Epic®, respectively). These products are registered for use on a number of different crops, including dicots and monocots, and are targeted for control of damage caused by fungal soil pathogens after seed-treatment fungicides have dissipated.

Mechanisms by which PGPR strains exhibit biological control against soil pathogens have been reported to include antibiosis through bacterial production of antifungal compounds (including antibiotics and hydrogen cyanide), competition for ferric iron, competition for infection sites, and production of lytic enzymes (Kloepper 1993). During the 1980s, work on mode-of-action of PGPR with biological control activity began to suggest that some PGPR strains may activate host defense systems, based on lack of direct antibiosis of the strains toward pathogens, or on correlation of biocontrol with plant growth promotion (Scheffer 1983, Voisard et al. 1989). In 1991, direct evidence supporting the conclusion that PGPR, which remain on plant roots, can induce resistance in plants to foliar or systemic pathogens was published independently for three pathosystems: cucumber and anthracnose (Wei et al. 1991), carnation and Fusarium wilt (van Peer et al. 1991), and bean and halo blight (Alström 1991).

"Systemic acquired resistance" (SAR) is a term first introduced by Ross (1961) to describe induction of resistance in tobacco by prior inoculation with tobacco mosaic virus. Since then, the term SAR has been commonly used in cases where induced resistance results from prior inoculation with necrotizing pathogens or application of chemical agents. Induction of SAR is characterized by an accumulation of salicylic acid (SA) and pathogenesis-related (PR) proteins (reviewed in van Loon et al. 1998). Some PR proteins (e.g., chitinases and glucanases) act directly on fungi by degrading cell walls, while the role of others has yet to be determined (Hoffland et al. 1995, van Loon et al. 1998, Hammerschmidt and Nicholson, this volume). SA has been implicated as a component of the SAR signaling pathway, but it does not appear to be the translocating, SAR-inducing signal (Vernooij et al. 1994, van Loon et al. 1998).

The term "induced systemic resistance" (ISR) is an alternative term sometimes used to denote induced resistance by non-pathogenic biotic agents, e.g., PGPR, in cases where SA signaling and accumulation of PR proteins do not occur (Pieterse et al. 1996, van Loon 1997, van Loon et al. 1998). PGPR-mediated ISR (PGPR-ISR) is similar to resistance induced by other agents (e.g., pathogens or chemical agents) in terms of disease suppression. However, ISR mediated by some PGPR is not associated with an accumulation of SA or PR proteins, and so may involve a distinct signal transduction pathway (see van Loon et al. 1998 for discussion and references on this topic). van Wees et al. (1997) demonstrated that SA is not required for the triggering of ISR in *Arabidopsis* by certain rhizobacterial strains. Pieterse et al. (1998) subsequently demonstrated that rhizobacterially-mediated ISR follows a novel signaling pathway in *Arabidopsis,* in which jasmonic acid and ethylene are involved in the signal transduction pathway. Once activated, PGPR-ISR is maintained for prolonged periods against multiple pathogens, even if populations of the inducing bacteria decline over time (van Loon et al. 1998).

Plant diseases caused by insect-transmitted pathogens are among the most difficult challenges in pest management, particularly in high-value vegetable production where loss of yield can drastically reduce profits. Effective control of insect-borne disease with insecticides is difficult or often impossible because most plant disease vectors are highly mobile insects that may colonize fields before growers are aware of their presence. In addition, even low numbers of insects may result in high field incidence of disease, as occurs with cucumber beetles and bacterial wilt disease (Yao et al. 1996). Plant diseases caused by viruses that are transmitted by aphids in a non-persistent manner, e.g., cucumber mosaic cucumovirus, are not effectively controlled by insecticides because incoming viruliferous aphids can inoculate plants in seconds before they are affected by insecticide exposure (Matthews 1991).

The application of PGPR for crop protection is relatively new. Therefore, rather than develop a review chapter on this topic, we chose to summarize the results of some of our experiments with cucumber and tomato that demonstrate the potential of PGPR as a crop protection tool. These efforts were directed towards PGPR-mediated induced resistance as an alternative strategy for management of three insect-transmitted diseases that have proven difficult to control with conventional methods.

PGPR-ISR Against Cucumber Beetles and Bacterial Wilt Disease

Bacterial wilt of cucurbits is a systemic disease caused by the xylem-inhabiting bacterial pathogen *Erwinia tracheiphila* (Smith) Holland. Yield losses in cucumber and muskmelon, the most susceptible host crops, can be as high as 75% (Sherf and MacNab 1986). The bacterial wilt pathogen is transmitted by diabroticite beetles, including *Acalymma vittata* (F.), the striped cucumber beetle, and *Diabrotica undecimpunctata howardi* (Barber), the spotted cucumber beetle. Early studies (Rand and Enlows 1916) demonstrated that primary infection occurs through feeding wounds made by beetles and subsequent transfer of the pathogen from the insect mouthparts or feces. Blua et al. (1994) provided serological evidence that herbaceous weeds served as overwintering hosts for the pathogen, but this evidence has been disputed (De Mackiewicz et al 1998). More recently, Fleischer et al. (1999) demonstrated that cucumber beetles serve as overwintering reservoirs for *E. tracheiphila*, and that beetle aggregation on host-plants facilitates delivery of a sufficient dose for infection to occur. Field studies in Alabama have demonstrated a positive, linear relationship between cucumber beetle density on cucumber plants and the incidence of bacterial wilt symptoms (Yao et al. 1996).

It is known that cucumber beetle feeding behavior is strongly influenced by cucurbitacins, a group of triterpenoid plant metabolites that occur in the plant family Cucurbitaceae (Chambliss and Jones 1966). Cucurbitacins act as a powerful feeding stimulant for cucumber beetles (Chambliss and Jones 1966, Metcalf 1986), and a strong, positive relationship has been established between cucurbitacin concentration and beetle feeding damage (Ferguson et al. 1983).

We first suspected that cucumber beetle feeding behavior was affected by PGPR treatment following cucumber field experiments in which PGPR afforded unexpected protection against bacterial wilt disease with large numbers of cucumber beetles present (Wei et al. 1995). We then initiated a series of experiments in cucumber to evaluate the effects of PGPR treatment on cucumber beetles and bacterial wilt disease.

Methods

Field Experiments. Field studies were conducted to assess the effects of PGPR treatment on populations of cucumber beetles, and to compare PGPR treatment with weekly applications of insecticide for control of cucumber beetles and bacterial wilt on cucumber (Zehnder et al. 1997a). For these experiments, PGPR strains were used that were shown previously to reduce disease incidence in cucumber caused by *E. tracheiphila* (Kloepper et al. 1993). Cucumber seeds were dipped in the pelleted bacterial cells or into distilled water (control) immediately before planting in plastic pots containing sterilized soilless planting mix. A dilute PGPR suspension (100 ml containing ~ 10^8 CFU/ml) was poured into each pot immediately after seeding. Seedlings ('Straight 8') were transplanted into the field at the 2nd leaf stage and grown in fumigated (methyl bromide + chloropicrin), raised beds with black plastic mulch and drip irrigation. Treatments in 1993 included the following PGPR: *Pseudomonas putida* strain 89B-61, *Serratia marcesens* strain 90-166, *Flavomonas oryzihabitans* strain INR-5, and *Bacillus pumilis* strain INR-7. Control treatments included an insecticide control (weekly sprays of esfenvalerate by backpack sprayer) and a untreated control. The 90-166 and INR-7 strains were re-evaluated in 1994 along with the insecticide and untreated controls.

Greenhouse Experiments. Greenhouse experiments were conducted to determine if resistance against feeding by cucumber beetles was a factor in PGPR-induced protection against bacterial wilt that was previously observed in the field (see Zehnder et al 1997b for complete details). In free-choice experiments, screen cages designed in a 'cross' arrangement with 4 arms (Fig. 1) were used to confine cucumber beetles on PGPR-treated (seed treatment and

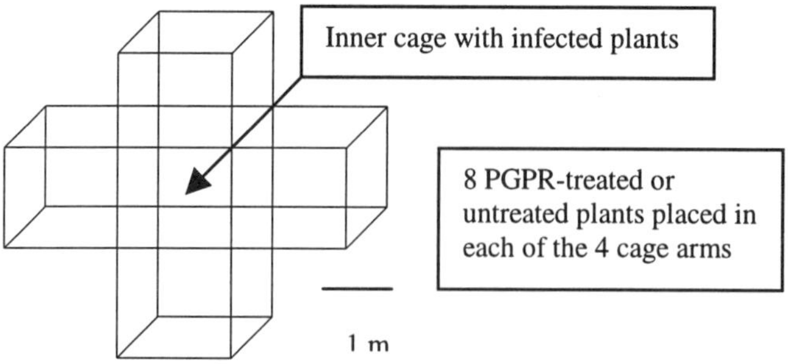

Figure 1: Diagram of screen cage used to confine cucumber beetles in free-choice experiments.

transplant drench with INR-7 strain) or untreated plants. PGPR-treated plants were placed in 2 arms/cage, and untreated plants in the other 2 arms/cage; 2 cages were used for each experiment (4 treatment replicates per experiment, 8 plants per replicate). Experiments were repeated twice. At the start of each experiment, 100 spotted cucumber beetles were confined on *E. tracheiphila*-infected cucumber plants in the center 'cage within a cage' for 48 h before doors were opened allowing beetles free access to all 4 cage arms. Data on beetle feeding damage and wilt incidence were recorded at 13 (experiment 1) or 17 (experiment 2) d after beetle release on uninfected plants. In separate no-choice experiments, beetle transmission of bacterial wilt was assessed in screen cages (1.0 by 0.5 by 0.5 m) where beetles were allowed to feed only on PGPR-treated or non-treated plants. In these experiments, 25 spotted cucumber beetles were released in each cage and allowed to feed on 3 *E. tracheiphila*-infected cucumber plants placed in the center of each cage for 48 h before 5 healthy PGPR-treated or untreated plants were introduced (see Zehnder 1997a for complete details). No-choice experiments were conducted separately for each of 3 cucumber cultivars; 'Poinsett' bitter (BI), 'Poinsett' non-bitter (bi), and 'Straight 8' (with low levels of cucurbitacin C). Wilt incidence was assessed by determining the percentage of wilted leaves per plant in each cage 17-23 d after the introduction of test plants into the cages with beetles.

Cucurbitacin Analysis. Cucurbitacin 'C', the putative sole cucurbitacin in cucumber (*Cucumus sativus*) (Rice et al. 1981), was detected in samples of fresh or frozen cotyledon leaves from PGPR-treated (INR-7 and INR-5 strains) or untreated plants using HPLC analysis (see Zehnder 1997b for analytical methods).

Results

Field Experiments. In both years, average numbers of cucumber beetles (spotted and striped species combined) were significantly lower in the PGPR

Table 1. Results of cucumber field experiments with PGPR for control of cucumber beetles and bacterial wilt.

PGPR treatment	Mean no. beetles/plant		Mean % wilted vines	Mean fruit weight/plot (kg.)	
	1993	1994	1994	1993	1994
89B61	0.61 cd	NT	NT	37.3 a	NT
90-166	0.44 d	2.34 c	2.61 c	35.9 a	28.1 a
INR-5	0.56 cd	NT	NT	32.7 ab	NT
INR-7	0.73 bc	2.96 bc	3.35 bc	37.1 ab	26.5 ab
Insecticide Control[1]	0.89 b	NT	11.48 b	25.6 b	21.9 ab
Untreated	1.73 a	5.42 a	24.56 a	29.4 b	20.8 bc

NT, not tested. Means within columns sharing a letter in common are not significantly different (P > 0.05; LSD test). Beetle and wilted vine means derived from 6 replicates; 10 plants per replicate. Beetle data averaged over 6 sample dates; wilted vines recorded on 24 June, 1994. Plants sprayed weekly with esfenvalerate insecticide at the rate of 0.05 lb (AI)/acre.

treatments compared with the untreated control (Table 1). In 1994, when bacterial wilt symptoms were observed, the incidence of wilted vines was significantly lower in the PGPR treatments than in the untreated control. In both years, yields in the PGPR treatments were higher (significantly only for some PGPR strains) than in the untreated controls. It is interesting to note that some PGPR strains provided significantly greater protection against cucumber beetles and bacterial wilt than the weekly applications of esfenvalerate, the recommended insecticide treatment.

Greenhouse Experiments. In free-choice experiments, beetle-feeding damage on cotyledons was greatly reduced on PGPR-treated plants compared with untreated plants, and damage to stems was also less severe on PGPR-treated plants (Fig. 2A and 2B). Wilt symptoms on test plants in the 4 cage arms were observed between 7 and 12 d after beetle release, demonstrating that beetles acquired *E. tracheiphila* from inoculated plants in the 'center cage', and then successfully transmitted the pathogen to healthy test plants. The average number of wilted leaves per plant ranged from 1.13 to 2.59 on untreated control plants, but only from 0 to 0.28 on PGPR-treated plants (Fig. 2C). In the no-choice experiments, the average percentages of wilted leaves in the 3 cultivar experiments ranged from 52.8% to 85.3% on the untreated plants, but only from 7.6 to 13.1% on the PGPR-treated plants (Fig. 3). Thus, spread of *E. tracheiphila* on both bitter and nonbitter cucumber cultivars was significantly reduced by PGPR treatment, even when beetles were restricted to feeding only on PGPR-treated plants for a prolonged period.

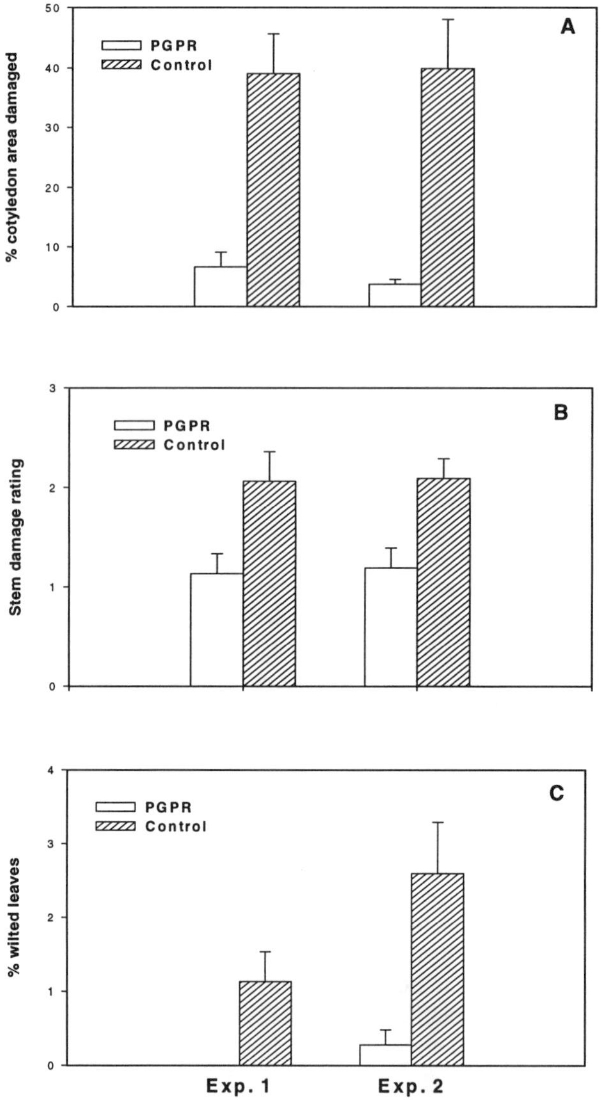

Figure 2: Cucumber beetle feeding damage and incidence of bacterial wilt symptoms on PGPR and untreated (control) cucumber plants. Infected cucumber beetles were released in greenhouse cages (see Fig. 1) and permitted to choose between treated and control cucumber plants (var. Straight 8, with low concentrations of cucurbitacin C). Experiments were repeated twice. (A) Mean percentage of cotyledon leaf area per plant with feeding damage. (B) Mean stem feeding damage rating per plant: 1 = <1/3 of stem from soil line to cotyledons damaged; 2 = 1/3 to 2/3 of stem damaged; 3 = > 2/3 stem with feeding damage. (C) Mean number of wilted leaves per plant.

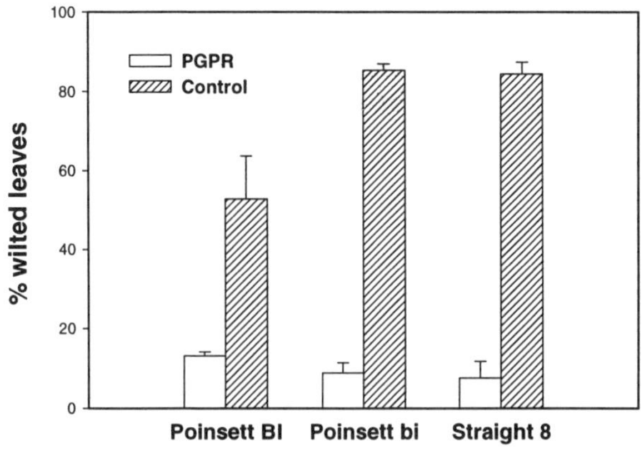

Figure 3: Comparison of mean percentage of wilted leaves per plant on 'Poinsett' bitter, non-bitter and 'Straight 8' (low concentrations of cucurbitacin C) cucumber plants in no-choice greenhouse cage experiments (see Fig. 1). Cucumber beetles infected with *E. tracheiphila* were released into cages with either PGPR-treated or untreated plants and allowed to feed for 17-23 days.

Table 2. PGPR-mediated reduction in cucurbitacin 'C' concentration.

	Mean cucurbitacin concentration (μg)	
PGPR treatment	Poinsett (bitter) cultivar	Straight 8 cultivar (non-bitter)
INR-7	117.3 b	27.1 c
INR-5	117.9 b	35.2 bc
Untreated	158.6 a	48.4 a

Means within columns sharing the same letter are not significantly different ($P > 0.05$; LSD test). Cucurbitacin 'C' values are μg cucurbitacin/g dry weight plant material. Means derived from 5 replicates per treatment. Data from Zehnder et al. 1997b.

Cucurbitacin Analysis. Cucurbitacin 'C' concentration was significantly lower in cotyledons of 'Poinsett' and 'Straight 8' cucumber treated with PGPR compared with untreated plants (Table 2). The greatest reduction occurred in 'Straight 8', with an average of 44% lower cucurbitacin in INR-7-treated plants compared with the untreated plants.

Discussion

The results of the field and greenhouse experiments demonstrated reductions in cucumber beetle feeding and spread of bacterial wilt were mediated by PGPR-induced resistance, and led us to hypothesize that the induction of resistance resulted in an unexpected physiological change in the plant: reduced levels of the feeding stimulant cucurbitacin 'C'. Cucurbitacin levels were reduced as much as 44% on PGPR-treated plants, providing support for our hypothesis (Table 2). The metabolic pathway for cucurbitacin 'C' biosynthesis involves squalene synthetase, an enzyme that catalyzes conversion of squalene epoxide to cucurbitadienol, the simplest tetracyclic triterpene with a cucurbitane skeleton (Balliano et al. 1982). Squalene is also a precursor of sesquiterpene plant defense compounds (i.e., phytoalexins, Tjamos and Kuć 1982).

Although not yet confirmed, PGPR-induced effects on cucurbitacin production may be the result of a metabolic shift, in which an increase in the production of plant defense compounds may result in deficiencies in other compounds requiring the same chemical precursors or intermediates (Karban and Kuć, Stout and Bostock, Felton and Eichenseer, this volume). Squalene used to manufacture cucurbitadienol may be diverted to the production of phytoalexins during PGPR-mediated induction of resistance, resulting in reduced cucurbitacin production.

We hypothesize that PGPR treatment provides two levels of protection against bacterial wilt in cucumber. First, the reduction in feedant (cucurbitacin) synthesis in PGPR-treated plants makes these plants less palatable to cucumber beetles, which may result in a lower proportion of beetles that acquire and successfully transmit the pathogen. Second, PGPR may elicit the induction of other plant defense mechanisms (i.e., phytoalexin production and other compounds involved in ISR) against the pathogen after it has been introduced into the plant.

PGPR-ISR Against Cucumber Mosaic Virus

Worldwide, cucumber mosaic cucumovirus (CMV) is one of the five most important viruses affecting production of field-grown vegetables (Tomlinson 1987). CMV is often the most prevalent virus in surveys of plant virus infections and it has been implicated in disease epidemics of fruit, vegetable and greenhouse crops (reviewed in Palukaitis et al. 1992). A recent epidemic of CMV in Alabama resulted in a 25% yield loss in the north-central tomato-growing region of the state (Sikora et al. 1998). CMV is difficult to control because of its extremely broad natural host range and the ability to be transmitted by more than 60 species of aphids (Zitter 1991). Bergstrom et al. (1982) demonstrated that previous inoculation of cucumber with *Colletotrichum orbiculare, Pseudomonas syringae* pv. *lachrymans*, or tobacco necrosis virus (TNV) could induce resistance against CMV. In greenhouse studies, Raupach et al. (1996) showed that two PGPR strains, which previously induced resistance in cucumber against certain fungal and bacterial diseases, also induced resistance in cucumber and tomato against CMV. The purpose of our study was to evaluate additional PGPR strains for activity against CMV on greenhouse-grown tomato, and to determine if PGPR-mediated induced resistance could be extended to tomato grown in the field using commercial production practices.

Methods

Greenhouse experiments were first conducted to evaluate the effects of 26 PGPR strains on the establishment of CMV infection in tomato. In these experiments, the PGPR strains (applied to seed as pelleted bacterial cells in densities of approximately 5 x 10^9 cfu/seed) were tested along with a disease control (CMV mechanical inoculation, no PGPR) and a healthy control (no CMV inoculation, no PGPR). Plants were examined daily for CMV symptoms (leaf distortion, mosaic patterns, general stunting of the plant). Based on these results, 4 PGPR strains were chosen for further evaluation in field experiments: *Bacillus pumilis* strain SE34, *Kluyvera cryocrescens* strain IN114, *B. amyloliquefaciens* strain IN937a, and *B. subtilus* strain IN937b.

Field experiments were conducted in 1996 and 1997 to evaluate these PGPR strains, a disease control and a uninoculated ("healthy") control. 'Mountain Pride' tomato seeds were mixed with pelleted PGPR cells resulting in approximately 5 x 10^9 cfu/seed. Tomato seeds in the healthy and disease control treatments were dipped in distilled water before planting. Control seeds were treated with 0.2M phosphate buffer. Tomato plants were transplanted into pots containing planting mix two weeks after seeding. PGPR suspension treatments (100 ml containing approximately 5 x 10^8 cfu/ml) were poured into each pot immediately after transplanting. Water or a buffer solution was applied to control plants. CMV-KM inoculum, originally isolated from tomato in North Alabama, was used throughout these studies. CMV-KM inoculum consisted of systemically-infected tobacco leaves ground in 50mM potassium phosphate, pH 7.5, with 10mM sodium sulfite. Tomato leaves were lightly dusted with carborundum and inoculum was applied and rubbed onto the first two-tomato leaves/plant one week after transplanting into pots; transplants were set in the field 3 days after inoculation. Tomato plants were grown on raised beds, fumigated with methyl bromide and chloropicrin and covered with black plastic mulch. There were 6 replications per treatment arranged in a randomized block design, each consisting of 15 tomato plants (single row plots). All plants in each treatment were examined weekly for virus symptoms using a rating scale from 0-10, followed by a calculation of disease severity (Tian et al. 1985). Marketable (non-damaged and mature) tomato fruit were weighed on 6 harvest dates during the season.

Results

In the initial greenhouse experiments, the number of plants exhibiting CMV symptoms was reduced in several PGPR strain treatments compared to the uninduced, challenged control (data not shown). The percentage of symptomatic plants in the 4 PGPR strains selected for further evaluation in the field ranged from 32 to 58%, compared with 88 to 98% in the uninduced, challenged disease control treatment. We also observed a delay in the onset of symptoms in PGPR treatments compared with uninduced control plants.

Table 3. Effects of selected PGPR treatments on cucumber mosaic cucumovirus (CMV) infection and yield in field tomato.

	AUDPC[1] value		Mean yield (kg/plot)	
Treatment	1996	1997	1996	1997
SE34	12.2 c	8.4 c	14.0 a	3.2 a
IN114	21.3 b	12.7 b	10.3 b	2.4 a
IN937a	9.9 c	9.1 bc	14.8 a	2.5 a
IN937b	11.1 c	10.7 bc	14.2 a	2.1 a
Uninduced, challenged control	24.8 a	18.3 a	9.5 b	2.0 a
Uninduced, unchallenged control	0.8 d	7.1 c	14.1 a	2.9 a

Means within columns sharing the same letters are not significantly different ($P > 0.05$; LSD test). [1]AUDPC, area under the disease progress curve.

In the 1996 field experiment, AUDPC values, indicating disease symptom progression over time, were significantly lower in all PGPR treatments compared with the uninduced, challenged control (Table 3). Enzyme-linked immunosorbent assay (ELISA) indicated that the percentage of infected plants in the uninduced, challenged control treatment was over 3-fold greater than in the IN937a and IN937b treatments (data not shown). Importantly, yields in the SE34, IN937a and IN937b treatments were significantly greater than in the disease control.

As in 1996, results of the 1997 field experiment indicated that AUDPC values were significantly lower in the PGPR treatments than in the uninduced, challenged control (Table 3). However, average tomato yields were not significantly different among treatments.

Discussion

These results suggest that specific PGPR strains can elicit induced resistance against CMV infection following mechanical inoculation onto tomato, and that protection can be maintained under field conditions. However, the level of PGPR-induced resistance observed was variable. In the 1996 field experiment, the incidence of CMV infection was significantly reduced and tomato yields were improved (relative to plants in the disease control) on PGPR-treated plants (strains IN937a, IN937b and SE34) mechanically challenged with virus before transplantation to the field. In 1997, AUDPC values were significantly lower in PGPR treatments than in the uninduced, challenged control, but the significant effects of PGPR on the incidence of infected plants and on tomato yields, as seen in 1996, were not evident. In 1997, 62.2% of unchallenged control plants tested positive for CMV infection by ELISA, compared with 4.4% in 1996. A possible explanation for the greater incidence of infection in 1997 is that the plants were subjected to lower levels of naturally transmitted CMV in 1996.

Aphid counts were not recorded in either year. We have previously observed a high incidence of natural CMV infection on PGPR-treated plants in a field trial on a north Alabama tomato farm where the level of CMV inoculum was known to be extremely high. Another explanation for reduced effectiveness of PGPR in 1997 could be that plants were naturally infected with a different strain of CMV, and that the PGPR strains tested were not as effective against the naturally occurring CMV strain. We have not yet conducted experiments specifically to evaluate PGPR on tomato for induced resistance against CMV by natural aphid transmission, or to measure the effects of changing abiotic factors on PGPR-induced resistance. The level of protection resulting from treatment by a given PGPR strain may vary from one cropping season to the next depending on existing conditions.

PGPR-ISR Against Tomato Mottle Geminivirus

Tomato mottle, caused by the tomato mottle geminivirus (ToMoV) poses a major threat for both transplant and field production of tomato in west-central and south-west Florida (Abouzid et al. 1992, Polston et al. 1993). ToMoV is transmitted by adult sweet potato whiteflies, *Bemisia tabaci*, biotype B (also known as the silverleaf whitefly, *Bemisia argentifolii*). Symptoms of ToMoV in field-grown tomatoes include chlorotic mottling and upward curling of leaflets, and an overall reduction in plant height as well as the number and size of fruit (Polston et al. 1993). Similar to the CMV pathosystem, traditional management of ToMoV has been very difficult. Tomato cultivars resistant to ToMoV are not yet commercially available, and foliar-applied insecticides have not provided effective management, in part because of the development of insecticide-resistant whitefly biotypes. Prompted by our findings that treatment of tomato with PGPR resulted in reduced symptoms of CMV infection, trials were conducted to evaluate some of the same PGPR strains for induced resistance in tomato against ToMoV.

Methods

Field experiments were conducted during the 1997 fall tomato-growing season at the University of Florida Gulf Coast Research and Extension Center in Bradenton, Florida. Tomatoes were exposed to high levels of natural whitefly infestation and ToMoV infection throughout the production season. Spores of PGPR strains IN937b and SE34 were produced in culture and formulated as both a seed treatment and a powder by Gustafson Corp. (Plano, Texas). The PGPR powder was diluted with water according to the manufacturer's recommendations and incorporated into the planting mix before seeding. At 40 days after planting, each plant in each treatment plot was rated for disease severity using a scale of 0 to 5.0, and all samples were analyzed for ToMoV DNA by nucleic acid dot blot analysis (Polston et al. 1993). Leaf samples for analysis of ToMoV DNA were also collected 80 days after planting. Tomatoes were harvested from all plots 80, 94 and 108 days after transplanting.

Results

The IN937b and SE34 PGPR treatments both resulted in reduced incidence of ToMoV and disease severity. Visual symptom ratings and the percentage of infected plants (based on dot blot analysis of leaves at 40 days after transplanting) indicated that the PGPR powder and powder + seed formulations were more effective than the PGPR seed formulations (Fig. 4 A, B). The severity of virus symptoms was significantly lower in all PGPR powder and powder + seed treatments than in the untreated control, but symptoms were not significantly different between the seed-only treatments and the control. There were no significant differences in symptom ratings between the PGPR powder-only and the PGPR seed + powder formulations. Contrast analysis of the percentage of plants testing positive for ToMoV DNA from leaf samples collected 40 days after planting generated similar results. By 80 days after planting, most plants in all treatments were infected with ToMoV, and differences in the percentages of infected plants among treatments were not significant (data not shown).

At the first harvest date (80 days after transplanting), tomato yields were higher in PGPR powder or seed + powder treatments than in the control or seed-only treatments; however, yield differences were statistically significant only between the IN937b powder treatment and the control (Fig. 4C). Analysis of tomato yield data from harvests at 94 and 108 days after transplanting did not indicate a significant effect of PGPR treatment on tomato yield. We suspect that PGPR-mediated resistance provided protection against ToMoV in the early stages of infection, but that continual whitefly infestation of plants (based on general observations, not direct insect counts) and inoculation of the virus eventually overcame the induced resistance response.

Discussion

The results with ToMoV in Florida demonstrate that specific PGPR strains can provide protection in the field against viruses in different groups with different insect vectors. The observed level of ToMoV disease symptom suppression resulting from PGPR treatment was encouraging given that the PGPR strains used in the Florida trial were selected based on screening for protection against CMV and not ToMoV. This illustrates the potential of PGPR to provide protection against multiple pathogens. Furthermore, vegetative cell treatments of PGPR were used in our previous experiments with CMV in tomato, and the PGPR spore seed and powder formulations used in the Florida trial had not previously been tested. The levels of whitefly infestation and ToMoV inoculum at the Bradenton field experiment site were greater than what typically occurs in commercial tomato fields (D. Schuster, personal

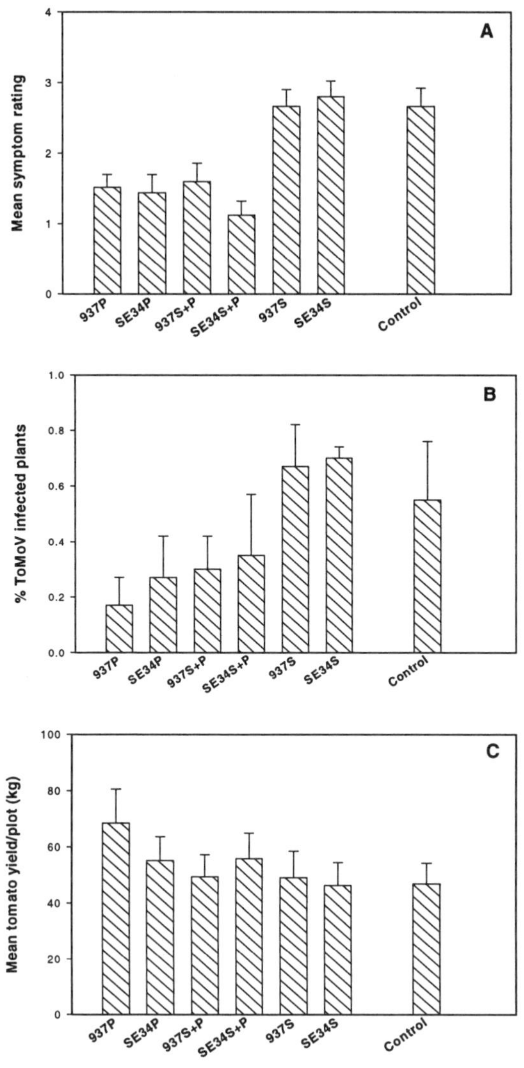

Figure 4: Evaluation of commercially prepared PGPR spore formulations IN937B and SE34 for induced resistance against tomato mottle geminivirus (ToMoV) in Florida field experiments in 1997. S, seed formulation; P, powder formulation; S+P, combination. (A) Mean virus symptom rating. (B) Mean percent of infected plants based on dot blot analysis. (C) Mean weight of marketable tomato fruit per plot on the first harvest date (Dec.1, 1997).

observation). Thus, conditions in the Florida trial provided a severe test for the PGPR treatments.

Results of the Florida experiment also indicate that the PGPR powder formulations more effectively induced systemic resistance than the treatments in which PGPR were applied directly to the seed. We do not know whether the occurrence and relative speed of PGPR germination and root colonization is correlated with higher levels of PGPR-induced resistance in tomato. If this is the case, it is possible that germination and colonization occurred earlier in the powder formulation treatments resulting in an earlier ISR response. Nonetheless, formulations of PGPR powder provide a practical delivery system for ISR in crops because powder can easily be added to planting mix, or mixed with water and applied as a transplant drench application.

Synthesis

Increased public demand to reduce reliance on chemical crop protectants combined with more stringent government regulations over pesticide registration (e.g., Food Quality Protection Act passed by the U.S. Congress in 1996) has focused more attention on the development of biological agents for pest management in agriculture. PGPR have practical applicability as components of environmentally-sound integrated pest management programs because they can provide safe, persistent and broad-spectrum protection that can be easily delivered to the crop.

To our knowledge, our results with PGPR-ISR in cucumber provide the first evidence that PGPR-induced plants are protected against insect herbivory and disease. The finding that the reduction in cucumber beetle feeding was associated with reduced levels of the secondary plant compound cucurbitacin, a beetle feeding stimulant, is of particular interest because it suggests that plant metabolism is affected by PGPR-ISR. Similar observations have been made for plants in which defense responses have been induced by other agents, including pathogens or chemicals (reviewed in Sticher et al. 1997) or herbivory (reviewed in Karban and Baldwin 1997).

PGPR appear to elicit both disease and insect resistance in cucumber. In contrast, other elicitors of induced resistance have been associated with increased insect feeding. Tallamy (1985) demonstrated an increase in cucurbitacins following herbivory or mechanical injury to zucchini, *Cucurbita pepo*. Apriyanto and Potter (1990) also showed that pathogen-induced resistance against anthracnose (*Colletotrichum lagenarium*) in cucumber was associated with increased cucumber beetle feeding, although cucurbitacin levels were not affected. Thus, it appears that different metabolic pathways are involved in induced resistance in cucurbits, and that the specificity of induced resistance differs in response to induction by pathogens, herbivores or PGPR (see also Stout and Bostock, this volume).

In a recent survey of Alabama tomato growers, a majority of respondents indicated that their most serious pest management problem was control of aphid vectors of plant viruses (Bauske et al. 1998). Recent work at Auburn University has shown that PGPR strains, previously shown to induce resistance against fungal and bacterial pathogens, also induce resistance against virus diseases in cucumber and tomato. Our results with tomato follow the work of Raupach et al. (1996) in which they reported a significant reduction in CMV symptoms on PGPR-treated tomato grown in the greenhouse compared with a untreated control. We have shown that PGPR-mediated induced resistance in tomato against two different viruses, CMV and ToMoV, can be maintained under field conditions; however, the level of virus disease protection may vary. We have not yet determined whether induced effects on aphid and whitefly feeding behavior and are involved with PGPR-mediated induced systemic resistance in tomato. It is conceivable that chemical signals elicited by PGPR-ISR in tomato may be translocated in the phloem and perceived by aphids or whiteflies during feeding, but this remains to be investigated.

Future Directions

Since 1985, with the first introduction of commercial PGPR products in the United States, the markets for PGPR have continued to increase. At present, 60-75% of the United States cotton crop is treated with *B. subtilis* bacteria targeted against soil borne pathogens (Backman et al. 1997). In China, 18 commercial PGPR strains or strain mixtures are sold, most of which are derived from the spore-forming genus *Bacillus* (Backman et al. 1997). *Bacillus* strains have been the most frequently exploited bacteria for commercial development because the resistant endospore produced by *Bacillus* spp. remains viable for long periods and is tolerant to extremes in temperature and pH, as well as to pesticides and fertilizers. However, the development of new formulation technologies has facilitated the use of biocontrol strains from other genera that have recently been introduced commercially (Backman et al. 1997).

In addition to improved formulation technology, the identification of effective combinations of bacteria for specific crop/pest systems is another research area that may lead to increased efficacy of PGPR products. Studies have shown that combinations of bacteria have been more effective than single strains for disease control and improved plant response (Pierson and Weller 1994, Sheng 1996, Raupach and Kloepper 1998). Another area of future development will undoubtedly include the genetic engineering of PGPR strains in which plant-beneficial genes are expressed, or in which such genes are expressed at an increased level. Genetically engineered PGPR may have an expanded spectrum of activity or level of control (Backman et al. 1997).

Another approach to achieve effective disease protection may be to use combinations of PGPR with other inducing agents that suppress diseases by complementary mechanisms (i.e., benzothiadiazole, Görlach et al. 1996, Tally et al., this volume). Additional studies in this area are needed to determine whether different inducers used simultaneously (e.g., PGPR and chemical inducers) enhance the induced defense response or act antagonistically (Karban and Kuć, Stout and Bostock, Felton and Eichenseer, this volume).

The enhancement of PGPR technology will be aided by a better understanding of the mechanisms involved with PGPR-ISR. Workers in the area of induced resistance are just beginning to compare and contrast the different mechanisms associated with plant resistance elicited by pathogens, chemicals, insects and non-pathogenic microbes (Stout and Bostock, this volume). In the case of PGPR-mediated resistance, it appears that different rhizobacterial strains induce resistance by different mechanisms (reviewed in van Loon et al. 1998). For example, bacterial production of salicylic acid is involved with ISR in some strains, but not in others (van Loon et al. 1998). van Loon et al. (1998) suggest that the mechanism by which a rhizobacterial strain induces resistance may vary depending on local conditions in the rhizosphere. Studies to elucidate the different mechanisms involved in the induction of systemic resisitance will enable us to develop and deploy PGPR products to their greatest advantage.

It remains to be determined if these strategies to optimize the effectiveness of PGPR as inducers of resistance will generate PGPR products that alone can consistently provide acceptable levels of disease protection. In the short term, at least, combinations of PGPR with other disease management tools (e.g., resistant or tolerant varieties, insecticides targeted against insect disease vectors, etc.) may prove to be the best strategy for the commercial use of PGPR. Furthermore, PGPR product development will be driven by economic considerations that may restrict its use to certain markets. PGPR represent a potentially valuable resource for IPM programs, particularly in high-value cropping systems like vegetables where regulations or lack of efficacy limit the availability of chemical crop protectants.

Acknowledgements

These projects were supported in part by grants from the USDA Southern Region IPM Program, and the USDA Pest Management Alternatives Research Program.

Literature Cited

Abouzid, A.M., J.E. Polston and E. Hiebert. 1992. The nucleotide sequence of tomato mottle virus: a new geminivirus isolated from tomato in Florida. Journal of General Virology 73:3225-3229.

Alström, S. 1991. Induction of disease resistance in common bean susceptible to halo blight bacterial pathogen after seed bacterization with rhizosphere pseudomonads. Journal of General Applied Microbiology 37:495-498.

Apriyanto, D. and D.A. Potter. 1990. Pathogen-activated induced resistance of cucumber: response of arthropod herbivores to systemically protected leaves. Oecologia 85:25-31.

Backman, P.A., M. Wilson and J.F. Murphy. 1997. Bacteria for biological control of plant diseases. Pages 95-109 *in* N.A. Rechcigl and J.E. Rechcigl, editors. Environmentally Safe Approaches to Crop Disease Control. Lewis Publishers, Boca Raton, Florida.

Bahme, J.B. and M.N. Schroth. 1987. Spatial-temporal colonization patterns of a rhizobacterium on underground organs of potato. Phytopathology 77:1093-1100.

Bahme, J.B., M.N. Schroth, S.D. Van Gundy, A.R. Weinhold, and D.M. Tolentino. 1988. Effect of inocula delivery systems on rhizobacterial colonization of underground organs of potato. Phytopathology 78:534-542.

Balliano, G., O. Caputo, F. Viola, L. Delprino and L. Cattel. 1982. The transformation of 10-alpha-cucurbita-5,24-dien-3-beta-ol into cucurbitacin C by seedlings of *Cucumis sativus* Cucumbers. Phytochemistry 22:909-913.

Bauske, E.M., G. W. Zehnder, E.J. Sikora and J. Kemble. 1998. Southeastern tomato growers adopt integrated pest management. HortTechnology 8:40-44.

Bergstrom, G.C., M.C. Johnson and J. Kuć. 1982. Effects of local infection of cucumber by *Colletotrichum lagenarium, Pseudomonas lachrymans*, or tobacco necrosis virus on systemic resistance to cucumber mosaic virus. Phytopathology 72:922-926.

Blua, J.J., F.E. Gildow, F.L Lukezic, S.J. Fleischer, and D. de Mackiewicz. 1994. Characterization and detection of *Erwinia tracheiphila* isolates (Abstract). Phytopathology 84:1370.

Broadbent, P., K.F. Baker, N.Franks and J. Holland. 1997. Effect of *Bacillus* spp. on increased growth of seedlings in steamed and untreated soil. Phytopathology 67:1027-1031.

Chambliss, O. and C.M. Jones. 1966. Cucurbitacins: specific insect attractants in Cucurbitaceae. Science 153:1392-1393.

De Mackiewicz, D., F.E. Gildow, M. Blua, S.J. Fleischer and F.L. Lukezic. 1998. Herbaceous weeds are not ecologically important reservoirs of *Erwinia tracheiphila.* Plant Disease 82:521-529.

Ferguson, J.E., E.R. Metcalf, R.L. Metcalf, and A.M. Rhodes. 1983. Influence of cucurbitacin content in cotyledons of cucurbitaceae cultivars upon feeding behavior of Diabroticina beetles (Coleoptera: Chrysomelidae). Journal of Economic Entomology 76:47-51.

Fleischer, S.J., D. de Mackiewicz, F.E. Gildow, and F.L. Lukezic. 1999. Serological estimates of the seasonal dynamics of *Erwinia tracheiphila* in *Acalymma vittata* (Coleoptera: Chrysomelidae). Environmental Entomology (in press).

Glick, B.R. 1995. The enhancement of plant growth by free-living bacteria. Canadian Journal of Microbiology 41:109-117.

Görlach, J., S. Volrath, G. Knauf-Beiter, G. Hengy and U. Beckhove. 1996. Benzothiadiazole, a novel class of inducers of systemic acquired resistance, activates gene expression and disease resistance in wheat. Plant Cell 8:629-643.

Hoffland, E., C.M.J. Pieterse, L. Bik and J.A. van Pelt. 1995. Induced systemic resistance in radish is not associated with accumulation of pathogenesis-related proteins. Physiological and Molecular Plant Pathology 46:309-320.

Karban, R. and I.T. Baldwin. 1997. Induced Responses to Herbivory. University of Chicago Press, Chicago.

Kloepper, J.W. 1993. Plant growth-promoting rhizobacteria as biological control agents. Pages 255-274 *in* F.B. Metting, Jr., editor. Soil Microbial Ecology. Marcel Dekker, New York.

Kloepper, J.W. and M.N. Schroth. 1978. Plant growth-promoting rhizobacteria on radishes. Pages 879-882 *in* Station de Pathologie Vegetale et Phytobacteriologie, editors. Proceedings of the Fourth International Conference on Plant Pathogenic Bacteria, Vol.2: INRA, Angers. Gibert-Clarey, Tours.

Kloepper, J.W. and M.N. Schroth. 1981. Relationship of in vitro antibiosis of plant growth-promoting rhizobacteria on potato plant development and yield. Phytopathology 70:1078-1082.

Kloepper, J.W., F.M. Scher, M. Laliberte, and I. Zaleska. 1985. Measuring the spermosphere colonizing capacity (spermosphere competence) of bacterial inoculants. Canadian Journal of Microbiology 31:926-929.

Kloepper, J.W. and C.J. Beauchamp.1992. A review of issues related to measuring colonization of plant roots by bacteria. Canadian Journal of Microbiology 38:1219-1232.

Kloepper, J.W., S. Tuzun, and J.A. Kuć. 1992. Proposed definitions related to induced disease resistance. Biocontrol Science and Technology 2:349-351.

Kloepper, J.W., S. Tuzun, L. Liu and G. Wei. 1993. Plant growth-promoting rhizobacteria as inducers of systemic disease resistance. Pages 156-165 *in* R.D. Lumsden and J.L. Vaughn, editors, Pest Management: Biologically Based Technologies, American Chemical Society Books, Washington, D.C.

Lazarovits, G. and J. Nowak. 1997. Rhizobacteria for improvement of plant growth and establishment. HortScience 32:188-192.

Loper, J.E., T.V. Suslow and M.N. Schroth. 1984. Lognormal distribution of bacterial subpopulations in the rhizosphere. Phytopathology 74:1454-1460.

Matthews, R.E.F. 1991. Plant Virology, 3rd edition. Academic Press. San Diego, California.

Metcalf. R.L. 1986. Coevolutionary adaptations of rootworm beetles (Coleoptera: Chrysomelidae) to cucurbitacins. Journal of Chemical Ecology 12:1109-1124.

Palukaitis, P. M.J. Roossinck, R.G. Dietzgen and R. B. Franki. 1992. Cucumber mosaic virus. Pages 281-348 *in* Advances in Virus Research. Academic Press, New York.

Pierson, E.A. and D.M. Weller. 1994. Use of mixtures of fluorescent pseudomonads to suppress take-all and improve the growth of wheat. Phytopathology 84:940-947.

Pieterse, C.M.J., S.C.M. van Wees, E. Hoffland, J.A. van Pelt and L.C. van Loon. 1996. Systemic resistance in *Arabidopsis* induced by biocontrol bacteria is independent of salicylic acid accumulation and pathogenesis-related gene expression. Plant Cell 8:1225-1237.

Pieterse, C.M.J., S.C.M. van Weese, J.A. van Pelt and L.C van Loon. 1998. A novel defense pathway in *Arabidopsis* induced by biocontrol bacteria. Mededelingen Faculteit Landbouwkundige en Toegepaste Biologische Wetenshappen Universiteit Gent. 63:931-940.

Polston, J.E., E. Hiebert, R.J. McGovern, P.A. Stansly and D.J. Schuster. 1993. Host range of tomato mottle virus: a new geminivirus infecting tomato in Florida. Plant Disease 77:1181-1184.

Rand, F.V. and Enlows. 1916. Transmission and control of bacterial wilt of cucurbits. Journal of Agricultural Research 6:417-434.

Raupach, G.S., L. Liu, J.F. Murphy, S. Tuzun and J. Kloepper. 1996. Induced systemic resistance in cucumber and tomato against cucumber mosaic cucumovirus using plant growth-promoting rhizobacteria (PGPR). Plant Disease 80:891-894.

Raupach, G.S. and J.W. Kloepper. 1998. Mixtures of plant growth-promoting rhizobacteria enhance biological control of multiple cucumber pathogens. Phytopathology 88:1158-1164.

Rice, C.A., K.S. Rymal, O.L. Chambliss and F.A. Johnson. 1981. Chromatographic and mass spectral analysis of cucurbitacins of three *Cucumis sativus* cultivars. Agricultural and Food Chemistry 29:194-196.

Ross, A.F. 1961. Systemic acquired resistance induced by localized virus infections in plants. Virology 14:340-358.

Scheffer, R.J. 1983. Biological control of Dutch elm disease by *Pseudomonas* species. Annals of Applied Biology 103:21-26.

Schippers, B.1988. Biological control of pathogens with rhizobacteria. Philosophical Transactions of the Royal Society of London 318:283-292.

Schippers, G., A.W. Bakker and P.A.H.M. Bakker. 1987. Interactions of deleterious and beneficial rhizosphere microorganisms and the effect on cropping practices. Annual Review of Phytopathology 25:339-358.

Schroth, M.N. and J.G. Hancock 1982. Disease-suppressive soil and root-colonizing bacteria. Science 216:1376-1381.

Sheng, H. 1996. Colonization and plant responses of soybean cultivars to *Bacillus subtilis* inoculants. M.S. Thesis, Auburn University, Auburn, Alabama.

Sherf, A.F. and A.A. MacNab. 1986. Vegetable diseases and their control. Wiley, New York.

Sikora, E.J., R.T. Gudauskas, J.F. Murphy, D.W. Porch, M. Andrianifahanana, G. Zehnder, E.M. Bauske, J.M. Kemble and D.F. Lester. 1998. A multivirus epidemic of tomatoes in Alabama. Plant Disease 82:117-120.

Sticher, L., B. Mauch-Mani and J.P. Métraux. 1997. Systemic acquired resistance. Annual Review of Phytopathology 35:235-270.

Suslow, T.V. 1982. Role of root-colonizing bacteria in plant growth. Pages 187-223 *in* M.S. Mount and G.S. Lacy, editors. Phytopathogenic Prokaryotes, Vol. I., Academic Press, New York.

Tallamy, D.W. 1985. Squash beetle feeding behavior: an adaptation against induced cucurbit defenses. Ecology 66:1574-1579.

Tian, W.H., S.X. Cao, Q. Sun, S.Q. Chang, X.H. Chang and P. Tien. 1985. Biological properties of CMV-S51 containing satellite RNA. Acta Phytopathologica Sinica 15:145-149.

Tjamos, E. and J. Kuć. 1982. Inhibition of steroid glycoalkaloid accumulation by arachidonic and eicosapentaenoic acids in potato. Science 217:542-544.

Tomlinson, J.A. 1987. Epidemiology and control of virus diseases of vegetables. Annals of Applied Biology 110:661-681.

van Loon, L.C. 1997. Induced resistance in plants and the role of pathogenesis-related proteins. European Journal of Plant Pathology 103:753-765.

van Loon, L.C., P.A.H.M. Bakker and M.J. Pieterse. 1998. Systemic resistance induced by rhizoshpere bacteria. Annual Review of Phytopathology 36:453-83.

van Peer, R., G.J. Niemann and B. Schippers. 1991. Induced systemic resistance and phytoalexin accumulation in biological control of Fusarium wilt of carnation by *Pseudomonas* sp. strain WCS417r. Phytopathology 81:728-734.

van Wees, S.C.M., C.M.J. Pieterse, A. Trijssenaar, Y. Vant Westende and F. Hartog. 1997. Differential induction of systemic resistance in *Arabidopsis* by biocontrol bacteria. Molecular Plant-Microbe Interactions 10:716-724.

Vernooij, B., L. Friedrich, A. Morse, R. Reist, R. Kolditz-Jawhar, E. Ward, S. Uknes, H. Kessmann and J. Ryals. 1994. Salicylic acid is not the translocated signal responsible for inducing systemic acquired resistance but is required in signal transduction. Plant Cell 6:959-965.

Voisard, C., C. Keel, D. Haas, and G. Défago. 1989. Cyanide production by *Pseudomonas fluorescens* helps suppress black root rot of tobacco under gnotobiotic conditions. Journal of the European Molecular Biology Organization (EMBO) 8:351-358.

Wei, G., J.W. Kloepper and S. Tuzun. 1991. Induction of systemic resistance of cucumber to *Colletotrichum orbiculare* by select strains of plant growth-promoting rhizobacteria. Phytopathology 81:1508-1512.

Wei, G., C. Yao, G.W. Zehnder, S. Tuzun and J.W. Kloepper. 1995. Induced systemic resistance by select plant growth-promoting rhizobacteria against bacterial wilt of cucumber and the beetle vectors (Abstract). Phytopathology 85:1154.

Weller, D.M.1988. Biological control of soilborne plant pathogens in the rhizosphere with bacteria. Annual Review of Phytopathology 73:463-469.

Yao, C., G. Zehnder, E. Bauske and J. Kloepper. 1996. Relationship between cucumber beetle (Coleoptera: Chrysomelidae) density and incidence of bacterial wilt of cucurbits. Journal of Economic Entomology 89:510-514.

Zehnder, G., J. Kloepper, C. Yao, and G. Wei. 1997a. Induction of systemic resistance against cucumber beetles (Coleoptera: Chrysomelidae) by plant growth-promoting rhizobacteria. Journal of Economic Entomology 90:391-396.

Zehnder, G., J. Kloepper, S. Tuzun, C. Yao and G. Wei. 1997b. Insect feeding on cucumber mediated by rhizobacteria-induced plant resistance. Entomologia Experimentalis et Applicata 83:81-85.

Zitter, T.A. Diseases caused by viruses. 1991. Pages 31-42 *in* J.B. Jones, J.P. Jones, R.E. Stall and T.A. Zitter, editors. Compendium of Tomato Diseases. The American Phytopathological Society, St. Paul, MN.

Commercial Development of Elicitors of Induced Resistance to Pathogens

Allison Tally, Michael Oostendorp, Kay Lawton, Theo Staub, and Bobby Bassi

Abstract

Novartis Crop Protection has been involved in basic research on induced plant resistance for many years. In this chapter we summarize our research efforts from the laboratory and field to develop CGA-245704 (benzo[1,2,3]thiadiazole-7-carbothioc acid S-methyl ester) as a commercial product. While finding a chemical that stimulated induced plant resistance was important, it is also important to continue to elucidate the mechanisms involved. Novartis, while actively developing a commercial plant activator, is also active in identifying genes and trying to elucidate signals in the plant, so that further improved plant protection can be achieved.

Introduction

Agri-chemical companies are always in search for the perfect control of unwanted pests and pathogens: environmentally friendly, broad spectrum, economical for the grower, and profitable for the manufacturer. Traditionally the search was for novel compounds that are directly toxic to major pest groups. An alternative approach was to develop screening methods to search for chemicals that stimulate the plants' own defense reactions. There are many mechanisms plants have evolved for their defense against pathogens. Novartis chose the multigenic, broad spectrum induction of systemic plant resistance to diseases, variously known as induced resistance and systemic acquired resistance (SAR) (Kuć 1982), as a biological model to follow. SAR was chosen because this defense response occurs in many, if not all, plant species and often conveys effective resistance against economically important pathogens. In principle, SAR presents some interesting new opportunities for the control of plant diseases and for enhancing the knowledge of disease resistance in plants - which could lead to improved crop yields.

In the mid-1970s, after hearing about SAR from Dr. Joe Kuć of the University of Kentucky, Novartis (formerly Ciba) became interested in pursuing commercial opportunities with SAR. Dr. Theo Staub spent a four month sabbatical with Dr. Kuć in order to learn more about the mechanism. Upon return to Basel, efforts were put forth to develop a screen that would find SAR inducing compounds. It was felt that three issues needed to be addressed: 1) find the natural SAR inducing signal, 2) develop screening methods and synthesize safe and effective synthetic compounds to stimulate these responses and 3) unravel the molecular and biochemical mechanisms underlying the SAR phenomenon (Den Hond 1998). We have successfully addressed the second issue and significant advances have been made on the first and third.

Searching For SAR

SAR is a natural defense response triggered by pathogens or by hypersensitive cell death in reaction to pathogen attack. Induction of SAR produces a systemic signal that is released that in turn activates a broad-spectrum resistance in the systemic tissue. Low molecular weight chemicals that are able to induce SAR would offer a great potential for disease control in economically important crops, therefore a screening method was developed that allowed for detection of compounds that mimicked the natural SAR activation. Similar to Dr. Kuć's basic biological model, the compounds were applied on the first leaves of cucumber plants, and the protection against diseases was examined on the subsequent leaves. In order to validate that a chemical did induce SAR, five criteria were established for screening (Kessmann et al. 1994):

- the treated plants are resistant to the same spectrum of diseases as those in which SAR is biologically induced
- no conversion of the compound *in vivo* into anti-microbial metabolites and lack of direct anti-microbial activity
- induction of the same biochemical mechanisms as seen in systemic plant tissues after biological induction of SAR
- no activity in plant mutants with a defective SAR signaling pathway
- induction time required to launch the resistance responses in the plant (typically 2-4 days between induction and resistance)

An isonicotinic acid (INA) derivative, 2,6-dichloroisonicotinic acid, was one of the first plant activators discovered in the cucumber tests to fulfill these criteria. INA does not exhibit any direct *in vitro* activity; it protects cucumber against the same spectrum of diseases as does biological induction of SAR (Métraux et al. 1991). Molecular studies with tobacco showed that INA induces the same set of genes as are induced in the non-infected tissue after local infection with tobacco mosaic virus. The two lead INA derived compounds, while inducing SAR, also had phytotoxic side effects which made them unsuitable for further commercial development. In 1989, CGA-245704 (benzo[1,2,3]thiadiazole-7-carbothioc acid S-methyl ester) was discovered (Fig.1). It is the first synthetic chemical developed and marketed that functions exclusively by activating the SAR response of plants (Ruess et al. 1996).

The actual signal for SAR has not yet been determined. It was thought at one time that it may be salicylic acid (SA), as SA is produced and accumulates during induction (see Hammerschmidt and Smith-Becker, this volume). To test this hypothesis, tobacco was transformed with the *nahG* gene from *Pseudomonas putida* which encodes salicylate hydroxylase, an enzyme that catalyzes the degradation of SA to the non-inducing metabolite catechol. The *nahG*-transgenic plants were shown to express the *nahG*-gene and did not accumulate SA or exhibit the SAR response following inoculation with SAR inducing pathogens. Thus, it appeared SA could be the systemically transmitted signal (Gaffney et al. 1993). However, results of reciprocal grafting experiments showed that SA is not the primary signal of SAR, but rather is required for the plant to respond to the systemic signal (Vernooij et al. 1994, Hammerschmidt and Smith-Becker, this volume). In these experiments, when *nahG* plants were grafted to wildtype rootstocks, SAR could not be induced in the *nahG* scion as expected. However, inoculation of *nahG* rootstocks unexpectedly resulted in activation of SAR in wildtype scions. Thus, *nahG* plants can release the systemic signal but cannot respond to it. Similar experiments using transgenic tobacco plants with inhibited SA synthesis confirmed these observations (Pallas et al. 1996). The search for the systemic signal that induces SAR continues.

As a result of activating the SAR signal transduction pathway(s), various plant genes begin production of enzymes and proteins that are thought to be involved in the inhibition of plant disease development (Fig. 2, Buell, this volume). These proteins have been termed SAR proteins and include the pathogenesis-related proteins (PR-proteins). Some of these proteins are known anti-microbial enzymes, i.e., acidic forms of chitinases and glucanases, or

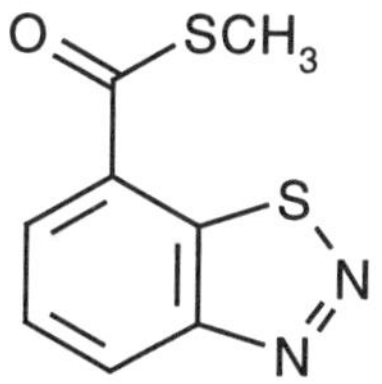

CGA 245704

Figure 1: Chemical structure of Plant Activator CGA-245704

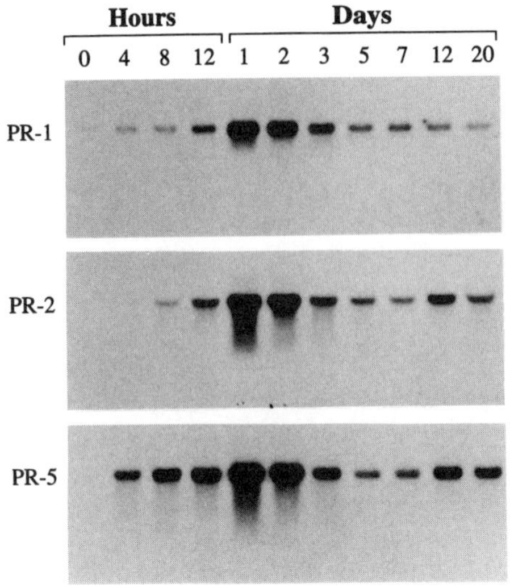

Figure 2: Induction of PR mRNA in *Arabidopsis* following a treatment of CGA-245704 (300 μM). Note the time lag of induction following treatment and the decrease after the peak expression.

thaumatin like proteins (see Hammerschmidt and Nicholson, this volume). Other SAR- and PR-proteins have antimicrobial properties of unknown mechanisms, i.e., SAR 8,2 and PR-1 (Neuenschwander et al. 1995).

Using the tools of biotechnology has allowed for a better understanding of how SAR works, i.e., how inducing agents (biological or chemical) can activate this plant disease resistance response. *Arabidopsis* mutants have been found in which defense responses cannot be induced biologically, via inducing pathogens, or chemically, via SA or INA (Delaney et al. 1995). Treatment of these *n*on-inducible *im*munity mutants (*NIM1*) with CGA-245704 does not result in reduction of disease. Since a functional *NIM1* protein is required for SAR activation by inducing pathogens, SA and CGA-245704, these results indicate that CGA-245704 activates the same plant response as do inducing pathogens and SA. Furthermore, experiments with the salicylic acid free transgenic plants (*nahG*), have shown that CGA-245704 is likely to be active as a functional analogue of salicylic acid, because SAR can be activated by CGA-245704 in these plants (Fig. 3) (Lawton et al. 1996).

How CGA-245704 Works

The new concept of chemically induced plant defense leads to product features that are intermediate between conventional pesticides and resistant plant

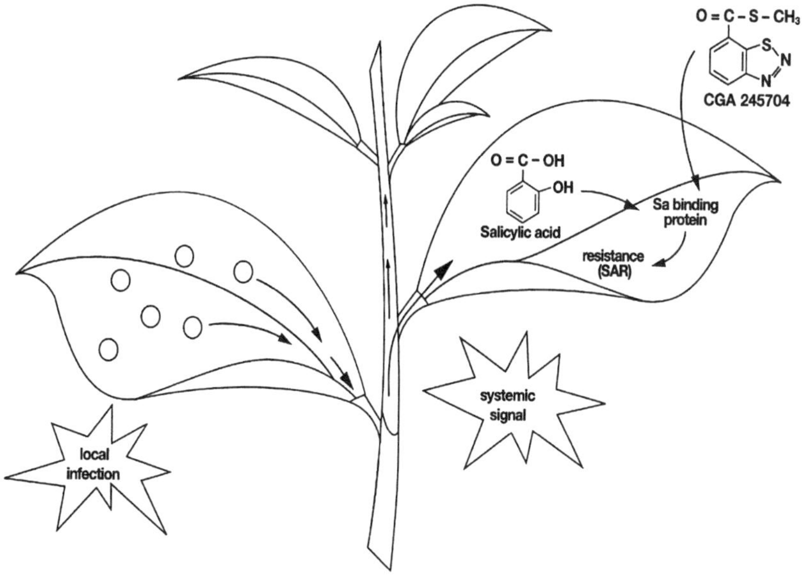

Figure 3: The Mechanism of Action of CGA-245704 in the Salicylic Acid Pathway. An infection, or chemical, induces an unidentified trigger that promotes the production of salicylic acid. Either the salicylic acid, or CGA-245704, activates other mechanisms in the plant which produce the enzymes and proteins which protect the plant from diseases.

varieties. Therefore, an understanding of the biology of SAR is needed in order to correctly use a SAR based product. Treatment with CGA-245704 elicits many responses in plants, similar to the defense responses seen after systemic activation of induced resistance by biological agents (Fig. 4). Many of them can also be observed upon unsuccessful pathogen attacks. Prominent among these defense responses are the production of PR-proteins, and a priming of the plant to react stronger or faster towards pathogen attack by papillae formation, hypersensitive cell death, oxidative burst, or other defense responses (Lawton et al. 1996, Görlach et al. 1996).

SAR requires active plant metabolism and a 2-4 day induction period under most conditions. Therefore, SAR based products work best when applied as a preventive strategy. Once activated, the defense mechanisms remain active even if the inducer is degraded. This is especially true in monocot plants, where defense responses remain active for many weeks. Consequently, intervals between application can be very long (e.g., banana) or only one application to

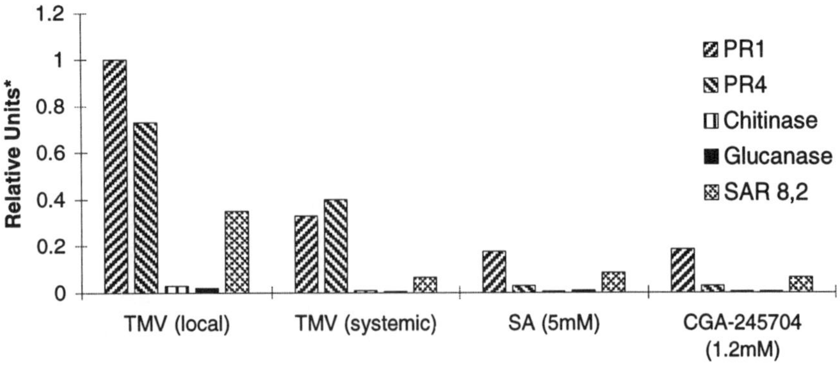

Figure 4: Induction of proteins and enzymes in tobacco by various inducing agents (*relative units – values normalized to maximum level of PR1 detected).

young plants may be recommended (e.g., rice, wheat). One theory is that since monocots only have one meristem, where all of the growth occurs, one treatment is sufficient for long lasting induction. In dicots, there are often many active meristems, with new ones becoming active, and so additional applications are needed to 'expose' the new cells to the inducing agent

Plants do not recognize biological systematics and use similar defense responses against seemingly unrelated pathogens. Consequently, a multitude of pathogens may be inhibited by SAR activation, since the response is non-specific (Karban and Kuć, Stout and Bostock, this volume). The main focus of SAR research has been the defense responses against fungal, bacterial, and viral pathogens, but inhibition of insects and nematodes has also been reported (Inbar et al. 1998, Owen et al. 1998, Stout and Bostock, this volume). Some of the induced PR - proteins are identical or similar to stress induced proteins, and an overlap with plant stress responses has been observed (Mauch et al. 1997). The spectrum of defense responses that can be observed in any given plant species is determined by the plant itself. Whereas some plants, like cucumber, exhibit a broad-spectrum defense response against many pathogens, SAR in other plants such as wheat is effective against only a few diseases (see also Stout and Bostock, this volume). Recent research resulted in increasing evidence that there are different inducible defense responses in many plants (Stout and Bostock, Staswick and Lehman, this volume). While the use of CGA-245704 is limited to activating the SA-dependent SAR pathway, other pathways may be employed against other pathogens (e.g., van Loon et al. 1998). CGA-245704 works solely via plant genes, making it more a complement to conventional disease control strategies than a replacement. It allows changing the resistance characteristics of a crop variety by the activation of additional resistance mechanisms, thereby broadening the farmers' options in cultivar selection.

Commercialization of SAR

CGA-245704, trade name BION™ in Europe and Actigard™ in the United States, is the first synthetic plant activator that triggers the plant to elicit the identical response as biological induction of SAR. It was registered in Germany in 1996 as a 'Pflanzenstaerkungsmittel' which loosely translates as 'plant tonic', since it is not a 'pesticide'. A full safety data package was developed for Europe and the United States and is currently in review by the respective authorities. The US EPA classified CGA-245704 as a Reduced Risk compound in April 1998. To obtain this classification, the active ingredient must be better or safer than other products in the areas of 1) environmental load 2) environmental effects 3) safety (including infants and children) 4) efficacy and 5) resistance management strategies in place as appropriate. The EPA's goal is to register Reduced Risk compounds within 18-20 months of submission. Registration is expected on tomatoes, leafy vegetables, and tobacco in the United States in late 1999 and on tomatoes, tobacco, and cucurbits in Europe in 2000. Other target markets are banana, mango, and rice.

Development of any new product for commercialization has challenges: defining the rate, defining the spectrum, and mitigating undesirable plant effects if any. Since CGA-245704 does not act directly on the pathogen, but via the plants, the challenges have been a little more intense.

Rates. Many trials have been conducted on rate titration. It appears that once the plant is 'turned on', higher rates of application do not improve disease protection or residual activity. Even finding the lowest rate has been difficult. There appears to be variability in the efficacy of SAR due to variability in the physiological state of plants (Guedes et al. 1980). Recommended application rates have gradually come down over the past few years. The initial rate for tomatoes was 100 g ai/ha (ai = active ingredient); it is now at 35 g ai/ha, while slightly lower may be adequate (Fig. 5).

Spectrum of activity. As pointed out earlier, CGA-245704 turns on various genes in plants that in turn produce specific proteins and enzymes. Consequently, each host/pathogen - plant activator interaction is unique. For example, in tomatoes, protection is afforded against late blight (*Phytophthora infestans*). In another solanaceous crop, potato, which is attacked by the same pathogen, there is little effectiveness from a CGA-245704 application. Thus, no specific assumptions can be made regarding the spectrum of resistance. The spectrum has to be worked out for each system. Table 1 summarizes the documented activity of the defense responses against a variety of pests. Plants activated with CGA-245704 are often able to defend themselves against a broad spectrum of diseases. Still, some effects are more pronounced or useful under practical conditions and it is anticipated that CGA-245704 will be labeled for activation against the

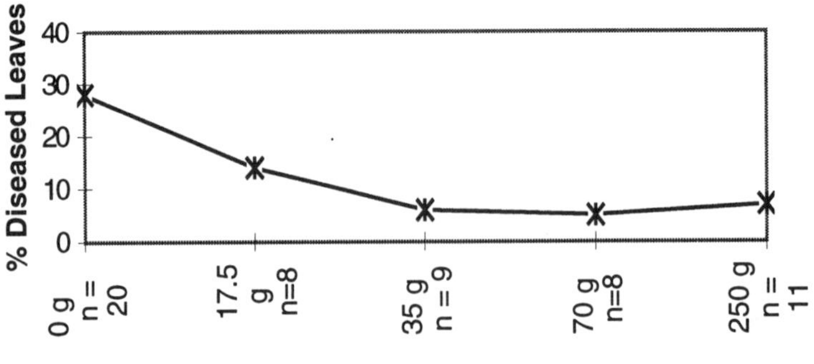

Figure 5: Rate response for protection against *Xanthomonas vesicatoria* on tomato (field trials in US).

diseases listed in Table 2. Fig. 6 illustrates the usefulness of an inducing agent for protection against plant diseases, and that inducing agents may enhance the activity of traditional fungicides. Conventionally viral diseases are controlled through plant breeding or through control of the insect vector. CGA-245704 (and SAR in general) offers a unique opportunity to reduce the effects of many viruses on the plant that have traditionally not been able to be controlled directly by pesticides (Fig. 7).

While a reduction of some sucking insect pests has been observed in field trials, the activity against insects does not appear to be up to commercial

Table 1. Spectrum of activity of CGA-245704 activated resistance on a variety of crops, against various plant antagonists.

Crop	Bacteria	Virus	Fungi	Insects/nematodes
Cereals		?	✓	
Rice	✓		✓	
Tobacco	✓	✓	✓	
Potato		✓		✓
Tomato	✓	✓	✓	✓
Vegetables	✓		✓	✓
Pome Fruit	✓		✓	
Stone Fruit	✓			
Mango	✓		✓	
Citrus	✓			
Grape				✓
Banana			✓	

Table 2. Diseases which CGA-245704 induced resistance effectively suppresses.

Crop	Disease	Pathogen
Bananas	Black Sigatoka	*Mycosphaerella fijiensis*
Chili peppers	Anthracnose	*Colletotrichum sp.*
Lettuce	Downy Mildew	*Bremia lactucae*
Tomato	Late Blight	*Phytophthora infestans*
Tomato	Bacterial speck	*Pseudomonas syringae* pv. *tomato*
Tomato	Bacterial spot	*Xanthomonas campestris* pv. *vesicatoria*
Rice	Blast	*Pyricularia oryzae*
Tobacco	Blue Mold	*Peronospora tabacina*
Wheat	Powdery Mildew	*Erysiphe graminis*

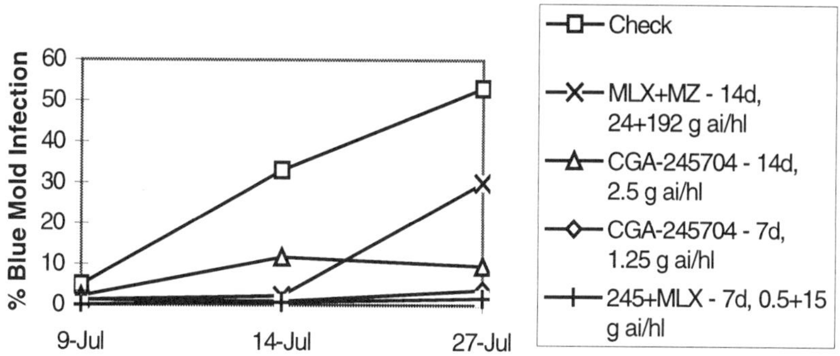

Figure 6: Field protection from *Peronospora tabacina* in tobacco (MLX = metalaxyl, MZ = mancozeb, d = indicates interval in days).

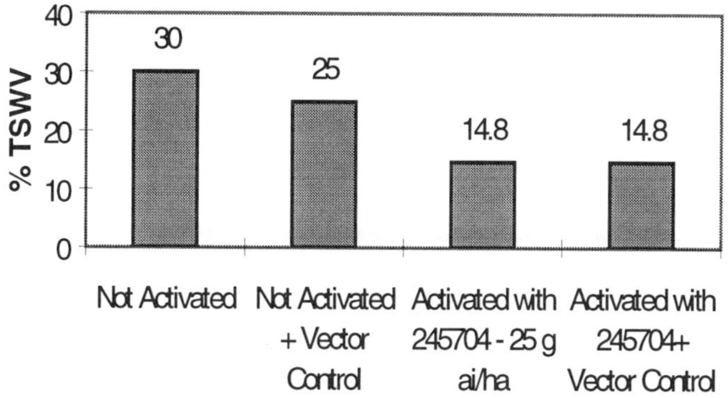

Figure 7: Field trials of tomato plants with insect vector control (pesticide) and application of CGA-245704 for control of Tomato Spotted Wilt Virus.

standards. However, there may be opportunities to combine inducing agents with insecticides to improve the activity at lower rates.

Plant Effects. Most of the early focus in the commercial development of SAR was on the evaluation of disease protection. What has become evident over time is that there are other plant effects of induced resistance that need to be understood. Novartis is currently undertaking more research into this area in order to more fully understand the benefits and limitations (see Agrawal, this volume) of the product that extend beyond disease protection. It appears that under certain circumstances, loading rate (the total amount applied per season) may result in undesirable effects in some plants (Table 3). For example, in green bell peppers, yields can be severely reduced (Fig. 8), but the product does not

Table 3. Effect of disease resistance induction (+ positive, -- negative) on the growth and development of a variety of crops.

SAR effect	wheat	tobacco	banana	pome fruit	seed potato
Leaf bronzing			--		
Leaf greening	+				
Stunting		--			
Up right flag leaf	+ /--				
Less flowering					+
Plant health	+	+	+	+	+
Starch/tuber size					+
Healthy leaves			+		

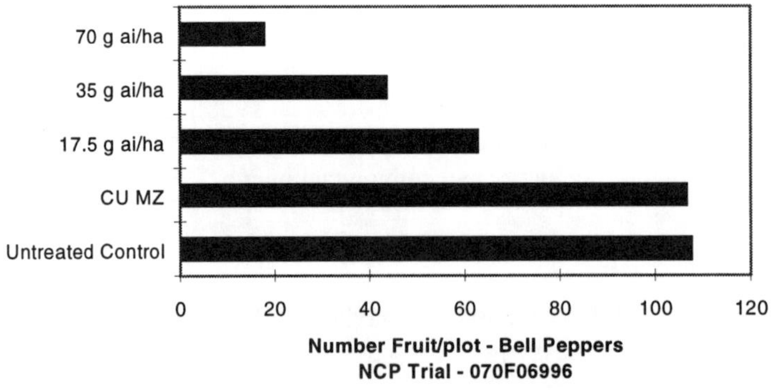

Figure 8: Negative yield effect of CGA-245704 applications on green bell peppers (rates per application with 8 applications/season; CU MZ = pesticide treatment: 2000 g ai/ha copper plus 1000 g ai/ha mancozeb).

Table 4. Contrast in terminology associated with resistance inducing agents (such as Actigard/Bion) and fungicides.

What an inducing agent does:	What a fungicide does:
• Activates defense responses	• Controls pathogens
• Defense mechanisms remain active for ... days	• Has residual activity for ... days
• Activated plants defend against....	• The compound is active against....

have this effect in chili peppers. In burley type tobacco, yellowing and stunting can sometimes result from repeated CGA-245704 treatments. However, there are numerous trials where no phytotoxicity was seen. This negative effect seems to be correlated, at least somewhat, with cool, cloudy conditions, but the physiological reason for this is unknown. The final use directions will reflect use patterns and crops where there is no negative impact or "cost" of induction.

There are also some very positive effects of SAR induction by CGA-245704. Trials on cabbage in Japan have shown non-treated plants with heads starting to crack open, while treated plants had intact heads. CGA-245704 treated hazelnut trees in Italy were protected against bacterial die-back and were much healthier than untreated controls.

Novartis does not intend to market Actigard/Bion as a pesticide, but rather as a plant health product. It is not a traditional pesticide nor is it a gene product in a transgenic plant. For this reason, new descriptions of activity must be used (Table 4).

Resistance. Several mechanisms appear to be activated simultaneously against the pathogen attack (Görlach et al. 1996). This feature seems to limit the risk of development of pathogens insensitive to the activated defense responses of SAR based products (see also Lyon and Newton, this volume). This may be a valuable tool in the management of resistance against conventional pesticides or in extending the life of cultivar resistance (Romero et al. 1998). SAR activation may also synergize the effects of fungicides (Molina et al. 1998). These aspects clearly will receive more scientific and commercial interest.

Synthesis and Future Directions

As in conventional chemical crop protection, research and development of SAR inducing compounds will focus on finding other molecules, with further adapted environmental and toxicological characteristics, even lower dose rates, alternative selectivity, and different crop spectrums. The increasing evidence for a multitude of inducible defense responses (Pennincks et al. 1996, Press et al. 1997, Karban and Kuć, this volume) creates an interest in finding chemicals that activate these pathways. These compounds may be able to induce resistance against other pathogens or selectively activate certain stress responses, such as drought stress responses. The search for natural elicitors of defense responses and for the endogenous signal molecules will continue to be an exciting area of research. Elicitors have been known for many years, and although none of them has been developed into a commercial product useful on many crops, this area of research has a high potential for discovering new compounds or potential lead structures for chemical development.

Research on the genetics of the plant signaling pathways in response to pathogens and other stresses has progressed rapidly in the last few years. Active research in this area has lead to the discovery of mutants that are unable to respond to the stress signaling molecules ethylene (i.e., *etr* and *ein* for *e*thylene *t*riple *r*esponse and *e*thylene *in*sensitivity), and jasmonic acid (i.e., *jar* for *j*asmonic *a*cid *r*esistant) as well as to SAR inducers (*NIM1*). The genes responsible for some of these mutant phenotypes (*etr1*, *ein2*, *ein3*, *NIM1*) have also been discovered (Buell, this volume). In order to fully exploit the potential of this technology, one future option might be the engineering of hyperinducible crops.

Finally, SAR switches on plant genes. Gene switches for the targeted expression of plant traits are very much in the center of the development of transgenic crops (Jepson et al. 1998). Not all switchable promotors participating in inducible plant defenses may be useful for switching engineered traits. The future research in inducible plant defense responses may result in interesting spin offs into the area of gene switch technology.

Literature Cited

Delaney, T.P., L. Friedrich, J. Ryals. 1995. *Arabidopsis* signal transduction mutant defective in chemically and biologically induced disease resistance. Proceedings of the National Academy of Sciences USA 92:6602-6606.

Den Hond, F. 1998. Systemic acquired resistance: A case of innovation in crop protection. Pesticide Outlook 9:18-23.

Gaffney, T, L. Friedrich, B. Vernooij, D. Negretto, G. Nye, S. Uknes, E. Ward, H. Kessmann, and J. Ryals. 1993. Requirement of salicylic acid for the induction of systemic acquired resistance. Science 261:754-756.

Görlach, J., S. Volrath, G. Knauf-Beiter, G. Hengy, U. Beckhove, K.-H. Kogel, M. Oostendorp, T. Staub, E. Ward, H. Kessmann, and J. Ryals. 1996. Benzothiadiazole, a novel class of inducers of systemic acquired resistance, activates gene expression and disease resistance in wheat. Plant Cell 8:629-643.

Guedes, M.E.M., S. Richmond, and J. Kuć. 1980. Induced systemic resistance to anthracnose in cucumber as influenced by the location of the inducer inoculation with *Colletotrichum lagenarium* and the onset of flowering and fruiting. Physiological Plant Pathology 17:229-233.

Inbar, M., H. Doostdar, R.M. Sonoda, G.L. Leibee, and R.T. Mayer. 1998. Elicitors of plant defense systems reduce insect densities and disease incidence. Journal of Chemical Ecology 24:135-149.

Jepson, I., A. Martinez, and J.P. Sweetman. 1998. Chemical-inducible gene expression systems for plants - a review. Pesticide Science 54:360-367.

Kessmann, H., T. Staub, C. Hofmann, T. Maetzke, J. Herzog, E. Ward, S. Uknes, and J. Ryals. 1994. Induction of systemic acquired resistance in plants by chemicals. Annual Review of Phytopathology 32:439-459.

Kuć, J. 1982. Induced immunity to plant diseases. BioScience 32:854-860.

Lawton, K.A., L. Friedrich. M. Hunt, K. Weymann, T.P. Delaney, H. Kessmann, T. Staub, and J. Ryals. 1996. Benzothiadizole induces disease resistance in *Arabidopsis* by activation of the systemic acquired resistance signal transduction pathway. Plant Journal 10:71-82.

Mauch, F., A. Kmecl, U. Schaffrath, S. Volrath, J. Görlach, E. Ward, J. Ryals, and R. Dudler. 1997. Mechanosensitive expression of a lipoxygenase gene in wheat. Plant Physiology 114:1561-1566.

Métraux J.P., P. Ahl-Goy, T. Staub, J. Speich, A. Steinemann, J. Ryals and E. Ward. 1991. Induced systemic resistance in cucumber in response to 2,6-dichloro-isonicotinic acid and pathogens. Pages 432-439 *in* Hnennecke and D.P.S. Verma, editors. Advances in Molecular Genetics of Plant-Microbe Interactions. Vol. 1. Kluwer Academic Publishers. Dortrecht, The Netherlands.

Molina, A.,M.D. Hunt, and J.A. Ryals. 1998. Impaired fungicide activity in plants blocked in disease resistance signal transduction. Plant Cell 10:1903-1914.

Neuenschwander U., K. Lawton, and J. Ryals. 1995. Systemic acquired resistance. Pages 81-106 *in* Stacey and Keen, editors. Plant-Microbe Interactions. Vol. 1 Chapman and Hall, NY.

Owen, K.J., C.D. Green, and B.J. Deverall. 1998. Systemic acquired resistance against root-knot nematodes in grapevines (Abstract). APPS Conference, Perth, September 1998.

Pallas, J.A., N.L. Paiva, C. Lamb, and R.A. Dixon. 1996. Tobacco plants epigenetically suppressed in phenylalanine ammonia-lyase expression do not develop systemic acquired resistance in response to infection by tobacco mosaic virus. The Plant Journal 10:281:293.

Pennincks, I.A.M.A., K. Eggermont, F.R.G. Terras, B.P.H.J. Thomma, G.W. de Samblans, A. Buchala, J.-P. Métraux, J.M. Manners, and W.F. Broekaert. 1996. Pathogen-induced systemic activation of a plant-defense gene in *Arabidopsis* follows a salicylic acid-independent pathway. The Plant Cell 8:2390-2323.

Press, C.M., M. Wilson, S. Tuzun, and J.W. Kloepper. 1997. Salicylic acid produced by *Serratia marcescens* 90-166 is not the primary determinant of induced systemic resistance in cucumber and tobacco. Molecular Plant-Microbe Interactions 20:761-768.

Romero, A.M., C.S. Kousik, and D.F. Ritchie. 1998. Systemic acquired resistance delays race shifts to major resistance genes in pepper (Abstract). Proceedings of the 1998 Annual Meeting of the American Phytopathological Society, Las Vegas, NV.

Ruess W., K. Mueller K. G. Knauf-Beiter and T. Staub. 1996. Plant activator CGA-245704: An innovative approach for disease control in cereals and tobacco. Pages 53-60 *in* Brighton Crop Protection Conference - Pests and Diseases.

van Loon, L.C., P.A.H.M. Bakker, and C.M.J. Pieterse. 1998. Systemic resistance induced by rhizosphere bacteria. Annual Reviews of Phytopathology 36:453-483.

Vernooij, B, L. Friedrich, A. Morse, R. Reist, R. Kolditz-Hawhar, E. Ward, S. Uknes, H. Kessmann and J. Ryals. 1994. Salicylic acid is not the translocated signal responsible for inducing systemic acquired resistance but is required in signal transduction. Plant Cell 6:959-965.

Implications of Induced Resistance to Pathogens and Herbivores for Biological Weed Control

Nina K. Zidack

Abstract

Pathogens and phytophagous insects constantly challenge plants, including weeds. Plants are also subjected to environmental stresses including atmospheric pollutants, physical wounding, and chemical exposure. All of the above scenarios are capable of inducing resistance (IR) to both plant pathogens and herbivores, agents that may be used in biological weed control. Biotic or abiotic elicitation of IR in the field may thwart efforts to use biocontrol agents. This chapter contains a review of the small body of literature dealing explicitly with defense mechanisms in weeds, and relates literature on IR in crop plants and model systems to biological weed control.

Introduction

Induced resistance (IR) is a plant response phenomenon that has been demonstrated in numerous plant/pathogen and plant/insect interactions. Since IR has implications for development of biorational disease and insect control measures, most studies have been either crop oriented or confined to model systems such as *Arabidopsis*. In addition to crop protection from pathogens and insects, IR has implications for weed control. Biological weed control systems which exploit plant pathogens and phytophagous insects rely on plant susceptibility to these attacks. Research addressing the specific problem of resistance to biocontrol agents has been limited, but many of the discoveries that have been made on crops and model systems can be extrapolated to weed systems. This information may be used to develop more successful biologically-based weed control.

Traditionally, biological weed control has been divided into two main areas, classical and inundative (often referred to as the bioherbicide approach) (Mortensen 1986, Hasan 1988). Classical biocontrol involves the introduction of a weed-specific phytophagous insect or pathogen from the weed's area of origin. Ideally, the establishment of a natural enemy against the invasive weed brings this weed into a more balanced position in the plant community. This method is primarily used to counter widespread infestations of alien weeds in forests and rangelands. The majority of the successes in this arena have been with insects, or insects that vector plant pathogens. Inundative biocontrol involves the use of endemic plant pathogens or insects. Plant pathogens are frequently candidates for this type of weed control. The ultimate goal of this approach is to replace or reduce chemical herbicide use, and/or provide controls for weeds that have no alternative controls, infest environmentally sensitive areas, or are closely related to the crop in which the weed is a problem.

Plants in nature are constantly facing challenge by pathogens, insects, and environmental stress. Their response to these challenges could have a significant effect on their susceptibility to artificially applied organisms intended for their control. Hoagland (1990) provided a comprehensive review of biochemical plant responses to pathogens and provided speculation on how some of this information might be used to improve the potential of biological weed control agents. This chapter presents a review of the small body of work that pertains directly to biological weed control, and also describes how some of the established findings in the field of IR can be applied to weed/pathogen and weed/insect relationships.

Hypersensitive Response

The hypersensitive response (HR) was defined by Stakman in 1915 as a defensive mechanism against infection by plant pathogens. It has been further characterized as a disorganization and granulation of cell cytoplasm at the site of pathogen attack and has been demonstrated for bacterial, fungal, and viral pathogens (Klement and Goodman 1965, Wood 1982, Goodman and Novacky 1994). More recently, the HR has been shown to be accompanied (though not always) by IR (Alvarez et. al. 1998).

Where weeds occur naturally, they are constantly challenged by a broad range of pathogens, and possibly even insects, which elicit the HR. It would be reasonable to speculate that this defensive mechanism would have a significant effect on pathogens and insects applied for biological weed control. An example of a potential inducer of the HR in weeds would be *Pseudomonas syringae*, a ubiquitous pathogenic epiphyte in nature. *P. syringae* pv. *syringae* causes a rapid induction of systemic resistance in cucumber, measured as increased peroxidase activity and reduced disease in plants challenged with *Colletotrichum lagenarium* (Smith et al. 1991). This particular pathogen has been demonstrated to epiphytically colonize plants within and outside its host range, and attain populations of greater than 10^6 - 10^7 colony forming units (cfu) per gram of leaf tissue (Hirano et al. 1987, Lindow et al. 1977). Populations of this magnitude are adequate for causing HR when conditions for infection are appropriate (i.e. driving rain, mechanical damage, etc.). This may induce a broad spectrum of resistance mechanisms in weeds and make them less susceptible to biological weed control agents. The expression of the HR has also been associated with phytoalexin production in *Arabidopsis thaliana* (Tsuji et al. 1992). A phytoalexin (3-thiazol-2'-yl-indole) was produced in *A. thaliana* in response to inoculation with a *P. syringae* pv. *syringae* strain pathogenic to wheat, and this accumulation was correlated with the restricted *in vivo* growth of *P. syringae* pv. *syringae*.

The HR has also been shown as a response to insects and is reviewed by Fernandes (1990). Some genotypes of *Brassica nigra* L., a cruciferous weed, resist attack by its common herbivore, *Pieris rapae*, by induction of the HR (Shapiro and DeVay 1987). Eggs laid by this Lepidoptera on leaves can elicit the formation of a necrotic zone around the site of the eggs, ultimately desiccating and killing them. The HR is also a resistance mechanism in tall goldenrod (*Solidago altissima)* to the gallmaker *Eurosta solidaginis* (Anderson et al. 1989). Significantly more larval mortality occurred in galls formed in the meristems of resistant clones than in the meristems of susceptible clones, with necrosis of gall tissue surrounding the larvae in resistant plants. Gall mortality of a leaf galling *Contarinia* sp. was also the result of the HR on its host plant *Bauhinia brevipes* (Fernandes 1998). The HR was considered the major mortality factor for over 90% of the galls, two years in a row.

Compatible Disease Reactions

Compatible disease reactions may also induce systemic resistance in weed plants. For example, many leaf spotting pathogens cause necrotic lesions which do not severely affect the plant. The formation of necrotic lesions by compatible pathogens induces resistance similar to the responses seen for the HR, and the resistance is non-specific with respect to both the inducing pathogen and the challenging pathogen (Kuć 1983, Karban and Kuć, this volume). Plant pathogen surveys on weed plants frequently identify pathogens such as *Alternaria* sp., *Septoria* sp., or *Pseudomonas* sp. which cause minor leaf spots and rarely cause plant death (Julien 1992, Zidack, unpublished results). These infections may induce resistance and render weed plants less susceptible to pathogen and arthropod biological weed control agents.

Another situation where IR may influence efficacy of biological control agents is in the use of endemic pathogens. In many cases, the endemic organism is already present in the environment and causing disease on the weed at non-lethal levels. These pre-inoculations may reduce the susceptibility of the weed to the applied pathogen and limit the potential for secondary cycling of the pathogen. This was demonstrated with *P. syringae* pv. *phaseolicola* (Psp), a candidate for biological control against kudzu (*Pueraria lobata*), an extremely invasive weed that is choking many natural ecosystems in the southeastern United States (Zidack et al. 1999). Psp causes halo blight on kudzu similar to the disease on common bean, *Phaseolus vulgaris*. Greenhouse and field studies were conducted using Psp as a bioherbicide on kudzu (Zidack and Backman 1996). While a percentage of plants in the greenhouse were killed by application of Psp, plant death was never achieved in the field. It was hypothesized that minor natural infection of kudzu by Psp may induce resistance against bioherbicide applications of the same pathogen. Leaf samples taken from young diseased and mature diseased leaves (in nature) had elevated levels of peroxidases when compared to asymptomatic leaves. Greenhouse studies showed systemic elevation of peroxidase in plants treated with Psp, and reduced disease in plants challenged with a second inoculation of Psp. This would indicate that this biological control agent may be negatively impacted by natural infections by Psp prior to the bioherbicide application. It also helps explain why secondary applications of Psp in the field did not achieve better results than one application alone.

Nickerson et al. (1993) demonstrated that peroxidases were induced in velvetleaf (*Abutilon theophrasti*) in response to infection by *C. coccodes* and the plant growth regulator, thidiazuron. This induction of peroxidase is presumably associated with IR. In addition, they measured soluble peroxidase activity in very young seedlings and found that certain defense related isoforms were not present in sufficient amounts to protect seedlings against pathogen attack. This growth stage corresponds with the window of maximum weed susceptibility to the control agent.

In another study, Weete (1992) showed IR in the weed sicklepod (*Cassia obtusifolia* L.) in response to infection by *Alternaria cassiae*. A spore suspension of *A. cassiae* was applied to the first compound leaf of sicklepod seedlings to induce resistance. Three days later, plants were challenged with the same pathogen and the resulting disease on the second leaf was reduced 43% when compared to the non-induced controls. A flavonoid phytoalexin (2-(*p*-hydroxy-phenoxy)-5,7-dihidroxychromone) has been detected in sicklepod in response to treatment with *A. cassiae* (Sharon and Gressel 1992). Plants also exhibited a similar development of resistance as a function of age. While Weete speculated that the induction of SAR would not limit the efficacy of *A. cassiae* as a bioherbicide, it does show why this bioherbicide is less efficacious on mature plants. Plants with minor infections in the field may be protected from bioherbicides.

IR may also interfere with biocontrol efficacy when endemic, weakly virulent isolates co-infect with a more virulent, biocontrol isolates of the same pathogen. In the crop oilseed rape, when a weakly virulent isolate of *Leptosphaeria maculans,* was co-inoculated with a highly virulent isolate of the same species, the average lesion size produced by the highly virulent isolate was significantly reduced (Mahuku et al. 1996). The authors concluded that when the weakly virulent isolate was inoculated 24 hours prior to the highly virulent isolate, it induced resistance against the highly virulent isolate.

Induced Responses to Herbivory

IR as a response to herbivory has not been studied with the objective of determining effects on biological weed control. But, crop plants and weedy plants have been the focus of many investigations and the results may be interpreted with biological weed control in mind. The literature suggests that classical biological control systems which utilize specialist phytophagous insects may not be adversely affected by IR, while biocontrol agents which are generalist feeders may be impacted. Agrawal (1999) showed that induction of the weed wild radish (*Raphanus sativus)* by chewing herbivores early in the season resulted in half the damage by later season chewing herbivores and 30% less colonization by aphids. A second study (Agrawal 1999) showed that induction of resistance in two species of wild radish (*R. raphanistrum* and *R. sativus*) by caterpillar feeding or treatment with the natural plant response elicitor, jasmonic acid, protected plants. Colonization of induced plants by leaf miners was reduced when compared to non-induced controls, and in field experiments, herbivory by grasshoppers was also lower on induced plants. In contrast, induced responses did not have a negative impact on the growth of specialist *Pieris rapae* larvae.

Insect response to induced factors in plants may also be a means of regulating populations of herbivores. This was demonstrated on a biological weed control system where the insect *Zygogramma suturalis* was introduced from the United States into the USSR to control ragweed (*Ambrosia artemisiifolia).* Reznik (1991) showed that *Z. suturalis* feeding on damaged ragweed caused a drop in subsequent oviposition intensity, and some of the females completely stopped ovipositing on damaged plants. Where insects were unable to find undamaged plants for several days, the females oviposited less and some entered diapause. This is probably an adaptive response of the insects, manifested as a decrease in the population to avoid starvation. This example provides quantitative evidence that even in the most successful biological control systems, eradication of the weed is not attained because the biological control agents do not fully exploit the plant.

Environmental and Wound Effects

Environmental factors, plant stress and physical wounding all have been shown to induce resistance. Ozone, a common atmospheric pollutant, has been shown to induce a number of defense-related genes that are also induced in the hypersensitive response (Sharma et al. 1996). This resulted in resistance of *A. thaliana* to virulent *P. syringae* strains. Wounding alone has been shown to induce resistance to herbivory. The weedy cruciferous plant *Sinapis alba* exhibited less feeding damage by the flea beetle *Phyllotreta cruciferae* after mechanical wounding of the cotyledons with a needle, or by induction through feeding damage (Palaniswamy and Lamb 1993). However, there were no measurable induced effects in two related plant species, *Brassica napus* and *B. rapa.* This indicates differential responses within a plant family and illustrates the need for specific studies with biological weed control targets.

Antagonism between Insect and Pathogen Weed Control Agents

In many biological weed control systems, insects and plant pathogens act synergistically and contribute to weed control (Wilson 1969, Hasan 1988). Much recent evidence points to a biochemical and ecological trade-off between resistance to herbivores and pathogens (Hatcher 1995, Thaler et al. 1999, Karban and Kuć, Stout and Bostock, this volume). This trade-off could lead to synergistic control of weeds, as herbivory may make plants more susceptible to pathogens, and vice versa; a dual challenge by herbivores and pathogens may overcome some of the plant's defenses. However, antagonistic relationships are also possible (i.e. cross-resistance: see Hatcher 1995, Stout and Bostock, this volume). In the case of the weeds *Rumex crispus* and *Rumex obtusifolius*, the beetle *Gastrophysa viridula* is negatively affected by the rust pathogen *Uromyces rumicis* (Hatcher et al. 1994). Eggs laid on leaves with rust infection had greater larval mortality and slower development than those raised on healthy leaves. While the greatest differences were attributed to a reduction in plant nutritional quality, the authors theorized that changes in feeding could have been brought about by chemical constituents such as phytoalexins.

Synthesis

The vast body of literature on IR in diverse plant families is evidence that weed plants are capable of deploying similar resistance mechanisms. Resistance factors in weeds may be induced by incompatible pathogens or insects through the hypersensitive response, compatible disease reactions, herbivory, or as part of a response to mechanical wounding or environmental effects. This information should not be used as an argument against the potential of biological weed control. Instead, it can be used to understand the successes and failures experienced in the field, and contribute to studies leading to more effective deployment of plant pathogens against weeds.

Future Directions

Studies on IR in weed plants should be incorporated into testing of biological weed control agents. Knowledge about the timing of induction of defense responses will aid researchers in optimizing treatment schedules. Also, specific information on the biochemical defenses employed by weeds could be used to develop treatments that incorporate biosynthetic inhibitors of defense compounds.

An important ecological question that should be addressed is whether induction of resistance mechanisms gives "weedy" plants a competitive advantage in the environment. In studies on wild radish, Agrawal (1998, 1999) showed that induced responses to herbivory increased plant fitness in the natural environment. Enhanced understanding of weed strategies for defense and competition in the environment will provide the basis for novel approaches to biological weed control.

Literature Cited

Agrawal, A.A. 1998. Induced responses to herbivory and increased plant performance. Science 279:1201-1202.

Agrawal, A.A. 1999. Induced responses to herbivory in wild radish: Effects on several herbivores and plant fitness. Ecology (In press).

Alvarez, M.E., R.I. Pennell, P.J. Meijer, A. Ishikawa, R.A. Dixon, and C. Lamb. 1998. Reactive oxygen intermediates mediate a systemic signal network in the establishment of plant immunity. Cell 92:773-784.

Anderson, S.A., K.D. McCrea, W.G. Abrahamson, and L.M. Hartzel. 1989. Host genotype choice by the ball gallmaker *Eurosta solidaginis* (Diptera: Tephritidae). Ecology 70:1048-1054.

Fernandes, G.W. 1990. Hypersensitivity: A neglected plant resistance mechanism against insect herbivores. Environmental Entomology 19:1173-1182.

Fernandes, G.W. 1998. Hypersensitivity as a phenotypic basis of plant induced resistance against a galling insect (Diptera: Cecidomyiidae). Environmental Entomology 27:260-267.

Goodman, R.N. and A.J. Novacky. 1994. The Hypersensitive Reaction in Plants to Pathogens: A Resistance Phenomenon. APS Press, St. Paul, MN.

Hasan, S. 1988. Biocontrol of weeds with microbes. Pages 129-151 *in* K.G. Mukerji and K.L. Garg, editors. Biocontrol of Plant Diseases. CRC Press, Boca Raton, FL.

Hatcher, P. E. 1995. Three-way interactions between plant pathogenic fungi, herbivores, and their host plants. Biological Reviews 70:639-694.

Hatcher, P.E., Paul, N.D., Ayres, P.G., and Whittaker, J.B. 1994. The effect of a foliar disease (rust) on the development of *Gastrophysa viridula* (Coleoptera: Chrysomelidae). Ecological Entomology 19:349-360.

Hirano, S.S., D.I Rouse, and C.D. Upper. 1987. Bacterial ice nucleation as a predictor of bacterial brown spot disease on snap beans. Phytopathology 77:1078-84.

Hoagland, R.E. 1990. Biochemical responses of plants to pathogens. Pages 87-113 *in* R.E. Hoagland, editor. Microbes and Microbial Products as Herbicides. ACS Symposium Series 439. American Chemical Society, Washington, D.C.

Julien, M.H. 1992. Biological Control of Weeds: A World Catalogue of Agents and their Target Weeds. CAB International, Canberra, Australia.

Klement, Z. and R.N. Goodman. 1965. The hypersensitive reaction to infection by bacterial plant pathogens. Annual Review of Phytopathology 6:17-44.

Kuć, J. 1983. Induced systemic resistance in plants to diseases caused by fungi and bacteria. Pages 191-221 *in* J. Bailey and B. Deverell, editors. The Dynamics of Host Defense. Academic, Sydney.

Lindow, S.E., D.C. Arny, and D.D. Upper. 1977. Distribution of epiphytic ice nucleation-active strains of *Pseudomonas syringae*. Proceedings of the American Phytopathological Society 4:107 (abstract).

Mahuku, G.S., R. Hall and P.H. Goodwin. 1996. Co-infection and induction of systemic acquired resistance by weakly and highly virulent isolates of *Leptospheria maculans* in oilseed rape. Physiological and Molecular Plant Pathology 49:61-72.

Mortensen, K. 1986. Biological control of weeds with plant pathogens. Canadian Journal of Plant Pathology 8:229-231.

Nickerson, R.G., T.J. Tworkoski, and D.G. Lusteer. 1993. *Colletotrichum coccodes* and thidiazuron alter specific peroxidase activities in velvetleaf (*Abutilon theophrasti*). Physiological and Molecular Plant Pathology 43:47-56.

Palaniswamy, P. and R.J. Lamb. 1993. Wound-induced antixenotic resistance to flea beetles, *Phyllotreta cruciferae* (Goeze) (Coleoptera: Chrysomelidae), in crucifers. The Canadian Entomologist 125:903-912.

Reznik, S.Y. 1991. The effects of feeding damage in ragweed *Ambrosia artemisiifolia* (Asteraceae) on populations of *Zygogramma suturalis* (Coleopetera, Chrysomelidae) Oecologia 88:204-210.

Shapiro, A.M. and J.E. DeVay. 1987. Hypersensitivity reaction of *Brassica nigra* L. (Cruciferae) kills eggs of *Pieris* butterflies (Lepidoptera: Pieridae). Oecologia 71:631-632.

Sharma, Y.K., J. Leon, I. Raskin, and K.R. Davis. 1996. Ozone-induced responses in *Arabidopsis thaliana*: The role of salicylic acid in the accumulation of defense-related transcripts and induced resistance. Proceedings of the National Academy of Sciences USA 93:5099-5104.

Sharon, A. and J. Gressel. 1992. Elicitation of a flavonoid phytoalexin accumulation in *Cassia obtusifolia* by a mycoherbicide: Estimation by aluminum chloride spectrofluorimetry. Pesticide Biochemistry and Physiology 41:142-149.

Smith, J.A., R. Hammerschmidt, and D.W. Fulbright. 1991. Rapid induction of systemic resistance in cucumber by *Pseudomonas syringae* pv. *syringae*. Physiological and Molecular Plant Pathology 38:223-225.

Stakman, E.C. 1915. Relation between *Puccinia graminis* and plants highly resistant to its attack. Journal of Agricultural Research 4:193-200.

Thaler, J. S., A. L. Fidantsef, S. S. Duffey, and R. M. Bostock. 1999. Tradeoffs in plant defense against pathogens and herbivores: a field demonstration using chemical elicitors of induced resistance. Journal of Chemical Ecology (In press).

Tsuji, J., E.P. Jackson, and D.A. Gage. 1992. Phytoalexin accumulation in *Arabidopsis thaliana* during the hypersensitive reaction to *Pseudomonas syringae* pv. *syringae*. Plant Physiology 98:1304-1309.

Weete, J.D. 1992. Induced systemic resistance to *Alternaria cassiae* in sicklepod. Physiological and Molecular Plant Pathology 40:437-445.

Wilson, C.L. 1969. Use of plant pathogens in weed control. Annual Review of Phytopathology 7:411-435.

Wood, R.K.S. 1982. Active Defense Mechanisms in Plants. Plenum Press, N.Y.

Zidack, N.K. and P.A. Backman. 1996. Biological control of kudzu (*Pueraria lobata*) with the plant pathogen *Pseudomonas syringae* pv. *phaseolicola*. Weed Science 44:645-649.

Zidack, N.K., S. Tuzun, and P.A. Backman. 1998. Biological Control of kudzu: Induction of systemic resistance and accumulation of peroxidases upon inoculation with *Pseudomonas syringae* pv. *phaseolicola*. Weed Science (in review).

Subject and Taxonomic Index